50 popular beliefs that people think are true

Praise for *Race and Reality:*
What Everyone Should Know about Our Biological Diversity

"This is a very important, profound, enjoyable, and enlightening book. It should go a long way in helping disprove man's most dangerous myth."

—Robert W. Sussman, professor of anthropology, Washington University,
editor of *Yearbook of Physical Anthropology*, and
editor emeritus of *American Anthropologist*

"A tour de force that conveys the current science on racial classification in a rigorous yet readable way. A book so clearly written, so elegantly crafted, so packed with nuggets that even those who think they know it all about race and racial classification will come away changed."

—David B. Grusky, professor of sociology,
director of the Center for the Study of Poverty and Inequality,
Stanford University

"Guy P. Harrison's well-written and passionate plea for eliminating the idea and ideology of race should be widely read. He has shown that the idea of race not only is contradicted by science but [also] is a social anachronism that should not be tolerated by society in the twenty-first century."

—Audrey Smedley, professor emerita,
Anthropology and African American Studies,
Virginia Commonwealth University

Praise for *50 Reasons People Give for Believing in a God*

"Deep wisdom and patient explanations fill this excellent book."

—James A. Haught, editor of West Virginia's largest newspaper,
Charleston Gazette

"Engaging and enlightening. . . . Read this book to explore the many and diverse reasons for belief."

—Michael Shermer, publisher of *Skeptic*

"A persuasive and frequently humorous book about an important topic. . . . [S]hould be read by religious practitioners, political leaders, and the general public."

—Nick Wynne, PhD, executive director of the Florida Historical Society

"[D]oesn't bully or condescend. Reading Harrison's book is like having an amiable chat with a wise old friend."

—Cameron M. Smith and Charles Sullivan,
authors of *The Top 10 Myths about Evolution*

Visit Guy's website at www.guypharrison.com
Contact him at guyfeedback@gmail.com

Guy P. Harrison

50 popular beliefs that people think are true

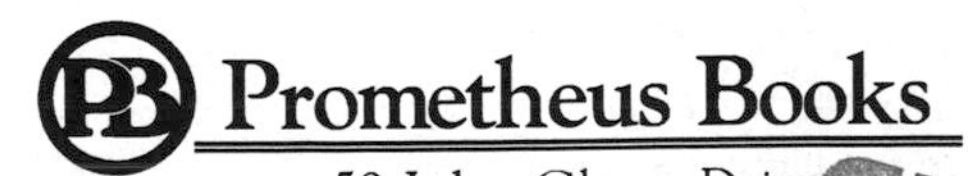

Prometheus Books
59 John Glenn Drive
Amherst, New York 14228–2119

2/16 18

Published 2012 by Prometheus Books

Cover design by Nicole Sommer-Lecht

Inquiries should be addressed to

Prometheus Books
59 John Glenn Drive
Amherst, New York 14228–2119
VOICE: 716–691–0133
FAX: 716–691–0137
WWW.PROMETHEUSBOOKS.COM

19 18 17 16 15 10 9 8 7 6

Library of Congress Cataloging-in-Publication Data

Harrison, Guy P.
50 popular beliefs that people think are true / by Guy P. Harrison; illustrations by Kevin Hand.
p. cm.
Includes bibliographical references and index.
ISBN 978–1–61614–495–1 (paper : acid-free paper)
ISBN 978–1–61614–496–8 (ebook)
1. Common fallacies. 2. History—Errors, inventions, etc. 3. Belief and doubt—Miscellanea. 4. Popular culture—Miscellanea. I. Title. II. Title: Fifty popular beliefs that people think are true.

AZ999.H37 2011
149'.73—dc23

2011032747

Printed in the United States of America on acid-free paper

For Natasha, Jared, and Marissa.
Always think before you believe.

ACKNOWLEDGMENTS

I am grateful to the following for their invaluable help in producing this book: Kevin Hand, Frank Ward, Neil deGrasse Tyson, Michael Shermer, Ian Birnbaum, Paul A. Offit, Steven L. Mitchell, Tariq Moosa, Benjamin Bradford, Sheree Harrison, Seth Shostak, Andrew Chaikin, Cameron M. Smith, Brian Dunning, Lewis "Nick" Wynne, Justin Uzell, Jefferson Fish, Coni Harrison, Andrea Roach, Victor Stenger, Hank Davis, Michelle Mead, Mark "Gunner" Lewis, Nancy White, Shawn Browning, Curtis Wienker, Kenneth Feder, Phil Plait, and Cayman Free Press.

CONTENTS

FOREWORD 13
Dr. Phil Plait

INTRODUCTION 17

MAGICAL THINKING

1. "I Believe in the Paranormal and the Supernatural." 23
2. "I Know There Is an Afterlife Because of All the Near-Death Experiences." 36
3. "A Psychic Read My Mind." 42
4. "You're Either Born Smart or You're Not." 51
5. "The Bible Code Reveals the Future." 58
6. "Stories of Past Lives Prove Reincarnation Is Real." 62
7. "ESP Is the Real Deal." 68
8. "Nostradamus Saw It All Coming." 74
9. "I Believe in Miracles." 81

OUT THERE

10. "NASA Faked the Moon Landings." 89
11. "Ancient Astronauts Were Here." 100
12. "UFOs Are Visitors from Other Worlds." 108
13. "A Flying Saucer Crashed Near Roswell, New Mexico, in 1947 and the Government Knows All about It." 121

14. "Aliens Have Visited Earth and Abducted Many People." 132
15. "Astrology Is Scientific." 139

SCIENCE AND REASON

16. "All Scientists Are Geniuses and Science Is Always Right." 147
17. "The Holocaust Never Happened." 154
18. "Global Warming Is a Political Issue and Nothing More." 161
19. "Television News Gives Me an Accurate View of the World." 167
20. "Biological Races Are Real." 180
21. "Biological Race Determines Success in Sports." 186
22. "Most Conspiracy Theories Are True." 192

STRANGE HEALINGS

23. "Alternative Medicine Is Better." 201
24. "Homeopathy Really Works, and No Side Effects!" 210
25. "Faith Healing Cures the Sick and Saves Lives." 218
26. "Race-Based Medicine Is a Great Idea." 226
27. "No Vaccines for My Baby!" 233

LURE OF THE GODS

28. "My God Is the Real One." 245
29. "My Religion Is the One That's True." 250
30. "Creationism Is True and Evolution Is Not." 259
31. "Intelligent Design Is Real Science." 273
32. "The Universe and Earth Are Fine-Tuned for Life." 279
33. "Many Prophecies Have Come to Pass." 283
34. "Prayer Works!" 290
35. "Religions Are Sensible and Safe. Cults Are Silly and Dangerous." 296

36. "They Found Noah's Ark!" 300
37. "Archaeology Proved My Religion Is True." 305
38. "Holy Relics Possess Supernatural Powers." 310
39. "A TV Preacher Needs My Money." 315

BIZARRE BEINGS

40. "Ghosts Are Real and They Live in Haunted Houses." 323
41. "Bigfoot Lives and Cryptozoology Is Real Science!" 333
42. "Angels Watch Over Me." 350
43. "Magic Is Real and Witches Are Dangerous." 354

WEIRD PLACES

44. "Atlantis Is down There Somewhere." 363
45. "I'm Going to Heaven When I Die." 368
46. "Something Very Strange Is Going on in the Bermuda Triangle." 372
47. "Area 51 Is Where They Keep the Aliens." 382

DREAMING OF THE END

48. "The Mayans Warned Us: It's All over on December 21, 2012." 391
49. "The End Is Near!" 397
50. "We're All Gonna Die!" 407

FAREWELL AND GOOD LUCK 417

NOTES 421

BIBLIOGRAPHY 443

INDEX 451

FOREWORD

Dr. Phil Plait

No one is born a skeptic.

Kids are natural scientists, though. They love to soak up knowledge, explore, experiment, name things (I can still remember my very young daughter, all those years ago, asking me to name the stars in the sky, one after another).

I suppose not all that is really science, though. Memorization and categorization are *important*, and the foundation of being able to understand relationships between objects, but they're not *science*. The basic property that makes science *science* is that it's self-checking. You don't just make an assumption; you test it. You see if it works the next time you use it. And you don't assume that just because it did, it always will.

And the most important thing, the one aspect of science that sets it apart from all other methods of knowing, is that science isn't loyal. You can rely on an idea for years, decades, but if something comes along that proves the idea wrong, boom! It gets chucked out like moldy cheese.

Well, not always. The other thing about science is that it builds on previous knowledge. If you learn something works pretty well, and then something else comes along that does better, a lot of the time you find out the second thing is just a modification of the first. Einstein didn't trash Newton; relativity *updated* Newton's mechanics, made it work better when objects are traveling near the speed of light, or where there's lots of gravity.

It was the accumulation of knowledge, of fact, that modified Newton's ideas. Hard-won, too, with experiments that contradicted centuries of "common wisdom." But that knowledge, when it's correct, builds over time. It all has to work, like a tapestry. And it *does*.

Still, it's hard to let go of an idea even when you know it's wrong. Sometimes the idea is stubborn (or its holder is). Sometimes it's comforting to have a warm, fuzzy idea. I bet that most of the time, though, it's ego, pure and simple. We identify with the ideas we keep, and if that idea is wrong, then that means some part of *us* is wrong. That's a difficult issue to deal with.

And that's why kids can be natural-born scientists, but terrible skeptics. And that's OK; sometimes kids need to just do stuff "because I said so," and you don't want them always questioning you. The real problem comes when they grow up and don't let go of that characteristic.

We all do it. Believing is easy. Being skeptical is hard. It's the road less traveled, rough-hewn and difficult. There are pitfalls everywhere, scary dark places, things that would be so much easier just to wish away when we close our eyes.

But reality, as author Philip K. Dick said, is what doesn't go away when we stop believing in it.

Reality doesn't care what you believe, what you do, for whom you vote. It just keeps on keeping it real. And since that's the case, isn't it better to see it for what it is? When you believe in something that's wrong, then other beliefs glom onto it, getting more complicated, getting harder and harder to balance and reconcile, like a pyramid built upside down. You build up more and more nonsense until the contradictions get so glaringly obvious, your only choice is to either completely ignore them, compartmentalizing your beliefs, or to let it all come crashing down.

You have to face reality.

In this book you will read about many such heels-over-head pyramids. Aliens. The Moon hoax. Bigfoot. Some are larks, fun little tidbits of silliness that on their own don't do much harm.

Others are dangerous. "Alternative" medicines that not only don't help, but keep people from seeking real medicine, making them sicker. Intercessory prayer, which is proven not to do anything, but which people sometimes employ instead of seeking real help. Self-proclaimed "psychics" who prey on the bereaved and grieving. And of course creationism, which shuts down curiosity and turns a blind eye to the true, and very ancient, nature of the world.

Science kicks over that pyramid, and sets it on its stable base. The best thing about science—and its multipurpose toolkit, skepticism—is that they show you how the universe really is. Yes, it can be scary, dark, and impersonal. But that's OK because it's also complex, deep, marvelous, profound, wondrous, magnificent . . . and above all, *beautiful*.

That beauty is out there. All you have to do is stop believing in it, and start understanding it.

INTRODUCTION

We all believe silly things. What matters is how silly and how many.

—Guy P. Harrison

Skepticism is the skill and the attitude that helps us navigate our way through an often-crazy world. Applied consistently and with force, skepticism can help us lead safer, happier, and more productive lives. It also helps to keep our minds healthy, sharp, and free by tossing aside much of the irrational junk that would otherwise obscure our view of an amazing universe. When crackpots, crooks, and fools try to lure us down costly or dangerous paths, skepticism is the shield we need to fend them off.

Some people think of skeptics as cynical, negative people with closed minds. Nothing could be further from the truth. Skepticism is really nothing more than a fancy name for trying to think clearly and thoroughly before making a decision about believing, buying, or joining something. It's about sorting out reality from lies and misperceptions. What's bad or negative about that? Embracing a skeptical attitude means approaching the world with open eyes, a switched-on brain, a willingness to ask the necessary questions, and the sense of humility that comes with knowing how easy it is for anyone to be fooled by things we see, hear, and think about. Being a skeptic means being honest and mature enough to seek answers that are based on evidence and logic rather than hopes and dreams. It also means being wise enough to accept that sometimes no satisfying answers are available.

James Randi, the world-famous skeptic and magician, has spent much of his life trying to save people from themselves by teaching how we can think our way clear of the relentless avalanche of kooky claims and irrational beliefs that bombard us all from childhood to the grave. Randi has been criticized for being too harsh toward irrational believers. But how gentle must one be when so much misery and waste comes from these unproven claims? Beyond the hundreds of billions of dollars thrown away on lies and fantasies, people literally suffer and die every day all around the world because of claims that any good skeptic could see right through in a few seconds. All irrational beliefs are not equal, yet they are all tied together in one gigantic cloud of danger. Believing in astrology may not be a direct health risk for an individual, for example, but the kind of faulty thinking that allows one to be impressed with astrology can be. Adopting weird beliefs without evidence is a dangerous game, warns Randi: "It can cost you money, emotional security, and it can cost you your life. I can think of a few exceptions, but almost any untruth or deception is bound to be a negative influence."[1]

Throughout this book I have tried my best to be positive and respectful. No doubt some readers will not see it that way. But I hope they will believe me when I say that my goal is not to win arguments or take away anyone's fun, happiness, or contentment. I fully understand that falling for weird, unproven beliefs is part of being human and happens to the best of us. It's part of the human condition. *We all believe silly things. What matters is how silly and how many.*

To push back against dangerous irrational beliefs, we have to pour every claim and every story through the filter of skepticism and science. Although it helps to know some history and science, education alone is not enough. Neither is exceptional intelligence enough. Many highly educated and highly intelligent people embrace some of the most ridiculous and baseless claims of all. No one should doubt that the lack of skepticism is a largely unrecognized global crisis.

I have traveled all over the world, and no matter where I found myself, I always saw money, time, and energy being squandered on beliefs that almost certainly were not true, and by people of every social and economic strata. This waste of energy and resources saddens me. We could do so much more and could possibly be so much better if only skepticism were more common. Still, frustrated as I may feel, I resist mocking, ridiculing, and dismissing those who cling to unproven beliefs. I prefer to be positive and offer help rather than condemnation to those who have yet to realize that not everything told to us in childhood is true.

I want readers to know my motivations for writing this book. I'm not scolding, lecturing, or preaching to make myself feel important. I'm only trying to encourage and inspire critical thinking and spread the word that skepticism is important. I want to help and to build, not condemn and tear down. Truth is, I really couldn't care less about what someone believes. It's only when I see unproven beliefs diminishing someone's life or causing harm to others that I feel obligated to speak up and offer a helping hand. If irrational beliefs weren't so often dangerous and such a drag on human progress, you would never hear a peep from me about anyone's beliefs. The way I see it, promoting reason and skepticism is a moral issue. It's about caring for your fellow humans.

Being a skeptic is the only way I can imagine living my life. It is part of a positive worldview that helps motivate me to get out of bed each morning. There are so many exciting experiences and important discoveries out there waiting on me that I don't want to be distracted or waste time believing things that are unlikely to be true. We should not be afraid to doubt and question, even when those around us do not. Skepticism is constructive, not destructive. It is a positive affirmation of being fully alive and mature enough to accept reality as it is rather than what we might like it to be or what somebody told us it is. The best kind of skeptic is not focused on rejecting and ridiculing. She or he embraces more of life, not less. Skepticism helps us to abandon astrology for astronomy, to see through the fog and find the stars, to stand up and exist fully as thinking human beings. Living a life as free from illusions and delusions as possible is to value that life and to understand that not one precious moment of it should be willingly sacrificed to a lie or an unproven belief.

Guy P. Harrison
Earth, 2011

MAGICAL THINKING

Chapter 1

"I BELIEVE IN THE PARANORMAL AND THE SUPERNATURAL."

It is error only, and not truth, that shrinks from inquiry.
—Thomas Paine

Every day we are confronted with paranormal, supernatural, or extraordinary beliefs. These claims find us no matter where we go. At the drugstore, homeopathic medicines are on the shelves right next to science-based treatments. The newspaper offers an astrological prediction of your future. A preacher promises that if you give him money, you will be rewarded one-hundred-fold via the supernatural hand of a god. TV commercials suggest that we can have better health with a pill or be better athletes if we wear a special bracelet. A friend swears she saw an alien spaceship in the sky last night. A family member tries to convince you that the end of the world is near. Is that strange noise you heard before falling asleep a ghost?

When weird ideas come along, we owe it to ourselves to pause and think before accepting them as real or true. Bad things can happen when people embrace beliefs for reasons no better than trust in authority or tradition, or because it "feels true." Countless people have died throughout history because they were not skeptical enough. Countless people who probably meant well have supported or participated in the exploitation, abuse, and even killing of fellow humans because they were not skeptical enough. Wherever and whenever skepticism is lacking, serious problems are sure to follow. Medical quacks and con artists cause great harm to people who don't know the difference between science and pseudoscience. How many times throughout history have unproven supernatural beliefs stood in the way of social and scientific progress? Where might we be today if we had rejected superstition five centuries ago? But the shortage of skepticism in the world today is not only a burden to advancement, it threatens to drag us back to the Dark Ages. Wait, did we ever really leave the Dark Ages? Even now, in the twenty-first century, witches are tortured and executed in some societies because people fear their magical powers. Many people still look to the stars and planets for insights into their personality and romantic prospects—even though the scientists who know more about the stars and planets than anyone say astrology is a preposterous concept. Millions believe that psychics read minds and the government is hiding extraterrestrial bodies at Area 51. As a species we are crippled by irrational beliefs. If we hope to ever shake off the costly and time-wasting habit of believing things that are almost certainly not true, then we have to embrace the scientific method and skepticism. Critical thinking skills must be appreciated and promoted widely. Progress depends on it.

Paranormal and supernatural beliefs—loosely defined as things that exist or occur outside the natural world—are not necessarily tied to intelligence or education. There may be some correlation between education level and the acceptance of a baseless claim such as tarot card reading or astrology, for example. But I warn against reading too much into that because we are all vulnerable. It is well established that intelligent and educated people can and do believe extraordinary claims that lack good evidence. Renowned scientist Jane Goodall is a Bigfoot believer, for example.[1] I once worked with a university-educated journalist who was convinced that a girl in Russia had X-ray vision that enabled her to see inside people and diagnose internal medical problems.[2] My colleague was reeled in, hook, line, and sinker, by an interesting but unproven claim. She isn't dumb, just short on skepticism and critical thinking skills. And she is hardly alone. When it comes to weird beliefs, accepting them seems to be more natural, or more human, than rejecting them.

According to a Gallup poll, three in every four Americans profess to hold at least one of the popular beliefs such as ghosts, astrology, and reincarnation.[3] This is important: Most people in the United States—and throughout the world, no doubt—are supernatural/paranormal believers. In America, ESP (extrasensory perception) leads with 41 percent, followed closely by haunted houses (37 percent) and ghosts (32 percent). Clairvoyance or the ability of psychics to read minds and know the future is real, according to 26 percent of Americans. Astrology's claims have convinced 25 percent, and 20 percent believe in reincarnation. More than half (57 percent) of all American adults have at least two paranormal beliefs, and 22 percent say they believe five or more.

In Great Britain, 40 percent of British people believe that houses can be haunted and 24 percent believe astrology works. In Canada, 28 percent believe in haunted houses and 24 percent believe that it's possible to communicate with dead people.[4] I didn't conduct a scientific survey, but my travels outside the United States leave me with no doubt that belief in claims that are unproven and unlikely to be true are immensely popular. Virtually everywhere I have visited—Africa, the Middle East, Asia, the Pacific Islands, the Caribbean—I came to the conclusion that an overwhelming majority of people believe in an assortment of paranormal and pseudoscientific claims. Without even including religious beliefs, I estimate that more than 90 percent of the world's people hold at least one paranormal belief. We are a believing species.

WHO CARES?

The easy reaction would be to just try to ignore all this irrational belief. After all, don't things like astrology, faith healing, and psychic readings make people feel good and give them a bit of reassurance in an often-confusing and scary world? Who am I to try to rob anyone of a source of comfort or amusement? It's none of my business what people decide to believe, right? What's the harm, anyway?

In my opinion, there is no choice but to speak out against irrational belief, if one has any concern and compassion for fellow humans. It doesn't require being mean or obnoxious about it, but silence is not an option. Belief in paranormal and pseudoscience claims is a chronic crisis that burdens us century after century. Those who do understand the damage caused by these beliefs every day around the world would be heartless monsters if they chose to do and say nothing. This is a matter of compassion for fellow humans and a belief that our world could be better if it were not so blinded and hobbled by superstition and unscientific thinking. I am not being mean and heartless when I explain to someone why alternative medicine is dangerous or how faith healers fool people. Keeping quiet would be mean and heartless. The proper question is not why skeptics protest, but rather how anyone can learn of "child witches" being murdered in Africa and feel no moral obligation to promote skepticism. Who can hear the story about an ill baby suffering and dying because stubborn parents treated her with homeopathic water instead of science-based medicine and not feel disgust toward all pseudoscience and medical quackery? We all share this world together and when an elected leader thinks Earth is six thousand years old or your neighbor believes that the position of a few stars determines what sort of day she will have, the stage is set for trouble. Dim thinking is dangerous thinking.

THE SHIELD OF SKEPTICISM

So how exactly does one wade through all the weird claims out there and make it to dry land safely? It's not as difficult as you might imagine. As readers will discover throughout this book, it often takes only one or two pointed questions to identify fatal weaknesses in claims that are unworthy of our belief. Constructive skepticism is compatible with open-minded curiosity, but it demands consistent vigilance and the courage to question anything and anyone.

It is important to always remember that the burden of proof is on those who make the claim. I would love for Bigfoot and the Loch Ness monster to be real, but I'm pretty sure there is no such thing as a ten-foot-tall bipedal primate running wild in the Pacific Northwest or an extinction-dodging plesiosaur in Scotland. I think this because, after all these years, no one has ever presented any convincing evidence, such as bones, DNA samples, or a body. If Bigfoot believers want me to believe, they need to show me proof. It's not my job to disprove the

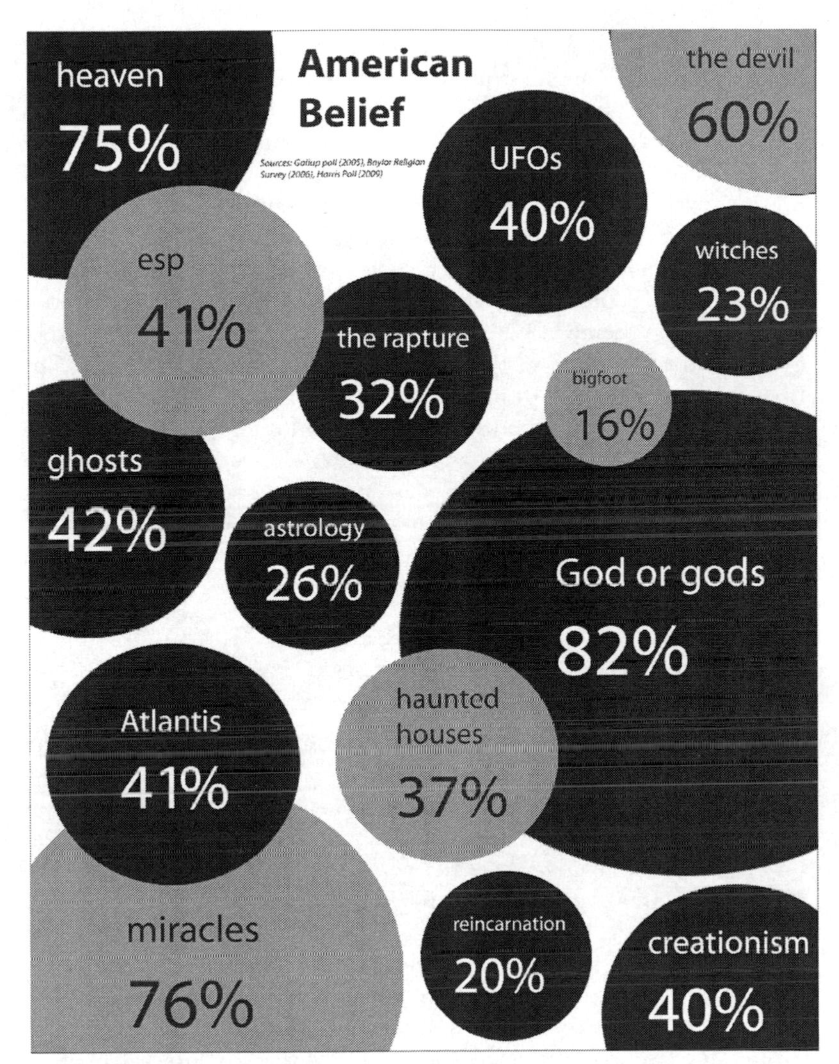

Figure 1. Illustration by the author.

existence of Bigfoot. How could I do such a thing, anyway? I can't look in every cave and behind every tree in North America.

Be on guard against stealth beliefs. These are partial truths that swell to include paranormal elements once you let them inside your head. For example, undoubtedly there are many cases of ancient coastal or island communities being devastated by earthquakes and tsunamis over the last several thousand years. But this is far different from the claim made by Atlantis believers who say a lost city or continent once ruled the world and was technologically advanced beyond even our time. Some UFO believers argue that intelligent life probably exists somewhere in the universe (a reasonable possibility) but then they seamlessly shift to the claim that extraterrestrials are visiting Earth regularly (unknown, unproven, and unlikely). We also have to be on the lookout for claims that are dressed up in science but are in fact pseudoscience. Just because someone—new age guru Deepak Chopra is a good example—frequently mentions "quantum mechanics" or other fancy science phrases does not mean that what they are promoting is necessarily valid or even scientific.

THE BIGGER THE CLAIM, THE BIGGER THE BURDEN OF PROOF

The smart skeptic adjusts the demand for evidence according to the scale of the claim being made. The nature of the claim being made—how outrageous or weird is it?—determines the degree of skepticism required. If my neighbors claim they saw a bird in their backyard yesterday, I'll probably give them the benefit of the doubt and believe it. No big deal. However, if they claim to have seen something far more unusual, say a thirty-ton dragon wearing leather pants and makeup, then I'm going to need to see high-definition video, footprints, and a DNA sample before I even consider believing it. Again, the quality and quantity of evidence should rise in conjunction with the claim. Although the quote did not originate with him, the late astronomer Carl Sagan popularized this important concept: "Extraordinary claims require extraordinary evidence." Keep those five words in mind whenever you think about ghosts, gods, astrology, psychics, intelligent design, UFOs, and other such beliefs.

Being a skeptic does not mean one is closed-minded or uninterested in everything that is weird and unproven. The history of science is filled with examples of bizarre ideas that turned out to be true. Germs were once a pretty strange idea, and were difficult to believe until Van Leeuwenhoek developed the microscope and helped establish the field of microbiology. Continental drift was difficult to accept until plate tectonics explained how it worked. The idea of rogue waves smashing ships under clear skies far out at sea seemed impossible, but we now know that they are real. What about meteorites? Rocks falling out of the sky? You must be joking—oops, it turns out that it really does rain rocks sometimes. The point is that good skeptics who understand how science works don't accept any wacky claim that comes along without evidence, but neither do they reject every wacky claim with absolute finality. The door is always slightly ajar, and if enough evidence comes forth, the door to acceptance opens.

When thinking about weird beliefs, it is important to be aware of how we perceive and assess the world around us. We know that humans are pattern-seeking creatures. Without even trying, we naturally attempt to "connect the dots" in almost everything we see and hear. This is a great ability if you are trying to catch a camouflaged bird in a tree for your dinner, trying to hear a potential mate's call amid a cacophony of distractions, or trying to spot your enemy hiding in the forest, hoping to ambush you. But pattern seeking also leads us to see things that are not there (see fig. 2), which might waste our time and maybe get us into trouble. Furthermore, our obsession with patterns doesn't stop at vision and hearing. We also have a tendency to automatically make connections and find patterns in our thinking. This is one reason that unlikely conspiracy theories are able to take root and blossom in the minds of so many people.

Former psychic-turned-skeptic Tauriq Moosa agrees that this pattern-recognition software in our heads is a primary reason irrational beliefs are so common. He saw it firsthand when his clients made absurd connections in order to support their prior belief that he was a genuine mind reader. Their minds did much of the work, making his job easier.

"We are by nature incredible at picking out patterns; but this also means we see patterns where there are, in fact, none," said Moosa. "This, to me, is the explanation behind *all* the supernatural or superstitious engagements people have, from UFOs to ghosts, from conspiracy-theories to astrology."[5]

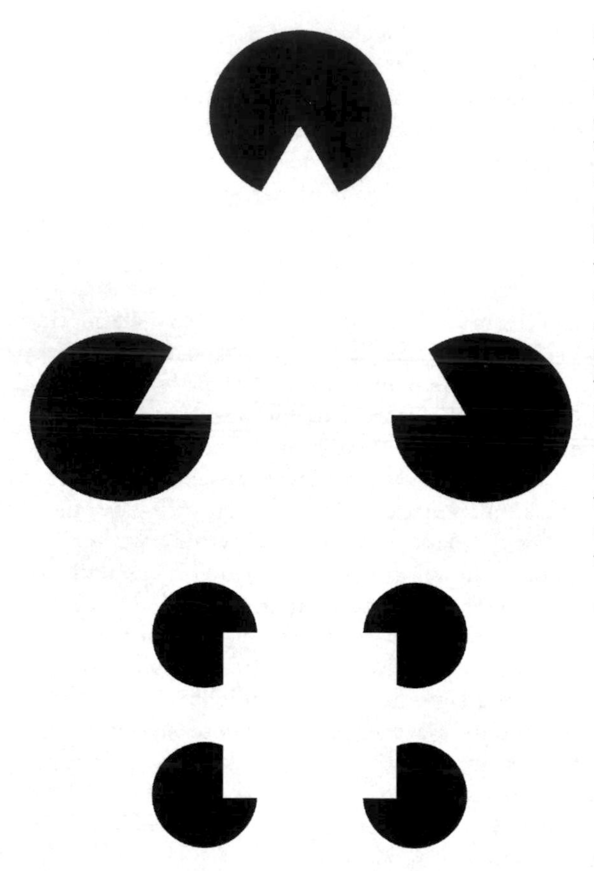

Figure 2. Do you see a triangle and a square? If so, you are seeing objects that are not there. Your mind creates the triangle and square based on its perception of the black objects. Remember this the next time someone claims to have seen an alien spaceship, Bigfoot, a ghost, or anything else extraordinary. It's possible that their mind misperceived visual input and created something that was never really there. Illustration by the author.

CONFIRMATION BIAS

One of the primary reasons that it can be so hard to dump a paranormal belief once it has set up camp inside your skull is that we all have a natural tendency to cheat. We just don't normally think about our beliefs objectively and honestly. Instead we tend to focus on and remember anything that confirms the belief, while missing, ignoring, and forgetting everything that contradicts or casts doubt on the belief. This is called *confirmation bias*, and it can lead the best of us astray, so be on guard.

"The confirmation bias is one of the most insidious and persuasive bits of software in your head," declares psychology professor Hank Davis, author of *Caveman Logic: The Persistence of Primitive Thinking in a Modern World.* "It is as much a part of being human as having two eyes, one nose, and two feet. To avoid evaluating the world through the confirmation bias, you have got to take conscious steps against it. Even then there is no guarantee you will succeed. If you allow your mental software to operate on its Pleistocene default settings, you will bring this bias into play."[6]

You have been warned.

SEEING IS BELIEVING AND BELIEVING IS SEEING

There should be little doubt that many claimed sightings of ghosts, UFOs, angels, and monsters by sincere witnesses are a result of the way our vision works. Contrary to what you may have assumed, we don't really "see" what we look at. What happens when you aim your eyes at something is that your brain "tells" you what you see. And your brain never tells you with 100 percent accuracy. It does this to be efficient and it really does help us function in a world with far too much detail and movement to take in. But sometimes this causes us to "see" things that were never there or at least not there in the form presented to us. It can also cause us to miss things that really are there. Sometimes things that our vision misses might have been the critical pieces of information that would have revealed to us that the UFO or ghost hovering out there is really just a bird or a patch of fog, for example.

FORGET WHAT YOU THINK YOU KNOW ABOUT MEMORY

What is currently known about human memory should send jarring waves of doubt up and down the spines of every paranormal enthusiast. Memory, researchers have discovered, does not work like a video camera. Many people think our eyes are like camera lenses and the brain is a hard drive that dutifully records all the images that enter. Not so. There is no such capturing of complete scenes before us and no such thing as rewind and playback. The reality is that our memories are *constructed* by our brains. Some things are left out. Some things that never happened are added. And sometimes the order of events is switched around. Weird as it seems, your brain produces the memory that it thinks is needed—100 percent accuracy be damned.

This means, of course, that all those stand-alone stories about sightings and encounters with everything from aliens to angels cannot be considered proof of anything because we know without a doubt that any honest, sane, and sober person can easily remember an event incorrectly through no fault of his own. Human memory is unreliable, which means something more than a story about a personal experience is called for when it comes to extraordinary claims. It has also been shown repeatedly by researchers that we all are highly vulnerable to suggestions and prior beliefs when forming our perception of reality and our recollections of what we experienced.

Michael Shermer, founding publisher of *Skeptic* magazine, has spent more than thirty years researching weird beliefs and talking to everyone from alien abductees to Holocaust deniers to psychics. He thinks that much of irrational belief comes down to "belief-dependent realism," the idea that we believe first and then come up with reasons for believing:

> We form our beliefs for a variety of subjective, personal, emotional, and psychological reasons in the context of environments created by family, friends, colleagues, culture, and society at large; after forming our beliefs, we then defend, justify, and rationalize them with a host of intellectual reasons, cogent arguments, and rational explanations. Beliefs come first, explanations for beliefs follow. I call this process belief-dependent realism, where our perceptions about reality are dependent on the beliefs that we hold about it. Reality exists independent of human minds, but our understanding of it depends upon the beliefs we hold at any given time.[7]

Like it or not, the fact is we can't be sure about everything we see, hear, think, feel, and remember. This has obvious implications for popular beliefs that so intrigue and entrance billions of people around the world. More people need to understand that our brains, wonderful as they are, are not very good at consistently sorting out reality from fantasy and illusion. Fortunately, we already have a system that is very good at doing just that.

SCIENCE HELPS US FIND OUR WAY

An appreciation for science goes hand-in-hand with constructive skepticism. It's unfortunate that so many people don't appreciate science's greatness and its importance to us. Science is not just a collection of facts and discoveries or some fringe pursuit of intellectuals. Science is the foundation of our modern world. Our civilization couldn't last one day without the products of the scientific method. However, science can be used by anyone for many different purposes, so no one should ever slip into naïve worship or deference to all scientists. Yes, science may generate cures and vaccines for diseases, feed billions through scientific agriculture, and reveal the working of the universe, but it is also the source of weapons with the potential to destroy humanity. Science may be necessary and wonderful, but it is only as good as the person using its methods.

More than anything, science is a *method* for figuring out and discovering things in our universe. It is also through science that we can best determine whether or not something is real. If something cannot be proven scientifically, it may still be real or true, but this would be a very good reason to have strong doubts until it is. We know science works because it has a record of success far superior to anything else. Even supernatural and paranormal believers rely on science and technology when they could try solutions more in line with their beliefs. People who need to communicate with someone far away don't use ESP, they pick up a phone. People who want to visit a place that is far away don't use astral projection, they board an airplane. Many believers in alternative medicine still turn to medical science when struck by a serious illness or injury. Why do most people who believe in a utopian afterlife fear death and avoid it at all costs?

A final point to keep in mind when thinking about paranormal, supernatural, and pseudoscientific beliefs is that letting go of them is not necessarily a sacrifice. Not only can thinking skeptically be safer and more economical over the course of a lifetime, it doesn't have to be any less fun, either. Whatever I may have lost by not believing in things like astrology and ghosts, I am confident that I more than make up for it by embracing reality with great enthusiasm. All scientific discoveries to date and all the mysteries still to be solved excite me, and I find plenty of reason for optimism and hope, even amid harsh realities. In my opinion, plenty of comfort and joy can be found in friends, family, romance, creative work, adventure, art, acts of kindness, nature, and fun in all its forms. I understand that it may feel comforting or stabilizing to believe that invisible forces influence us, but it can also be comforting and stabilizing to realize that as humans we are smart enough and strong enough to face up to the universe as it really is and get on with our lives.

GO DEEPER . . .

Books

- Barker, Dan. *Maybe Yes, Maybe No: A Guide for Young Skeptics*. Amherst, NY: Prometheus Books, 1990.
- Buonomano, Dean. *Brain Bugs: How the Brain's Flaws Shape Our Lives*. New York: W.W. Norton, 2011.
- Carroll, Robert Todd, ed. *The Skeptic's Dictionary*. Hoboken, NJ: John Wiley and Sons, 2003.
- Davis, Hank. *Caveman Logic: The Persistence of Primitive Thinking in a Modern World*. Amherst, NY: Prometheus Books, 2009.
- Dunning, Brian. *Skeptoid: Critical Analysis of Pop Phenomena*. Seattle, WA: CreateSpace, 2008.
- Dunning, Brian. *Skeptoid 2: More Critical Analysis of Pop Phenomena*. Seattle, WA: CreateSpace, 2008.
- Dunning, Brian. *Skeptoid 3: Pirates, Pyramids, and Papyrus*. Seattle, WA: CreateSpace, 2011.
- Frazier, Kendrick. *Science under Siege: Defending Science, Exposing Pseudoscience*. Amherst, NY: Prometheus Books, 2009.
- Hines, Terrence. *Pseudoscience and the Paranormal*. Amherst, NY: Prometheus Books, 2003.
- Kurtz, Paul, ed. *Skeptical Odysseys*. Amherst, NY: Prometheus Books, 2001.
- Nickell, Joe. *Adventures in Paranormal Investigation*. Lexington: University Press of Kentucky, 2007.
- Park, Robert. *Voodoo Science: The Road from Fraud to Foolishness*. New York: Oxford University Press, 2000.
- Pigliucci, Massimo. *Nonsense on Stilts: How to Tell Science from Bunk*. Chicago: University of Chicago Press, 2010.

- Radford, Benjamin. *Scientific Paranormal Investigation: How to Solve the Unexplained Mysteries*. Corrales, NM: Rhombus, 2010.
- Randi, James. *An Encyclopedia of Claims, Frauds, and Hoaxes of the Occult and Supernatural*. New York: St. Martin's Griffin, 1995.
- Ruchlis, Hy. *How Do You Know It's True? Discovering the Difference between Science and Superstition*. Amherst, NY: Prometheus Books, 1991.
- Sagan, Carl. *The Demon-Haunted World: Science as a Candle in the Dark*. New York: Random House, 1995.
- Schick, Theodore, and Lewis Vaughn. *How to Think about Weird Things*. New York: McGraw-Hill, 2011.
- Shermer, Michael. *The Believing Brain: From Ghosts and Gods to Politics and Conspiracies—How We Construct Beliefs and Reinforce Them as Truths*. New York: Times Books, 2011.
- Shermer, Michael. *The Borderlands of Science: Where Sense Meets Nonsense*. New York: Oxford University Press, 2002.
- Shermer, Michael. *Why People Believe Weird Things*. New York: MJF Books, 1997.
- Smith, Jonathan C. *Pseudoscience and Extraordinary Claims of the Paranormal: A Critical Thinker's Toolkit*. West Sussex, UK: Wiley-Blackwell, 2010.

Other Sources

- *Skeptic* (magazine).
- *Skeptical Inquirer* (magazine).
- Skeptic (website), www.skeptic.com.
- Skeptical Inquirer (website), www.csicop.org.
- James Randi Educational Foundation, www.randi.org.
- Point of Inquiry (podcast), www.pointofinquiry.org.
- Skepticality (podcast), www.skepticality.com.
- The Skeptic's Guide to the Universe (podcast), www.theskepticsguide.org.
- Skeptoid (podcast), www.skeptoid.com.
- *Here Be Dragons: An Introduction to Critical Thinking* (DVD), CreateSpace, 2008.

Chapter 2

"I KNOW THERE IS AN AFTERLIFE BECAUSE OF ALL THE NEAR-DEATH EXPERIENCES."

The boundaries which divide life from death are at best shadowy and vague.
Who shall say where one ends, and the other begins?
—Edgar Allan Poe

Frank Ward was dying. He had a leg infection that turned extremely nasty and spread to his bloodstream. Septic shock caused his blood pressure and heart rate to plummet. "I was basically in a coma," Frank explained. "My heart was beating just four to six beats per minute. My brain wasn't getting enough oxygen to do anything or to perceive anything. I was dying—or already dead."

Frank says that at some point during the process of dying, he left his body. Frank bases this on what he remembers seeing in the hospital room.

"I saw the doctor praying. He was bent over at the edge of the bed with his hands together. He wasn't on his knees, but he was bent over. I saw him from another place in the room. It was like I wasn't looking at him from the bed where [my body] was. I wasn't in my body. That's how it appeared to me, anyway."

Frank recovered from what he calls a near-death experience and later discussed it with his doctor.

"When I came to later, I asked him if he had been praying for me in the room," Frank said. "He turned about thirteen different colors. He couldn't believe that I knew. I saw things in the room, details that I couldn't possibly know. That tells me it wasn't just a hallucination."

Unlike many near-death experiences, Frank's did not include a tunnel of light, seeing loved ones, or catching a glimpse of heaven or hell. "I didn't see fire and brimstone—that's a good sign. But I didn't see pearly gates or angels either. Let's face it, when we die we want something to be on the other side. I just can't subscribe to the theory that there's nothing. As human beings we don't want to end up just being a bunch of bones in a box for maggots to eat. I saw the doctor praying and he says he was. But did I have some otherworldly experience beyond that? No. I wish I had, but I didn't. The way I see it, neither science nor religion can answer everything. I don't know what to say; I just know that something happened to me in that hospital."[1]

Frank is hardly alone. He is a bright, thoughtful, and honest person who experienced a very strange event at a time when his life was in jeopardy. Variations of Frank's story have been reported by many people around the world. As many as eighteen million Americans may have had a near-death experience.[2] Even more common are so-called out-of-body experiences that are very similar to near-death experiences. Something as simple and non-life-threatening as fainting can cause them. There is no doubt that people really do have these strange experiences. It's just not reasonable to suspect that millions of people are lying. The key questions, of course, are what causes the events and do they really prove that there is a soul and afterlife as many claim.

I learned firsthand in childhood that sometimes weird things happen when your body is in crisis. Pain and panic make perfect sense to an injured kid, but peace and tranquility? That I didn't expect. One day during my childhood, I was out emulating Huckleberry Finn or whatever and slipped off of a large, elevated sewer pipe that spanned a canal. My head slammed into the metal pipe, and a protruding bolt left a deep gash on my shin. Fortunately I landed on the bank of the canal and not in the water. It was a semiserious accident by any standard, but from a child's perspective, it qualified as downright catastrophic.

A funny thing happened on the way to my suffering and doom, however. I felt no pain. There was no screaming and no tears. I just lay there in the dirt, feeling nothing but extraordinary peace and calm. It was more than the absence of pain. I felt *incredibly good*. It seemed like I was detached from everything. I can't say that I felt like I left my body or hovered high above the scene, but I did experience a strange detached feeling. It wasn't quite flying, but it felt like I was floating and had lost contact with the ground. Weird, to say the least. Eventually my brain rebooted and I got back on my feet. Then the pain came, and I faced a long limp home.

Today I don't think of that childhood accident as a near-death experience or even an out-of-body experience. It just doesn't seem quite dramatic enough, in my view. But it did feel great. If I could push a button and get that feeling anytime, I'd never leave my house. What is significant about the event is that it gave me a tiny taste of what people likely feel when they go through out-of-body or near-death experiences. I should have felt terrible and been screaming in agony, but instead I felt wonderful and at peace with the universe. As a result of that event, I find it easy to understand when people say they were profoundly happy or calm at the brink of death or even that they think they floated above their own bodies. It is also not difficult for me to accept claims of powerful visions, seeing long-lost relatives or friends, and even visiting heaven. After all, the mind can take us anywhere. I once dreamed that I was somewhere far away in outer space. It was all very realistic at the time, but I'm pretty sure I never left Earth. Scientists know well that the human brain is better than a Hollywood studio when it comes to producing images and stories. CGI (computer-generated imagery) is all the rage in moviemaking now, but it's nothing compared to *mind-generated images*. The dreaming mind; the stressed-out, tired mind; and the dying mind can conjure up virtually anything and make it convincing. This is not proof of life after death, the existence of heaven, or souls, not when the same experiences happen to people who black out in high-speed jets or during g-force testing in centrifuges. Maybe those things are real, but near-death experiences and out-of-body experiences have alternative explanations that are far more down-to-earth and more likely to be true.

The strange sensation of leaving one's body is not, it turns out, all that strange. Nothing supernatural is required to explain it. A confused brain can even do the trick. It's not the same as a dying brain, but it does show how this sort of thing can happen by natural means. The feeling of separation from one's body has been induced by simple experiments with mirrors. Brain researchers have achieved out-of-body experiences in 75 percent of subjects in one experiment that involves a person sitting at a desk with a large mirror in front of them and only their upper body visible. A second person then walks toward the seated person. That second person would in most cases experience a brief out-of-body sensation.[3] By replicating another similar experiment with myself as the test subject, I induced what could be called an out-of-hand experience. With my left hand hidden behind a mirror, I could only see my right hand and its reflection, I was able to confuse my mind (even though I knew what was going on) to the point where my left hand felt dead and vaguely detached from my body. Weird as it was, however, I saw no reason to believe that the left hand of my soul had vacated my body.

Those who are convinced that near-death experiences prove supernatural claims often point to the fact that they are reported by people in different cultures. This, they say, means it can't just be some specific belief-centered or culturally derived quirk. If it happens to all kinds of people in all sorts of places, then these events must be real as described, right? Well, no. More likely these events are of a similar nature because people in Japan, Nairobi, Damascus, and New York City, and everywhere else, all have one important thing in common: they have human brains that function in the same way. Tunnel vision, for example, is common in near-death experiences. Sure, it might be the road to heaven, as many say, but a simpler explanation is that vision is restricted during oxygen deprivation and tunnel vision results. "When not enough blood is pumped to the head," explains neurologist Kevin Nelson, "the eyes fail first, causing tunnel vision before the brain fails and unconsciousness occurs."[4] It's as simple as that.

As for the sensation of leaving one's body and floating around above it, it may seem like an exceptional and rare event, but it is surprisingly common. One study of more than thirteen thousand Europeans found that nearly six percent of them said they have had an out-of-body experience. Many millions of Americans have as well.[5] While many people believe this is evidence or even proof of something supernatural, such as "astral projection," the existence of souls, or heaven, there is a much more simple explanation available once again.

Scientists have induced out-of-body experiences in people by stimulating a specific region of the brain with electricity, casting doubt on the need to invoke supernatural explanations. "It takes just a small trickle of electrical current to induce such an experience," explains neurologist Nelson. "Other factors, such as a temporary lack of blood or oxygen to the brain, may also interfere with sensory integration in the temporoparietal region and cause out-of-body experiences."[6] And, of course, many people have induced out-of-body experiences by taking drugs. Drugs that do this are not "magical soul-extraction potions"; they are merely hallucinogens that cause hallucinations. The point is that if we can induce these experiences by tinkering with brain biology, then isn't it likely that all such experiences can be attributed to brain biology rather than supernatural causes?

UK psychologist Susan Blackmore had a dramatic out-of-body event that motivated her to research near-death experiences. Although she had hoped and expected to confirm that something paranormal or supernatural was going on, Blackmore eventually concluded that evidence for the dying brain, or at least the oxygen-starved brain, as the cause of these events is "overwhelming."[7]

There is also the challenge of addressing very basic questions about how a soul, whatever that is, maintains its structure and moves around. How does it retain thoughts, memories, and personality? What is it made of? California Institute of Technology physicist Sean M. Carroll says we cannot just conveniently forget hard-won knowledge about how reality works when we think about souls and life after death:

> Claims that some form of consciousness persists after our bodies die and decay into their constituent atoms face one huge, insuperable obstacle: the laws of physics underlying everyday life are completely understood, and there's no way within those laws to allow for the information stored in our brains to persist after we die. If you claim that some form of soul persists beyond death, what particles is that soul made of? What forces are holding it together? How does it interact with ordinary matter?
>
> Everything we know about quantum field theory says that there aren't any sensible answers to these questions. Of course, everything we know about quantum field theory could be wrong. Also, the Moon could be made of green cheese.
>
> Among advocates for life after death, nobody even tries to sit down and do the hard work of explaining how the basic physics of atoms and electrons would have to be altered in order for this to be true. If we tried, the fundamental absurdity of the task would quickly become evident.[8]

The bottom line here is that we know beyond any doubt that natural events occurring inside one's skull can cause all the sensations repeatedly associated with out-of-body and near-death experiences. Human minds often see and hear things that are not real. They also frequently misremember events. And this is under the best of conditions. During extremely stressful situations—such as dying—our minds are even more likely to distort reality. Knowing this, how can we justify taking the gigantic leap necessary to involve souls, heaven, and other supernatural causes for these events?

Neurologist Nelson adds:

> People like to say that these experiences are proof that consciousness can exist outside the brain, like a soul that lives after death. I hope that is true, but it is a matter of faith; there is no evidence for that. People who claim otherwise are using false science to engender false hope and I think that is misleading and ultimately cruel.[9]

Frank Ward, the man who had the near-death experience described at the beginning of the chapter, contacted me a few weeks after our discussion. He told me that sharing his story prompted him to think more deeply about it and do some research. As a result, he said he has changed his mind about what likely occurred that day. "Now I'm convinced it can be explained by the natural dying process," he said. "My brain was starved of oxygen, and it made me hallucinate. It was an amazing experience, but after doing some reading about it, I think the scientific explanation makes the most sense."

GO DEEPER . . .

- Blackmore, Susan. *Dying to Live: Near-Death Experiences.* Amherst, NY: Prometheus Books, 1993.
- Woerlee, G. M. *Mortal Minds: The Biology of Near-Death Experiences.* Amherst, NY: Prometheus Books, 2005.
- Zimmer, Carl. *Soul Made Flesh: The Discovery of the Brain—and How It Changed the World.* New York: Free Press, 2005.

Chapter 3

"A PSYCHIC READ MY MIND."

> **I know, as much as I know anything, that there are no psychic powers or spiritual realms. Anyone who performs psychic or divination tricks is lying, either to themselves or to their clients.**
>
> —Tauriq Moosa, former psychic

"Look into my eyes," I say. "Pull back the curtain that hides your thoughts. Let me in. . . . Yes, I'm starting to feel your thoughts. You are doing great. Just relax."

She's definitely unsure about this, but she hasn't laughed or protested yet, so I'll keep going.

"You have a strong mind. I'm beginning to sense some of the things that are troubling you. I can help."

Is she buying it? I force a smile, worried that she knows I'm faking. I figure it's a long shot for me to be able to convince someone that I am a psychic, but I'm going to give it a try. No, I'm not charging her money, pretending to speak to her dead relatives, or anything quite so terrible as that. I simply want to see if I can pull off something called a "cold reading." Before continuing with the story of my first attempt at mind reading, however, let's explore the concept of a cold reading.

Experienced skeptics have long known that cold readings are a very effective way to deceive people. Once you understand how the process works, it's easy to see how professional psychics are able to fool so many people. It's much like playing the old game called "twenty questions," only with a thick layer of supernatural silliness applied to it. A cold reading is performed with no prior specific knowledge about the individual being "read." The psychic thinks of some information about people in general that probably hits the mark and then delivers it as if it is a unique and personal insight for that person alone. Knowing something about common concerns and behaviors according to gender and age is useful. Being aware of health and lifestyle statistics can help, too. A critical part of cold readings is gleaning information from simple observations of the subjects' reactions to questions and guesses during the reading. The cold reader starts with general comments and then watches for a positive response to something that may indicate a productive path to pursue.

Many psychics use a shotgun tactic. They blast a flurry of comments and questions rapid-fire at the subject, and if one of them hits the mark, they claim success. All the wrong guesses are ignored as if they don't count. For example, a psychic may say, "I see sadness; you have lost someone close. A mother, father, grandparent, uncle . . ." If they get a nod or verbal agreement for one of the choices, the psychic then pretends to have known that specific fact all along. While this may seem ridiculous, it does work.

One reason cold readings are so effective is that people tend to remember the hits and forget the misses. A correct guess or prediction grabs our attention and lights up our brains with excitement. But bad guesses and failed predictions are boring and forgettable. This is that problem of confirmation bias rearing its troublesome head again. Those who already believe in psychics and mediums are predisposed to having their belief strengthened no matter what happens when they sit down with one. We all have a natural tendency to look for and remember what we expect or hope to find, while ignoring or rejecting that which contradicts our expectations and beliefs. If someone goes into a reading convinced that psychics have real powers, then they are far more likely to forgive and forget numerous errors and remember the rare correct comments. This is how psychics can be wrong most of the time yet still convince so many people that they can read minds. Primed believers also are more likely to connect vague generalizations ("You will be lucky today.") with specific events ("I found $10 this afternoon!") and then conclude that it was an accurate prediction. It may be difficult for some to accept that this is all it takes to convince many people, but it does. I know because I pulled it off.

At the start of my first attempt at fake mind reading, I picked up two key things about the lady I was giving the psychic reading to. She was in her early thirties, from a Spanish-speaking society, and not married. I remembered from personal experience that some Spanish-speaking girls I knew in high school and college were very family oriented. They seemed far more concerned about their parents' wishes and much more connected with extended families than other girls I knew. The idea that marriage was a big deal in Spanish culture popped into my head as well. I had a spontaneous hunch that she might be feeling some pressure from her family for not being married yet. Of course, this was all based on generalizations about Spanish-speaking people, a diverse population numbering in the hundreds of millions. When doing a cold reading, however, a psychic can afford to take shots with crude stereotypes and hunches because the misses don't matter anywhere near as much as the hits. It's like throwing darts in the dark. The more darts you throw, the better your chances of hitting the dartboard—and no one sees all the darts stuck in the wall that missed the target, because it's dark.

"I feel your deep concern about something very personal and important," I said in the most serious tone I could manage. Again, all I had done was put two and two together in my head: she was thirty-something, single, and probably in a family that placed a high priority on marriage and family.

"You worry about the fact that you have not yet found the right man to marry. It frustrates you because you know that you are an attractive woman and a good person. There is no good reason that you should not be married yet, but still you are not. It doesn't help that your family keeps asking you about this and pressuring you. Sometimes you wonder if you will ever get married."

I half-expected her to laugh and tell me to shut up at that point.

But a remarkable thing happened. Her eyes widened. She leaned forward. I sensed that she was hooked. *Wow!* I thought, *This really works! She thinks I'm reading her mind!*

My confidence soared. I instinctively knew to follow this path at full speed. Without missing a beat, I grimaced and rubbed my forehead. "All you want is a good man who treats you like you deserve to be treated. He should be a good father, too. You feel frustrated because you don't think this is too much to ask. And you are fed up with all the guys you meet who turn out to be losers or only interested in one thing."

Based on her facial expressions, it was obvious that she was eating up my trite comments. But then I got overconfident. I changed course and squandered my momentum. Rookie mistake.

"I also sense that you worry about money all the time. You have a good job, you work hard and save, but you are still concerned about bills and having enough. You want to be rich."

She did nod in agreement but it didn't seem sincere. I stayed in character but could tell that I had lost her. Apparently money didn't matter very much to her and I had blown it. I tried my best to recover and finish strong. "Listen, I want you to know that things will work out for you one way or the other," I added. "You will find a nice man and get married or you won't. But either way, you will be happy. You are a good person and your life is going to be great. You will be a success in life."

Cold reading fundamental: keep it positive and pile on the compliments. People are more willing to believe good things about themselves than anything else. This is why mediums always say that the grandfather loves everyone from the great beyond and has nothing but hugs and kisses for all. Psychics never seem to report back that dear old grandpa says he hates his family and couldn't be happier to finally be free of his annoying grandchildren.

I felt I had bungled my cold reading at the end. I cringed, expecting ridicule or anger for daring to waste her time with such an absurd charade. She couldn't possibly believe I was a psychic after that performance, could she? Fortunately, the encounter ended with no harsh words. She chuckled, rolled her eyes, and left. Lesson learned, I figured, cold readings are for professionals, and I'm clearly an amateur. I felt stupid for even trying. But that was not the end of the story.

I felt embarrassed when I saw the woman a few days later. I could only assume that she must have thought I was some sort of weirdo for pretending to be a psychic and talking about her personal life. But then she hit me with one of the biggest surprises of my life. She drew close and whispered: "How did you do that the other day? It's amazing. How long have you been a psychic? You knew so much about me."

I was stunned. Based on a cold reading that I fumbled through in five minutes, the woman was awestruck and completely believed that I possessed psychic powers. Simply by throwing out a few guesses I convinced her that I could read minds. Undoubtedly I could have sat down with her right then and there, conducted another reading, and this time billed her $300 for it. But I wouldn't do that, of course. I'm a decent human being with a conscience. In fact, I was already feeling a few stabs of guilt for having tricked her at all.

I immediately confessed that I was just pretending and have no psychic powers. I explained that I hadn't realized that she believed it when we parted ways that day. If I had known, I would have told her then. Even though I justified the encounter to myself by thinking of it as a harmless experiment, I admit feeling a bit creepy for doing it. After that experience, I find it deeply disturbing that professional psychics around the world make a living doing that to trusting people all the time.

Tauriq Moosa, a young South African intellectual, has lived on both sides of the fence. Years ago he was lured into the world of psychics. Convinced he had special powers, Moosa found it easy to convince others—and get paid. Today, however, he is a graduate school student who writes about skepticism and science topics frequently. He offers his story as a case study in how good minds can stumble.

Uncomfortable with the religion of his parents, Moosa says he sought spirituality elsewhere and found it in "Norse mythology and divination." Soon he was convinced that he was gifted at reading minds, seeing things others could not, and giving life advice. Friends were eager to believe he was gifted.

"I began practicing on the very friends who proclaimed my talents," he said. "Of course, I got amazing results. I began practicing on friends of friends. It was my thing at parties to talk to girls. I then took it to the streets, with a cardboard sign. I offered free readings at first, but then I obtained regular customers who wanted private sessions. I started charging."

Moosa eventually recognized that things didn't add up. One of his clients was in awe of his powers because he hit on an obvious guess about her personal life. She told others that he was amazing, but Moosa had doubts.

"I knew then that something wasn't quite right. I began recalling all the other spot-on readings I had made. To my disappointment, they were all exactly the same—guesswork. I carried on for about a week, but eventually I confessed to one of my clients that all I really was to her was a cheap, unqualified therapist.

"I retained some aspect of the spiritual for some time. I thought perhaps my medium had become tainted. I tried reading palms and auras and other nonsense. Then I enrolled at the University of Cape Town, studied psychology, and had all pretentions of unfalsifiability and superstition slowly beaten out of me."

Looking back, Moosa says it was extraordinarily easy for him to convince people that he had supernatural powers, even though he didn't.

"I think it's easier to convince people that you have magical powers than it is to convince them otherwise. Most people, it seems, think that the universe has some underlying or external realm which we're connected to. The idea that the universe is as it is, as science explains it, is difficult, since it means that we are part of those mechanisms. It means we are not special. Working on that hunger to be special, on having forces external to us yet working through us, and so on, only taps in to what many people *want* to believe. Couple this powerful desire with the human mind's ability to misfire its pattern recognition—seeing faces in mountains, men in shadows, and so on—and you have the template for any quack, huckster, and self-appointed guru. If you know even a little bit about human psychology, you too can become a successful psychic."[1]

The enduring popularity of psychics is a strange phenomenon, especially considering the fact that they are wrong so much more often than they are right and that they never produce or contribute anything of value to society. Imagine if psychics really could read minds and see into the future. Wouldn't it be great if they could predict earthquakes, fires, tornadoes, and tsunamis so nobody had to die from those things? Wouldn't it be wonderful if they could prevent and solve crimes, give advance warnings about terrorist attacks, predict winning lottery numbers for charities, and let us all know with 100 percent accuracy if it will rain this weekend? But, of course, they can't do any of that. Based on the absence of compelling evidence—and very good counter explanations for their performances—all psychics are most likely either dishonest or self-deluded cold readers.

Is it possible that some people really do have psychic powers? I suppose it could be true. But until someone can demonstrate it scientifically under controlled conditions, it makes no sense to believe it. Unicorns might be living on the far side of the Moon, too. But until we catch one or some fossils turn up, it would be a pretty silly belief, right? If psychics can do what they say they can, then show us. But that's not all. Besides proving that people can read minds or hold conversations with dead people, believers also need to come up with a theory that explains *how* it works.

So why don't professional psychics seem more concerned with proving to the rest of us that they are anything but fakes? They earn a living by reading minds and communicating with the dead. The general lack of enthusiasm about letting science confirm or reject their claims just doesn't add up. Imagine how many more customers they would attract and how much more money they would earn if they had scientific approval. Tellingly, however, most of them seem uninterested in any serious analysis or testing of their work. Professional psychic/medium James van Praagh actually warns people against thinking and analyzing "too much" because it can hinder psychic ability.[2] This is the typical warning sign often associated with irrational beliefs that everyone should be aware of. Watch out whenever someone discourages questions, downplays evidence, and tells you that it's better to just trust them than to think for yourself. Never let your guard down around people who stand to profit from an extraordinary claim but have little or no interest in credible testing to see if the claim is even valid.

Millions of people continue to believe in psychic phenomenon despite the fact that none of the usual psychic claims have been proven to the satisfaction of the world's psychologists and neuroscientists (the people who would be the first to know if there were anything to this). A 2009 CBS News poll found that a majority of Americans (57 percent) believe in "psychic phenomena."[3] In Britain, more than half of respondents say they believe in "psychic powers," with 43 percent claiming to have read someone else's mind or had their own mind read by a psychic.[4] The fact that millions of people around the world believe this stuff when it's very unlikely to be true might not seem like that big of a deal until one considers the money and time wasted. Billions of dollars and countless hours are wasted, decade after decade, on psychic readings, supposed conversations with dead people, books, and television programs. It's not just a belief, it's big business. Visit a website of one of the more prominent psychics and see for yourself. If having too much money is a problem for you, then just sign up for a few private readings with one of them and he or she will happily relieve you of a couple thousand dollars. Or, if you prefer, you can make money vanish into thin air by buying a course that teaches you things like how to "receive messages," meet your "healing guides," and "understand why you are here." Maybe not all, but many of the world's psychic customers, people who don't just believe but also spend hard-earned money on this stuff, must be able to find better things to devote their energy and resources to. What about family, friends, recreation, exercise, education, or charity work? In my opinion, purchasing time with a self-proclaimed psychic/prophet who accepts Visa®, and MasterCard®, so he or she can tell you things you already know about yourself and offer vague predictions and recommendations that could apply to millions of other people is not a sound investment. There is one alternative to giving money to a psychic that makes a lot of sense to me: one could always just subscribe to *Skeptic* and *Skeptical Inquirer* and develop an immunity to such silliness.

IT'S JUST NOT THAT HARD TO FOOL PEOPLE

The next time you see a psychic in person or on television, just remember that the author of this book was a psychic too. With no practice or intense preparation, I aced my debut. It took me about five minutes to convince an intelligent adult that I could read minds. My half-hearted and clumsy effort was a spectacular success. Imagine if I spent years perfecting my cold-reading skills. How good could I be if I devoted time to memorizing key trends and statistics linked to age and gender in order to make better guesses? What if I was intensely motivated to master this technique because I wanted a lot of money or saw it as a route to fame? Could I fool you? I hope not.

GO DEEPER . . .

- Fine, Cordelia. *A Mind of Its Own: How Your Brain Distorts and Deceives*. New York: W. W. Norton, 2006.
- Kida, Thomas. *Don't Believe Everything You Think*. Amherst, NY: Prometheus Books, 2006.
- Randi, James. *The Truth about Uri Geller*. Amherst, NY: Prometheus Books, 1982.
- Rowland, Ian. *The Full Facts Book of Cold Reading: A Comprehensive Guide to the Most Persuasive Psychological Manipulation Technique in the World*. London: Ian Rowland Limited, 2008.
- Shermer, Michael. "Learn to Be Psychic in Ten Easy Lessons!" www.skeptic.com/downloads/10_Easy_Psychic_Lessons.pdf.
- Stenger, Victor J. *Physics and Psychics: The Search for a World beyond the Senses*. Amherst, NY: Prometheus Books, 1990.
- Wiseman, Richard. *Deception & Self-Deception: Investigating Psychics*. Amherst, NY: Prometheus Books, 1997.
- Wiseman, Richard, and Robert L. Morris. *Guidelines for Testing Psychic Claimants*. Amherst, NY: Prometheus Books, 1995.

Chapter 4

"YOU'RE EITHER BORN SMART OR YOU'RE NOT."

The worst kind of blame, and the most common, is on one's own biology. This is the great final irony of genetic determinism: the very belief of possessing inferior genes is perhaps our greatest obstacle to success.

—David Shenk, *The Genius in All of Us*

Intelligence is definitely not what most of us had imagined.

—*New Scientist* editorial

While watching the last Super Bowl with my son, we got to talking about what it takes to be an NFL quarterback. I explained that it's one of the most demanding positions in any team sport. It takes a very special skill set to be able to read defenses, react appropriately to rapidly changing conditions, select a receiver, and deliver accurate passes—all while dodging very large, quick, and violent defensive players. My son asked me if I had ever played quarterback. I explained that I never even played football in school. I jokingly added that I could have ended up playing in the NFL—*if* I had been born with perfect vision, lightning-fast reflexes, and an elite throwing arm. Yeah, apart from those minor factors, I could have made it, no problem. The underlying message was that I couldn't have been an NFL quarterback because I was shortchanged at birth—no fault of mine.

Only later did it dawn on me that I had played the coward's card of genetic limitations in order to dodge the key fact that I had never even tried to play quarterback at any level anywhere. If I had participated in Pop Warner, middle school and high school football, attended off-season camps and clinics, obsessed over success, prayed to a Joe Montana shrine in my bedroom, ate my Wheaties®, *and actually played the position*, then maybe I could make some kind of a judgment about my genetic potential to throw spirals under pressure. If I had tried, I might have clawed my way into the NFL. I doubt it, given the odds, but who knows? The *genetic* potential to be a pro quarterback may be real, and *inherited* limitations may indeed exist, but we are not capable of identifying them in childhood except perhaps where severe disabilities are present. Given countless factors that come into play on the road to NFL quarterback stardom, it is impossible to know who could *not* have made it. We can know with certainty who has the right stuff only *after* they make it. Far more important than the ability to throw a football well is potential intelligence, and this too is unknowable in the same way.

Unfortunately, many people believe strongly in the existence of identifiable limits that fence in human minds. The reality, however, is that while one's current intellectual *ability* can be measured, intellectual *potential* cannot. Your brain is not cast in stone in the womb, in the first few years of life, and not even in adulthood. It can change dramatically, based on what it does and what is done to it. *Neuroplasticity* is a fancy word that refers to the brain's ability to rewire or physically restructure itself throughout life. Much, if not most, of our intellectual destiny is what we make of it and what our environment makes of it. When it comes to intelligence, we are not the passive passengers popular opinion often suggests we are.

A MOST DANGEROUS BELIEF

One of the most destructive myths of all is the one that tells us intelligence is innate and fixed. It has been widely believed for centuries, and still is, that a person is either born smart or not, that education, opportunity, motivation, and hard work can only carry one so far because most intellectual ability is tightly confined by inherited or genetic restraints. Worse, these limits are believed to be identifiable by a test, a few report cards, or maybe by nothing more than a mere glance at one's physical appearance. Alfred Binet, the French psychologist who developed the intelligence quotient (IQ) test in the early 1900s, would surely be one of its most vocal critics today if he were still alive. He created the test specifically for the purpose of identifying children who suffered from severe mental problems or learning disabilities so that they could receive special attention early on. Binet never intended or imagined that his test, and its descendants such as the SAT and ACT, would be seen as valid ways to measure innate intelligence and grounds for make sweeping assumptions about the innate intelligence of large groups of people. Yes, these types of tests can measure current ability in some forms of intellectual activity, and they may also do a pretty good job of predicting future success with schools or jobs. But this is light-years away from measuring someone's overall intelligence, and it's even farther away from determining the inherent potential of a healthy human brain.

Attached to racial identity and gender, the myth of knowable and measurable inherited potential intelligence has inspired outrageous crimes of neglect and abuse against countless numbers of people. It places an unjustified cloud of doubt above us all as individuals as well. The myth also provides a handy excuse when things don't go as well as we would like. For example, how many students attempt to explain poor grades in a particular subject by claiming they "just aren't smart enough"? The implication is that their genes are to blame and it can't possibly have anything to do with nutrition, sleep, prior education, the quality of teaching, personal motivation, or study habits. Based on what is now known about mental development and the impact of environmental factors, it is ridiculous for anyone to presume to know the *inherent* or *genetic* limits of a person's intellect.

If everything transpired perfectly for an individual from the moment of conception forward (though in reality it never can, of course), we might then be able to glimpse the upper limit of that person's mental development. But even then we could not be sure where the real limits for that individual lie because intelligence is more of a process than a specific measurable thing. A genius in math might have been an even more brilliant pianist or sculptor—had he or she pursued playing the piano and sculpting. A mediocre musician might have been a great juggler. A dim person might have been a bright person, given a few different opportunities and life choices here and there. Who can say that Isaac Newton and Albert Einstein would not have been even smarter and more productive if their childhood nutrition, social stimulation, opportunities, and academic encouragement had been different? The same applies to every human who has ever lived. School systems all over the world routinely write off children at young ages, consigning them to less demanding and less stimulating educational tracks designed for those students who "just aren't smart enough for higher learning." How much intellectual potential does our species waste by giving up on the brains of millions of children year after year?

The brains of infants who are deprived of adequate affection, security, stimulation, and nutrition do not develop as well as babies who have those things. Children who hear fewer words per day from their parents, for example, do not develop in the same way as children who hear more words. Some cultures place more emphasis on education than others. Some parents demand more academic work from their children and help them do it. Some parents do neither. Some children go to terrible schools; some go to great schools. Some groups of people have more or less opportunities and more or less confidence than others due to culturally embedded prejudice and discrimination. Even though all these factors should be obvious by now, many people still believe that nothing overrides genetic destiny when it comes to intelligence.

Here's a simple thought experiment to illustrate the absurdity of claiming to know an individual's genetic potential based on current appearance or ability. To visualize this, consider a physical example. Imagine a massive, vein-popping bodybuilder standing next to an extraordinarily skinny man. One of them looks like freshly shaved King Kong and the other looks like the slightest gust of wind might knock him over. At first glance, most people probably would assume that the large, muscular man has the genetic advantage for success in competitive bodybuilding over the skinny man. But we can't assume this to be the case. It could easily be the skinny guy who had been born with the superior genes for bodybuilding. But maybe nothing in his environment activated those genes. Maybe he was deprived of something in childhood—good nutrition perhaps—and those genetic gifts were muted. Maybe he grew up in a place that had no gyms and he never had the opportunity to train with weights. Or, maybe he thought bodybuilding was weird or too difficult so he never even tried.

Now imagine a woman with a doctorate in physics who writes books about cosmology. She is standing next to a woman who dropped out of high school and now cleans bathrooms for a living. Can we look at their résumés and determine which of them was born with the superior genetics for intelligence? We cannot. If we are honest and reasonable, all we can do is make a judgment about their current intellectual *abilities*.

While writing this chapter I decided to seek out the champion of smarts from my high school days. Debbie was the classic teen goddess: beautiful, varsity cheerleading captain, and valedictorian of my graduating class. She would have been so easy to hate, but nobody did because her personality matched everything else about her. I remember Debbie as a great person but also as a bit intimidating. To me she seemed like some kind of academic terminator who never stopped, never ate, never slept; she just kept making As. In fact, Debbie never made a single B during four years of high school. But did her brain have classroom success etched into it from birth? Looking back, she doesn't think so.

"Some As were easy; some I had to work for," Debbie explained. "Math took more effort. Mostly it was dedication to the goal of being valedictorian. It seemed like once I decided I wanted that, I just found a way. I was *not* one of those geniuses who could sleep through a class, though, and still get an A. I made a long-term commitment and didn't feel I could let my guard down ever. It was as much about endurance as ability for me."

Debbie says family support and expectations were critical to her success in school.

"Expectations were high. We were always told that it was expected we would make whatever grade was within our reach, and in our parents' mind that meant As. Close family was key for just basic support. Plus, my brothers were so accomplished, I wanted to keep up."

Keeping up with her brothers could not have been easy. They were both outstanding students. They also excelled outside the classroom. One was a member of a World Series championship team, and the other was a PGA golf pro for ten years. But don't assume that this all adds up to prove that she merely coasted on the good fortune of being born into a genetically gifted family. Debbie was adopted.

GROUP INTELLIGENCE?

Beyond individuals, what about inherited intelligence shared among vast groups of people, such as races? It has long been a common belief that races can be ranked by innate intelligence. Can they? I did a lot of research on racial differences in IQ scores and academic achievement while writing my book *Race and Reality: What Everyone Should*

Know about Our Biological Diversity. Along the way I discovered how glaringly obvious the likely reasons for these much-hyped differences are. While some people focus on average group scores on "intelligence" tests and cite genetic limitations as the cause—something they cannot possibly know—others point to the fact that the groups that score higher on tests and perform better in school simply *work harder*.[1] Yes, doing homework and studying correlates nicely with success in school. Amazing, huh? But belief in genetic destinies leads many to the unwarranted conclusion that all those test scores and grades were predetermined by genes handed out at conception. The reality is that what our genes do and do not do for us is determined by the infinitely complex way they are influenced by our environments. David Shenk describes this in his book *The Genius in All of Us*:

> [G]enes are not like robot actors who always say the same lines in the exact same way. It turns out that they interact with their surroundings and can say different things depending on whom they are talking to. This obliterates the long-standing metaphor of genes as blueprints with elaborate predesigned instructions for eye color, thumb size, mathematical quickness, musical sensitivity, etc. Now we can come up with a more accurate metaphor. Rather than finished blueprints, genes—all twenty-two thousand of them—are more like volume knobs and switches. Think of a giant control board inside every cell of your body.
>
> Many of those knobs and switches can be turned up/down/on/off at any time—by another gene or by any miniscule environmental input. This flipping and turning takes place constantly. It begins the moment a child is conceived and doesn't stop until she takes her last breath. Rather than giving us hardwired instructions on how a trait must be expressed, this process of gene–environment interaction drives a unique developmental path for every unique individual.[2]

LIMITS LIKELY EXIST, BUT WE SHOULD NOT PRETEND TO KNOW THEM

It is wrong to imagine ourselves and others as prisoners fenced in by our genes. Science is now showing us more clearly than ever that whatever genetic limitations we may have, the specific boundaries of these limitations are unknown to us. They may be forever unknowable, given the infinitely complex interaction between genes and environment. Therefore, it makes no sense to pretend that we know these prison walls and then navigate our lives accordingly. We know what it takes to develop healthy minds, and this is where our attention and efforts ought to be. Genetic inheritance is a huge factor in intellectual potential, of course. Sure, people really are born with brains that will never produce Nobel Prize–winning achievements no matter how hard they work or how many opportunities they have. I don't think for a second that every brain in every newborn baby is equal or has equal potential. Some people are born with more or less innate potential than others. But how exactly do we know who falls into which category? All we can do is measure current ability. We cannot determine intelligence that *might have been* or intelligence that *could be*. Therefore, we must never let the lie of known genetic limits to our intelligence hold us back either as individuals, as groups, or as a species.

GO DEEPER . . .

- Fish, Jefferson, ed. *Race and Intelligence: Separating Science from Myth*. New York: Routledge, 2001.
- Medina, John. *Brain Rules: 12 Principles for Surviving and Thriving at Work, Home and School*. Seattle: Pear Press, 2008.
- Murdoch, Stephen. *IQ: A Smart History of a Failed Idea*. Hoboken, NJ: Wiley and Sons, 2007.
- Nisbett, Richard E. *Intelligence and How to Get It*. New York: W. W. Norton, 2009.
- Shenk, David. *The Genius in All of Us*. New York: Doubleday, 2010.

Chapter 5

"THE BIBLE CODE REVEALS THE FUTURE."

It is easy to be wise after the event.
—English proverb

We humans rarely meet a pattern we can resist. As mentioned in chapter 1, our minds have evolved to be very good at detecting things like animal tracks in a thick forest or spotting a camouflaged bird perched on a branch. While this obviously has been an invaluable skill for hungry humans over the millennia, it comes with a price. Sometimes we can be so good at identifying patterns that we "see" things that aren't really there. That's how we end up with people seeing the Virgin Mary on a slice of toast, Allah's name written in clouds, and ghosts in the shadows.

The Bible code, also known as the Torah code, is the astonishing claim that says the Abrahamic god embedded messages into the text of the Bible. According to believers, the coded words include not just the names of important people and events but also accurate predictions about the future. The interest and excitement that this has generated has been remarkable. I think we are primed not only to look for and recognize patterns, both real and imagined, but also to experience joy or some form of emotional reward when we find one. This may explain some of the popularity of all those word-seek puzzles—and the Bible code. It's like the *Where's Waldo?* books, but this time with supposed life-and-death implications for all of us. It's not difficult to see why millions of people find it irresistible. But resist they should, because there is nothing to substantiate this claim.

SEEK LONG ENOUGH AND HARD ENOUGH AND YE SHALL FIND

It is difficult, if not impossible, to read through the Bible and find numerous disparate words that can be linked together in a semisensible way to make predictions about the future. However, thanks to computers, it is possible to plug in an extraction formula and let it run through the text to produce "hidden messages." The way it's done is by using something called the "equidistance letter search." First you choose a letter—let's say *B*, for example. Next you choose a number—5, for example. Then you plug this sequence into a computer program that will find every fifth letter that falls after every *B* throughout the text (also before, above, or below the letter, if you wish). Some of the letters generated will spell out words. Some of the words will have meaning. How much meaning is in the eye of the beholder, as we shall see.

Using this technique, say believers, the Bible accurately predicted historical events such as the Holocaust, the assassination of John F. Kennedy, and the assassination of Israeli Prime Minister Yitzhak Rabin thousands of years before they happened. Wow, if true, this would appear to be very strong evidence that the Bible really is much more than a mere book written by humans. This code might even be proof for the existence of the Judeo-Christian God. Once again, however, a claim for the miraculous and the supernatural has a rather simple down-to-earth explanation.

It is not difficult to explain how Bible-code proponents come up with names like "Kennedy" and "Rabin." The Bible contains many thousands of letters. Given enough chances, the creation of words that can be strung together so that they *seem* to have meaning is bound to happen. There certainly are plenty of chances, as one can pick different letters for starting points as well as choose different number sequences to select letters after, before, above, below, and diagonal from the starting-point letters. Dave Thomas, a physicist and mathematician, found "Roswell" and "UFO" in just one verse of Genesis using the same sort of extraction process Bible-code believers rely on.[1] Thomas doesn't see anything magical in the Bible code: "The promoters of hidden-message claims say, 'How could such amazing coincidences be the product of random chance?' I think the real question should be, 'How could such coincidences not be the inevitable product of a huge sequence of trials on a large, essentially random database?'"[2]

Bible- and Torah-code claims have been around for many years. However, the claim surged in popularity and gained widespread media attention in the 1990s when Michael Drosnin's book *The Bible Code* hit the *New York Times* bestseller list. In response, skeptics explained repeatedly that names and "predictions" can be found "encoded" in any book. Drosnin fired back, promising he would believe the critics if they found the assassination of a world leader encrypted in *Moby Dick*. So some skeptics did just that, multiple times: "Lincoln" and "killed" turned up in one passage of the 1851 novel. "Prepare for death" and "M L King to be killed by them" were also decoded in *Moby Dick*. So too was "Kennedy," "shot," and "head" in close proximity to one another.[3] Sadly, the professors responsible for this demonstration had to issue disclaimers because some people took their demonstration as evidence that Herman Melville, the author of *Moby Dick*, must have had supernatural powers too.

Experienced skeptics know to be on the lookout for *postdictions*. These are *predictions* that are discovered and lauded *after* an event occurs. Nostradamus believers love them. It's much harder, of course, to go on record predicting something before it happens. Drosnin tried it using his Bible-code method, and it didn't work out too well for him. He wrote in his 2002 follow-up book, *Bible Code II*, "The Bible Code clearly states the final danger in modern terms—'atomic holocaust' and 'World War' are both encoded in the Bible. And both are encoded with the same year, 2006."[4] Clearly that was a miss. Drosnin's code-breaking efforts also came up with phrases in the Bible that seemed to describe the assassination of Yasser Arafat, the former leader of Palestine. "Assassin will assassinate," "the ambushers will kill him" and "shooters of Yasir [*sic*] Arafat."[5] Missed again. Arafat died of illness at age seventy-five. Bible-code believers must ask themselves why a god would encode incorrect predictions.

Jews and Christians who believe that the Bible code proves something significant about the Bible/Torah might give some thought as to why the Koran code doesn't prove that the claims of Islam are correct. Yes, that's right, there is a Koran code, and, according to many Muslims around the world, it proves once and for all that the Koran is divinely inspired and Islam is the one true religion.[6] For some reason, however, Jews and Christians do not seem to be very impressed by the Koran code. It seems that they're just too skeptical to fall for something like that.

GO DEEPER . . .

- Dunning, Brian. *Skeptoid: A Critical Analysis of Pop Phenomena*. Laguna Niguel, CA: Skeptoid Media, 2007.
- Wheen, Francis. *How Mumbo Jumbo Conquered the World: A Short History of Modern Delusions*. New York: Public Affairs, 2004.

Chapter 6

"STORIES OF PAST LIVES PROVE REINCARNATION IS REAL."

Within your aura is all the information needed for life, including your genetic and ancestral lineage, a record of your past lives, and the karmic contract and lessons that you intend to resolve in this lifetime.

—James van Praagh, *Heaven and Earth*

This would be a strange afternoon, even by my standards. A small group of transvestites surrounded me on a Mumbai sidewalk. They were an impressive sight in their colorful saris, each of them adorned with a pound or two of gleaming jewelry. The one who did most of the talking was a dead ringer for the late Charles Bronson, only she was slightly more masculine than the late Hollywood action star. Squinting in the afternoon sun, she politely explained that she and her friends felt that I was a "most handsome man" and that it would be "most wonderful" if I would spend the day with them. As she spoke, one of her comrades squeezed my arm and purred. Yes, I swear, she purred. I couldn't be sure if this encounter was a sincere grasp at a love connection or just another routine attempt to extract money from a stupid tourist. As interesting as a date with the Magnificent Seven may have been, I looked deep into the leader's yellow eyes and respectfully declined. I smiled and retreated back into the herd of pedestrians. And then the day became weird.

India is the most intense, irritating, colorful, exhausting, stimulating, bizarre, and beautiful country I have ever experienced. During my time there I saw an amazing assortment of juxtaposed images. On one side of a street I see stunningly gorgeous women strutting, while on the other side my eyes find hideously deformed beggars scurrying about on crooked limbs and twisted spines. One moment I'm entranced by the breathtaking architecture of the Taj Mahal. A brief walk later, I'm staring at human corpses rotting in the mud of the adjacent riverbank. One of the biggest surprises for me during my visit was that India inspired me to think deeply about life, death, and even reincarnation. At times I even felt close to what many might describe as spiritual—minus the actual spirit, of course. Something about confronting extreme beauty and ugliness side by side led me to think of things other than what I would have for dinner later. One thing I allowed myself the freedom to wonder about was what it would be like to have lived previous lives and to experience more lives after this one ends. I don't remember anything from before my birth, but I do feel oddly comfortable when I wander through faraway lands for the first time. Could it be because I have been there before? I also seem to have a knack for getting along with people in societies very different from the one that spat me out. Could it be that I once lived in that culture long ago? I doubt it, but it is a fascinating idea. Hundreds of millions of people today, mostly Hindus and Buddhists, believe it is more than an idea. They think it is reality.

A central belief in Hinduism, one of the world's oldest and most popular religions, is that we all experience many lives in succession. It varies depending on which Hindu you ask, but the general claim is that these lives are opportunities to learn and improve. Eventually, if all goes well, our soul attains perfection and we can finally relax. Belief in reincarnation is not exclusive to Hindus, of course. Many millions of non-Hindus around the world think it is true as well. This includes many Christians, which is interesting given the contradiction with Christian dogma. A 2005 Gallup poll found that 20 percent of Americans believe in "the rebirth of the soul into a new body after death."[1] That's at least fifty million American believers in reincarnation. One can certainly understand the attraction. Death is scary. Believing that it's not final and that we all get a few more turns at it can be comforting. But appeals to emotion aside, is there anything to it?

Whenever I think of reincarnation, I recall a basic law from physics class: energy cannot be created nor can it be destroyed. It may change form or location, but it never vanishes. Certainly there is energy in my body and my brain. Many people think of that energy as the soul. So when I die my energy (soul?) may change form and relocate, but it won't vanish from the universe entirely. Interesting, but this is where we run into the critical problem with reincarnation belief. No one has ever been able to show that human energy equates to a soul with thoughts, personality, and memories. There simply is no good evidence to support the claim that something containing our thoughts and memories survives our physical death. This is the gigantic obstacle looming in front of reincarnation claims. Shouldn't the basic question of the existence of souls be answered first before we even consider accepting the claim that these souls leave dying or dead bodies and enter new bodies?

Some who believe in multiple lives have told me that they think reincarnation is true because it "makes sense" or "feels right." But *every* reincarnation believer I have ever spoken with brings up stories of people knowing things about their past lives, things they could not possibly have known if they had not been reincarnated. Strangely, however, believers never seem to know many key details, or if these stories were analyzed and verified by credible researchers. Nonetheless, they cite them as proof anyway. Is this good enough? Are vague stories enough to confirm reincarnation belief? Of course not.

Another interesting aspect to this is that nobody seems curious or concerned about how souls maintain themselves through physical death or how they power their way through the air when seeking the next body to inhabit. Some reincarnation believers say that human souls only inhabit human bodies. If so, does one soul inhabit millions of bodies across millennia? Or do many souls share each body? I suppose they would have to share because it is estimated that more than one hundred billion people have lived, but there are only about seven billion people alive today.[2] That's a lot of souls with relatively few bodies to accommodate them. In fairness, I suppose some believers would say that the animal kingdom accommodates those excess human souls. Or maybe the success rate for soul progression is high and all those extra souls have achieved nirvana and escaped the cycle.

ANECDOTES ARE NOT SCIENTIFIC EVIDENCE

If stories of past lives are the best evidence of reincarnation, then what are the best stories? Reports of children who "remember" past lives probably offer the most compelling evidence of reincarnation. The reason child stories are so appealing is that people don't think a child would know compelling details about some historical setting and they would be less likely to perpetrate a hoax. But why should anyone think that? Just like adults, children can fantasize, misinterpret, perform as coached, and lie. They can also learn things and repeat them. Take the case of James Leininger, a boy some believe is a reincarnated pilot who was shot down over the Pacific during World War II. On the surface, the story seems like iron-clad proof of reincarnation. But is it?

According to a *Primetime* report, James showed an interest in airplanes very early and knew things about them that most people are unaware of. For example, his mother said he knew the difference between a bomb and a drop tank. He also began having nightmares about a plane crash. The concerned parents sought help for the boy, not from a psychologist or psychiatrist, however, but from a "past-life therapist" who specializes in helping people "remember" their past lives. Not surprisingly, it was determined soon after his sessions began that James is indeed a reincarnated fighter pilot. According to the *Primetime* story, the boy also named a specific pilot and ship that he flew off of.[3] Is this conclusive proof that reincarnation is real? Not even close.

It turns out that the boy's father took him to an aviation museum when he was around two years old and that James was fascinated with the World War II planes on display there. Shortly after that museum visit, James began having nightmares about planes crashing. He also wanted to play with planes and look at books about them, an interest his parents indulged.[4] It seems far more reasonable that the planes in the museum made an impression on James and this inspired his deep interest in aviation. Thinking a lot about military airplanes and then having nightmares about a plane crash is not so strange, nor is it odd for a child to pick up a lot of information about planes or anything else if he or she is deeply interested in them. I happen to have gone through a childhood phase of World War II–aviation obsession myself. I knew much more about WWII fighters and bombers than 99.9 percent of adults. Researching and retaining information can be effortless and fun when the subject fascinates you. Just like James, I knew the difference between a bomb and a drop tank when I was a child. As far as James naming a pilot and a ship, we can't be sure about that either because the name he gave was "James," his own name and a common one at that. Naming the ship is interesting, but isn't it possible that he saw the name in a book, at the museum, or on TV? Kids pick stuff up everywhere. Maybe he simply said something out of the blue that the adults matched to a ship by chance. After all, there were many US Navy ships with many names in that war. However, when parents and a "past-life therapist" attach meaning to it, another reincarnation story is born.

I have no reason to suspect that the Leiningers taught their son about WWII aviation in order to intentionally fool people, but that sort of thing is always a possibility with these kinds of stories. Reincarnation proponent J. Allen Danelek suggests in his book, *The Case for Reincarnation*, that children can't easily be coached to say and do things that might convince people they had past lives.[5] Oh really? I saw a tiny child in China transform herself into a living pretzel while balancing spinning plates on three sticks. Have you ever seen one of those toddler beauty pageants on television? If parents want their children to do something badly enough, they will find a way to squeeze just about any desired performance out of a kid. Getting a child to describe a fictitious past life would be easy. There is also the possibility that parents or some other adult could innocently and unconsciously plant the idea of a past life and then supply supporting information to the child in order to make it seem real. All these possible explanations are much more likely to be true than the extraordinary claim that an old soul inhabits the body of a child.

There are also interesting stories about people speaking a language they never learned. When you first hear this, it sounds like proof too. How could it be possible unless the person had a previous life? It turns out, however, that whenever somebody takes the time to investigate such claims, they always seem to fall apart. For example, a person who was said to be able to speak Bulgarian under hypnosis, despite not knowing Bulgarian, was not speaking that language at all. The person apparently just made up words with an accent that the hypnotherapist guessed was Bulgarian.[6] Despite the stories, there are no confirmed cases of a person speaking a language unknown to him or her. Everyone, including skeptics, would sit up and pay attention if a four-year-old living in Detroit suddenly spoke a tribal New Guinea language that went extinct last century. But nothing like that has ever been confirmed by credible investigators.

Whether I am feeling "spiritual" during a solo journey across India or reading about various claims of past lives, I always end up sensing an abundance of hope amid the absence of evidence. I cannot fault people for desiring more than one life. Optimism and dreams of something better are positive human traits. But hope alone is not proof.

Reincarnation defender Danelek writes, "Reincarnation is the mechanism through which we may live the very life we've always wanted—or relive the one we've always loved—upon a stage from which we may act out a million possibilities, dream a billion dreams, and live on throughout eternity."[7]

I agree. The claim of reincarnation is exciting to imagine and a wonderful thing to hope for. Unfortunately, however, nothing suggests that it is true.

GO DEEPER . . .

- Baker, Robert A. *Hidden Memories: Voices and Visions from Within*. Amherst, NY: Prometheus Books, 1996.
- Edwards, Paul. *Reincarnation: A Critical Examination*. Amherst, NY: Prometheus Books, 1996.
- Harris, Melvin. *Investigating the Unexplained*. Amherst, NY: Prometheus Books, 2003.

Chapter 7

"ESP IS THE REAL DEAL."

Numerous ESP studies conducted over the years point to one overriding conclusion. Studies supporting ESP consistently lack proper controls, and studies with proper controls consistently find no support for ESP.

—Thomas Kida, *Don't Believe Everything You Think*

You are reading this book right now thanks to the most magnificent and amazing three-pounds of matter in the known universe. Your brain is made up of about one hundred billion nerve cells called neurons that keep you alive by doing things like telling your body to breathe while making love and to duck when someone throws a brick at you. These tiny cells also work together to produce complex thoughts and new ideas. They can imagine both possible and impossible things. For example, my brain has taken me on journeys far beyond our galaxy, through a spectacular nebula, and even inside a terrifying black hole—and I didn't even have to drink or use drugs before departure. All I did was close my eyes and imagine.

For all its flaws that often trip us up when trying to distinguish reality from make-believe, the human brain is undeniably special, very special. Obviously far apart in form and function from hearts, lungs, kidneys, and other organs, your brain is *you*—everything else is basically Tinker Toys™ and plumbing. Human brains are big too. You know those giant-headed aliens we see in old science-fiction movies? Well ours are more impressive than you might think because what we have is essentially a size-9 brain that evolutionary pressures over time have creased, folded, and layered in order to fit inside a size-3 skull. This is necessary because of restrictive birth-canal issues that come with bipedalism. If baby heads were any larger, I suspect very few women would be willing to get pregnant.

Far more interesting and important than physical dimensions, however, is the brain's ability to think, to analyze difficult problems and come up with novel solutions, to dream in great detail of things we cannot touch or see. I certainly can't say I think that the human brain is beautiful because, after seeing one up close, I know better. But I am in awe of it and drawn to it nonetheless. I have a plastic model of a brain on my desk, and I often imagine my own brain humming away inside my skull, doing its thing. Yes, I think about my brain thinking.

Measured by its creative potential, the human brain is larger than the universe itself. By that I mean we can imagine and think about the limits of our universe and beyond. Grand ideas, such as string theory and multiverses, come from our brains. Given the admiration and strong feelings I have for the brain, why don't I believe in extrasensory perception, or ESP? If the brain is so complex, so wonderful, and capable of so much, isn't it possible that it could communicate with other minds without need of sight, touch, or sound, and perhaps even know the future?

Yes, it is possible that the human mind is capable of feats that could only be thought of as magical or paranormal by today's standards. But ESP, or "psi" as researchers often call it, has been studied for many years now, and still no researchers have managed to produce an experiment that can be replicated by others and confirmed. A small body of enticing data is overshadowed by mountains of negative results. While this doesn't disprove ESP, it certainly means it remains unproven. We can't ignore the fact that mind readers, card readers, and every other kind of ESP practitioner has failed to survive scientific scrutiny.

"Many parapsychologists have adopted a 'heads I win, tails you lose' approach to their work, viewing positive results as supportive of the psi hypothesis while ensuring that null results do not count as evidence against it," explains UK psychology professor Richard Wiseman.[1] The lack of positive experiments and good evidence should trouble ESP believers deeply, for if there were something to this, how hard could it be to have somebody consistently identify cards they can't see at a better-than-chance rate? If ESP is real, then why can't test subjects consistently tell when people are staring at them or accurately report what another person is thinking? I'm willing to believe, but not until somebody proves it. Not everyone is as picky I am, however.

A Gallup survey of Americans found that ESP is very popular, with 41 percent of adults professing to believe in it. ESP topped Gallup's list on that survey of paranormal beliefs that did not include traditional religious claims. Belief in haunted houses was second with 37 percent.[2] The US government apparently believed as well, having spent many millions of dollars on efforts designed to exploit paranormal powers like telepathy and remote viewing during the Cold War to spy on the Soviet Union.[3] Despite claims of success by some, nothing came of it and the project was cancelled after being reviewed by the American Institutes for Research in the 1990s. It was finally recognized to be a waste of time and money. "In no case had the information provided ever been used to guide intelligence operations," states the report. "Thus, remote viewing failed to produce actionable intelligence."[4]

Like most other paranormal and pseudoscientific claims, ESP's popularity depends largely on anecdotes, nothing more than stories about a vision that correctly predicted a future event, a sense of impending doom just before something bad happened, a phone call from someone who was just thought of, and so on. These kinds of stories may be compelling to people, "but the only way to find out if the anecdotes represent a real phenomenon or not is controlled experimental tests," warns scientist and *Skeptic* magazine publisher Michael Shermer. "Psi phenomena have now been subjected to rigorous scientific experiments for over a century . . . and the results are unequivocal: psychic power is a chimera."[5]

Shermer adds that good evidence is not all that is needed to make the case for ESP: "The deeper reason scientists remain skeptical of psi—and will even if more significant data are published—is that there is no explanatory theory for how psi works. Until psi proponents can explain how thoughts generated by neurons in the sender's brain can pass through the skull and into the brain of the receiver, skepticism is the appropriate response. If the data shows that there is such a phenomena as psi that needs explaining (and I am not convinced that it does), then we still need a causal mechanism."[6]

BUT WHAT ABOUT PSYCHIC DETECTIVES?

Believers often cite the success of psychic detectives as proof of ESP. The problem, however, is that psychic detectives aren't any good. Their track record is abysmal. There have been many claims made by psychics and their fans, of course, but look closer and one finds little or nothing of significance. Virtually every case can be explained as coincidence, the exaggerated value of vague tips, or the gullibility of a law enforcement person who gave undeserved credit to a psychic. For example, saying things like, "the victim knew the killer" or "the victim was killed with a knife and buried in the woods" are just guesses. And, thanks to confirmation bias, ESP believers will latch onto the rare hits and forget the numerous misses. Skeptics have analyzed many claims over the years, and they always find problems. Regardless of what you may have heard, there is no great number of cases out there somewhere that were solved by psychics. It's a lie.

A typical example of how people are misled would be a psychic predicting that the body of a missing crime victim would be found "in the woods," "near water," or "near a highway." These are smart guesses, because the bodies of missing persons who were murdered are often found in rural areas near a pond, lake, river, or ocean because almost everywhere is "near" water. "Near a highway" is also an excellent guess because many places are "near a highway." Certainly more missing bodies are found in the woods near a highway than, say, in a telephone booth or in a theater. It is a standard cold-reading technique to make vague predictions that can easily be spun to look like direct hits. Psychics have an even better chance to claim success if they throw out multiple predictions, which they almost always do because they know people will remember and focus on the "correct" prediction while forgetting all the wrong ones.

Here is the most telling fact of all to consider when assessing the validity of psychic detectives: the police don't use them. Some misguided law enforcement people somewhere may call on these people sometimes, but the vast majority of police departments and detectives do not bother with psychics because they know they are a waste of time. Detectives may listen to a tip from a psychic because they have to listen to all tips. But that doesn't mean they pin their hopes on these people. If psychic detectives really could "see" crimes that have already occurred and find missing people, this would be well known to all by now. Word would spread very fast and criminal investigators would jump to exploit their powers. There would be a full-time psychic on the payroll of every police department in the world. But this is not the case, which says a lot about the real abilities of psychics as crime solvers.

KEEPING AN OPEN MIND ABOUT THE MIND

Those who are skeptical of ESP are often accused by believers of having closed minds. Jefferson M. Fish, professor emeritus of psychology at Saint John's University in New York, is unconvinced but says he is open to accept it if new evidence ever justifies it. "I think it is possible, though unlikely, that there is something there." He says that there has not yet been a convincing case made for ESP, but neither has it been completely discredited. Fish explains that there are two lines of evidence for ESP claims:

> (1) Many experiments showing a slight but statistically significant deviation from chance, and additional results that make sense in terms of psychological principles—e.g., believers scoring above chance and disbelievers scoring below, when everyone should score at chance. Unfortunately, there are objections to all these experiments, and while some have been replicated, none are consistently reproducible.
>
> (2) Many isolated instances of amazing unexplained individual or shared experiences. There are, of course, alternative possible explanations for all of these events.
>
> As a result, I'm inclined to be skeptical, but don't consider the case closed.[7]

It appears that an open mind is typical of many ESP skeptics, despite what ESP believers often charge. I certainly am open-minded on the issue, and no serious skeptic I know would turn away from compelling evidence if it were produced. Terrence Hines, a professor of psychology and neurology, reviewed a batch of studies and found a pattern of paranormal believers being "rigid and unchanging" in their beliefs compared with skeptical nonbelievers, who are more willing to revise their conclusions when shown contrary evidence.[8]

I think I understand why some people would want to maintain a tight grip on their belief in ESP. I feel some of what they feel because I too believe in a magical mind. The only difference is that the kind of "magical mind" that I believe in is the one that has been revealed by science. This mind doesn't require supernatural or paranormal elements to impress and excite. The mind I am in awe of has no need for unproven claims and gross exaggerations of its power and value. The real mind is impressive enough.

GO DEEPER . . .

- Blackmore, Susan. *In Search of the Light: The Adventures of a Parapsychologist*. Amherst, NY: Prometheus Books, 1996.
- Charpak, Georges, and Henri Broch. *Debunked! ESP, Telekinesis, and Other Pseudoscience*. Baltimore: Johns Hopkins University Press, 2004.
- Horstman, Judith. *The Scientific American: Brave New Brain*. San Francisco, CA: Jossey-Bass, 2010.
- Horstman, Judith. *The Scientific American: Day in the Life of Your Brain*. San Francisco, CA: Jossey-Bass, 2009.
- Hyman, Ray. *The Elusive Quarry*. Amherst, NY: Prometheus Books, 1989.
- Nickell, Joe. *Psychic Sleuths: ESP and Sensational Cases*. Amherst, NY: Prometheus Books, 1994.
- Van Hecke, Madeleine. *Blind Spots: Why Smart People Do Dumb Things*. Amherst, NY: Prometheus Books, 1997.

Chapter 8

"NOSTRADAMUS SAW IT ALL COMING."

> **Credulous France, what are you doing, hanging on the words of Nostradamus? . . . Don't you understand that this dirty rascal offers you only nonsense? . . . One must ask, in the end, who is sillier, this evil charlatan or you, who accept his impostures?**
>
> —From a letter written by a Nostradamus critic in the year 1555

It's not difficult to understand why many people are drawn to Michel de Notredame, the sixteenth-century French medical doctor, poet, and astrologer. Who doesn't want to know the future? If the world is going to end tomorrow, for example, then I would love to know because it would mean I can skip taking out the trash this evening. What is difficult to understand, however, is how Nostradamus ever became history's undisputed champion of supernatural predictions. It's not like there isn't plenty of competition for the title.

There was no shortage of astrologers and soothsayers around in Nostradamus's day, and there are still many around today who claim to be able to predict the future. So how has Nostradamus remained king of the hill, more than four hundred years after his death? He must have some amazing predictions to his credit, right? One might think so, but in fact he does not. His record is no better than the astrologers and TV preachers of our time. The popular image of Nostradamus as a mysterious man who had the ability to see far into the future rests entirely with the vague nature of his prophecies that can be loosely interpreted and twisted to mean just about anything anyone wants them to. Nostradamus only continues to be a popular and convincing prognosticator because those who believe in him don't think critically about the predictions.

In addition to being a renowned skeptic and elite magician, James Randi is a Nostradamus scholar. He looked deep into the extraordinary claims and found no extraordinary evidence to back them up. He couldn't even find any *weak* evidence to back up the Nostradamus phenomenon. Randi's book *The Mask of Nostradamus: The Prophecies of the World's Most Famous Seer* is a brilliant analysis and critique of the Nostradamus prophecies. It clearly shows that there is nothing substantial to be found in the hype. Nostradamus left behind no remarkable, uncanny, or spooky predictions that defy explanation. The only thing that happened is that five centuries ago, a European astrologer wrote a bunch of poetic gibberish that others have rewritten, reinterpreted, and spun to appear like accurate predictions of future events.

Some beliefs, no matter how hollow, are difficult to kill. Especially when some people put so much effort into selling the lie. It seems that scarcely a month goes by, for example, without the History Channel pushing Nostradamus belief on a new generation of unsuspecting innocents with yet another slick pseudodocumentary. These days they include impressive special effects as well as compelling soundbites from Nostradamus and paranormal experts who lay out "overwhelming evidence" to make the case.

Before we pull the curtain back and expose Nostradamus's legacy for what it really is, let's review who the man was. He was born in Saint Remey de Provence, France, in 1503. According to Randi, Nostradamus was a bright student who graduated from school early and went on to have a successful medical practice. He specialized in treating the plague, the disease that killed his first wife and two of his children.[1] Nostradamus's astrology and prophecy work was no great stretch for a doctor at the time, as medicine during the sixteenth century was more a mix of art and hocus-pocus than science.

The Nostradamus prophecies that continue to impress people today are contained in quatrains (four-verse poems). In my opinion, this in an important element to the attraction because it seems that many people are far more impressed by predictions embedded in ambiguous collections of obscure wording than a straight, no-nonsense message. The Bible code is another example of this. If Nostradamus simply wrote, "The world will end at noon on December 1, 1927," it would not be as dramatic or mysterious. It would also hurt his chances to impress multiple generations, because then it would be easy for everyone to know exactly what he meant and it would be clear when he was wrong. The Nostradamus industry would sink fast if his predictions could be given an unambiguous pass/fail assessment.

Before we go further, we need to acknowledge that it's not really fair to judge Nostradamus based on his quatrains. Maybe he really did know the future and maybe he really did write down accurate predictions of major events to come. The fact is, we can't know for sure because there are no original Nostradamus manuscripts! Yes, that's correct. All the noise coming from Nostradamus believers is based on their interpretations of original writings that no longer exist.

"The very first editions of his prophetic writings are lost and we must depend upon the accuracy and integrity of those who transcribed them [whomever that may have been more than four centuries ago]," explains Randi. "Numerous known forgeries have been published, some to prove points not originally intended by the seer and others merely to take financial advantage of a public hungry for anything in any form bearing the Nostradamus name. Typographical errors, transpositions, changed italicization, punctuation, and capitalization, altered spellings and 'improvements' on his writings have bastardized his works to the point where proper scholarship is difficult."[2]

There is also the colossal problem of meaning. Even if we were to assume that existing versions of Nostradamus's predictions are somewhat fair representations of his original work, it is clear that the wording is too imprecise to be useful. Colorful language might be great for poetry, but it is terrible for assessing predictions. Like any well-written horoscope, his quatrains can be interpreted in multiple ways which, of course, increase the chances of them being "correct." Randi shows this by identifying four very different interpretations of the same quatrain (1-57) from four Nostradamians. One believer said the quatrain predicted a political revolution and dead king. A second said the same quatrain was a "clear and forthright prediction" of the attack on Pearl Harbor. A third expert claimed it was a warning about the rise of Adolph Hitler. Finally, a fourth said it foretold an earthquake.[3] Could the interpretations of one small batch of words be any more different? This is the same way horoscopes in newspapers are able to impress so many people. No matter what happens, they are seen as accurate by true believers.

In his book *The Mask of Nostradamus*, James Randi provides a helpful list of guidelines for anyone who would like to become a great prophet:

> Make lots of predictions, and hope that some come true. If they do, point to them with pride. Ignore the others;
>
> Be very vague and ambiguous. Definite statements can be wrong but "possible" items can always be reinterpreted;
>
> Use a lot of symbolism. Be metaphorical, using images of animals, names, initials. They can be fitted to many situations by the believers;
>
> Cover the situation both ways and select the winner as the "real" intent of your statement;
>
> Credit God with your success, and blame yourself for any incorrect interpretations of His Divine messages;
>
> No matter how often you're wrong, plow ahead. The believers won't notice your mistakes and will continue to follow your every word;
>
> Predict catastrophes; they are more easily remembered and more popular by far;
>
> When predicting after the fact, but representing that the prophecy preceded the event, be wrong a few times, just enough to appear uncertain about the exact details; too good a prophecy is suspect.[4]

WILL THE REAL ANTI-CHRIST PLEASE STAND UP

One of the more amusing flaws within the Nostradamus industry is the repeated naming of the Anti-Christ. The biblical villain was Napoleon, according to confident Nostradamus scholars in the nineteenth century. When the French emperor failed to deliver the end of the world, however, new predictions were "discovered" that clearly and obviously pointed to Adolph Hitler as the man destined to bring down the final curtain.

If you read a book or watch a pseudodocumentary that promotes the Nostradamus myth, you are likely to encounter the name "Hister." Hmmm, sounds a little like "Hitler," right? Sure, but "Hister" also sounds exactly like "Hister," which was the name given for the lower portion of the Danube River on maps at the time Nostradamus lived. That, as well as the fact that the line immediately preceding the line containing "Hister" mentions "swimming," would seem to indicate that Nostradamus was referring to a river and not the twentieth century German dictator.[5]

I am old enough to remember when Ayatollah Khomeini, Muammar Gaddafi, and Saddam Hussein were each at various times declared to be the Anti-Christ that Nostradamus had warned about. Of course, none of them worked out too well, so believers were quick to embrace Osama bin Laden as the guy Nostradamus really meant when the 9/11 attacks occurred in 2001. But when Bin Laden was killed by Navy SEALS in 2011, another Anti-Christ candidate bit the dust.

An interesting example of how Nostradamus nonsense can spread occurred in the wake of the 9/11 attacks. Millions of people around the world, myself included, received this Nostradamus quatrain in a chain e-mail shortly after the World Trade Center towers fell:

> In the City of God there will be a great thunder,
> Two Brothers torn apart by Chaos,
> While the fortress endures, the great leader will succumb.
> The third big war will begin when the big city is burning.

Wow, that's an attention grabber. "Two brothers torn apart by chaos" sounds like it could be the World Trade Center towers, right? And during the tense days after 9/11, who could ignore this line: "The third big war will begin when the big city is burning"? There is, however, a problem with this Nostradamus prophecy—Nostradamus didn't write it. I mean he *really* didn't write it this time, not even in a fake way. It was the work of a university student in Canada. The prediction was included in the student's 1997 essay as an example of a vague prediction that could later be applied to a wide variety of events. The student points out in his pre-9/11 essay that "city of god" could be attributed to many cities; "a great thunder" could mean anything from war to a storm to an earthquake; and "brothers torn apart" could apply to many things (a split between nations, a government, two populations, two former friends or allies, or two actual brothers). The funny part of the story is that this wise student asks readers of the essay to imagine how this brief 1997 prediction might be viewed if it were put aside for centuries and then matched with some event that it seemed to predict. The kid was right on the mark about everything except the time frame. His prediction was just vague enough to be tied to the 9/11 attacks only four years later. By the way, just as it has been the case with Nostradamus's writings, someone dishonestly added "The third big war will begin when the big city is burning," to give it even more punch. As this non-Nostradamus quatrain circulated throughout the world via e-mails, other lines were added as well. Sadly, millions of people were probably influenced by this bogus e-mail to believe that there must really be something to the Nostradamus claim. Four days after the 9/11 attacks, *Nostradamus: The Complete Prophecies*, was the number one bestselling book on Amazon. Five other Nostradamus books made the top twenty-five list.[6]

As irrational beliefs go, this is not one of the more dangerous ones. Believing that a sixteenth-century astrologer accurately saw the distant future is, however, a symptom of a greater problem that is risky indeed. Where one irrational belief creeps in, others may follow. People in a democracy, for example, who can't recognize the utter emptiness and failure of Nostradamus claims are probably at greater risk of being fooled by corrupt politicians, dishonest marketing efforts, and bogus medical treatments. I think it's important to confront irrational beliefs, both superficial and serious. One doesn't have to be rude or overly aggressive about it, of course, but we shouldn't silently accept any of them in a society that places any value on reason.

James Randi certainly never surrendered to nonsense, but he does seem resigned to accept the enduring popularity of the great seer of France. "The legend of Nostradamus, silly as it is, will survive us all," he writes. "Not because of its worth but because of its seductive attraction, the idea that the Prophet of Salon could see into the future will persist. An ever-abundant number of interpreters will pop up to renew the shabby exterior of his image, and that gloss will serve to entice more unwary fans into acceptance of the false predictions that have enthralled millions in the centuries since his death. Shameless rationalizations will be made, ugly facts will be ignored, and common sense will continue to be submerged in enthusiasm."[7]

Unfortunately, that's one prediction I can believe.

GO DEEPER . . .

- Randi, James. *The Mask of Nostradamus*. Amherst, NY: Prometheus Books, 1993.

Chapter 9

"I BELIEVE IN MIRACLES."

The supernatural is a failure of human imagination and an insult to the majesty of the real.

—Edward Abbey, *Confessions of a Barbarian*

It is likely that unlikely things should happen.

—Aristotle

On August 5, 2010, a cave-in trapped thirty-three Chilean miners more than two thousand feet deep in the earth. Many observers assumed that death was likely, if not certain, for the men. However, after a $20 million, sixty-nine-day, round-the-clock rescue effort, all of them were saved. An estimated one billion people watched as the men were pulled up, one by one, in a slim metal cylinder. It was not surprising, of course, that the word *miracle* was immediately attached to the event and repeated often whenever people spoke about the rescue. Was it a miracle? That depends on how miracles are defined.

Traditionally a miracle has been, for most people, an unusual event that seems to violate or transcend the normal workings of the natural world. Most people who believe in these kinds of miracles link them to their religion's specific god or gods. They never seem to credit the god or gods of a rival religion. In this way, miracles are seen as verification of one religion or another, determined by who is claiming the miracle. But it's even more complicated these days because in common, everyday speech, people toss the word *miracle* around without much thought. The bar has never been lower. A late comeback by a sports team is sure to be called a miracle. A winning hand in black jack might qualify too. By that standard, I believe in miracles. Every time my son eats all his vegetables at dinner, for example, I feel that I have witnessed a profound miracle. But do I believe in the kind of miracles that are supposed to contradict the laws of nature? No, and here's why.

Miracles that cannot be explained as naturally occurring events depend on ignorance. I don't mean "ignorance" in a mean-spirited or condescending way. Let's take myself as an example. There are so many things I don't know about the universe that it would be ludicrous for me to witness something that I can't explain and, because it stumps me, declare that it must be a magical or miraculous event. Wouldn't it be far more likely that I simply don't know enough to explain what is going on by natural means? The only sensible and honest way to react to something you do not understand is to admit that you do not understand it. Filling in a blank with a made-up answer is intellectually shallow and dishonest. It seems to me that we all would do better to simply admit that many so-called supernatural miracles are mysteries. Sometimes things happen that we cannot explain. Of course some people can't stand to leave loose strings dangling so they pretend to know by saying it was the act of a god. I suggest we embrace the phrase, "I don't know." It seems to get a bad rap, but "I don't know" is a respectable answer when one *doesn't know*.

My travels and encounters with diverse cultures taught me an important thing about miracles. I learned that the less people know about basic science, the more they talk about miracles. In places where there is little awareness of astronomy and medical science, for example, one hears much talk of miracle eclipses and healings from minor illnesses and injuries that most people recover from. In societies with higher levels of science literacy, I still heard claims of miracles, but it was less frequent and almost always limited to unusual events, such as people surviving a plane crash or the rescue of some lost hiker. The correlation is clear: more understanding of the natural world means less reliance on miracles to explain events. This can be seen in history as well. Centuries ago, things we now understand were thought to be unexplainable and therefore supernatural. It is likely that this trend will hold true in the future. Today's miraculous occurrence will probably be tomorrow's routine occurrence, thanks to future generations' greater understanding of how the universe works.

A 2009 Harris Poll found that 76 percent of Americans believe in miracles.[1] I think they believe for three primary reasons. The first is that miracles are closely associated with religious belief. Claims of miracles are abundant and important within Christianity, and most

Americans are Christians. Therefore, it should be expected that most Americans would be heavily predisposed to believe in miracles.

The second reason so many believe in miracles is that few people ever slow down and think critically about miracles. Even a tiny bit of skeptical thinking can easily bring the concept of miracles down to Earth. Just because an event is rare, conveniently timed, or can't readily be explained is no justification to jump to the conclusion that there must be something supernatural going on. Many miracles are almost certainly nothing more than random events and coincidences that *must* occur in a busy world filled with seven billion people.

The third reason miracle belief is reported by so many people is that referring to unusual (but clearly nonmagical) events as miraculous has become entrenched in our culture. To say the Chilean miners were saved by a miracle in the supernatural sense is clearly not a fair assessment of what really happened—unless one somehow defines a miracle as the efforts of hundreds of rescue workers, many tons of heavy machinery, sixty-nine days of intense and nonstop hard work, and $20 million spent. The loose way in which the word *miracle* is applied to almost any event makes believing in miracles a default position for many people.

Let's analyze events that seem too unlikely to happen naturally and for this reason alone are often called miracles. Imagine an unusual occurrence that is so rare it only happens to one person out of one million in the entire world during an entire twenty-four-hour period. Most people would agree that hitting one-in-a-million odds is special, many would even say miraculous. But wait, there are approximately seven billion people alive on Earth today. This means that our one-in-a-million event would happen seven thousand times per day! That's 2,555,000 times per year. Even a once-in-a-billion event would still happen seven times every day. Suddenly long shots don't seem quite so miraculous, do they? It's a numbers game; if something can happen, it will happen at some appropriate rate. Thinking in realistic ways about statistics and probabilities does not come naturally to us. This is how casinos are able to drain billions of dollars from their customers year after year.

British mathematician John Littlewood had some fun by crunching some numbers to show just how common "miracles" can be. He proposed that a typical person experiences one thing per second while awake, say twelve hours per day. These "things" experienced would include everything from driving a car to looking at a table to feeling an itch on your toe. According to Littlewood, it adds up to more than one million experiences every thirty-five days. This means every person on Earth should experience those rare, one-in-a-million "miracle" events roughly once every month.[2]

There also is something called the "gambler's fallacy" that leads most of us to miscalculate the chances of a particular thing happening. For example, imagine if you flipped a coin five times and got heads every time. Would the odds favor tails coming up on the next flip? Many people would say yes, but the correct answer is no. The odds would still be even at fifty-fifty because all the flips before have nothing to do with the next flip. Coins do not have an internal memory system that is linked to a guidance control mechanism that allows them to adjust their landing positions based on previous flips. Las Vegas was built with money from people who didn't understand this.

It can be a challenge, but when faced with something that feels like it could not possibly have been a coincidence, try to think of the big picture. If the dream you had last night perfectly predicted something unusual that actually happened today, put it in proper context. How many dreams have you had in your lifetime that failed to predict events accurately? You also have to think of all the other people on Earth who had a dream last night that didn't come true. The odds are that somebody's dream would hit the mark just by chance. It's like the lottery: the odds of winning may be low, but *somebody* does win. So, if billions of people are having dreams every night, shouldn't it be expected that some of them will "come true" just by chance? If millions of people buy lottery tickets, don't we expect that someone out of the crowd will pick the winning numbers? If we want to think clearly, we have to accept the fact that coincidences happen all the time. In isolation, many of them can seem eerie and supernatural. Placed into proper context, however, they usually seem inevitable more than anything else. If enough things are going on all the time—like seven billion people scurrying about on this planet—then unusual and unexpected coincidences will occur. They *have* to occur. I've had several weird things happen over the course of my lifetime, but I don't think magic or gods were involved. I once ran into an acquaintance from the Cayman Islands on a sidewalk in New York City where both of us just happened to be visiting at the same time. What are the odds of running into each other in a city that big? Even weirder, I ran into the same guy on a crowded day at Disney World while on vacation the following year. The longer you live, the more weird things should happen to you—*must happen* to you. If I went my whole life without experiencing a few unusual and unexpected events, *then* I would have reason to be suspicious that something very strange is going on in the universe.

The primary reason people have long believed in miracles is that sometimes things happen that people can't explain. By this reasoning, the unknown is defined as a miracle. But that makes no sense because our ignorance is not proof of anything. It's unreasonable to expect that we can explain everything. We are a young species, and it was only very recently that we started using the scientific method to figure out the world and universe around us. We have come far, but there is still a long way to go. It's laughable to imagine that we should have an answer for everything that happens, so anything we can't readily figure out must be supernatural. What if an uneducated person from the year 900 somehow visited our time, sat down in front of a laptop with voice-recognition technology, and started asking random questions to Google. What if she were filmed and then shown the footage of herself on TV? What if this visitor from the Middle Ages were given a cell phone and talked to someone on the other side of the world? It likely would be very difficult to convince her that she was seeing anything other than magic and miracles. We know, given our location in history, that none of those experiences are supernatural. But she would not know this immediately, and her first temptation likely would be to call them miracles. Many people today do the same thing when confronted with the unknown.

Regardless of what you may have been led to believe, it's OK to say, "I don't know" when faced with a genuine mystery. Ignorance should be embraced openly as motivation to learn and discover. When we come up short and don't have answers for something, admitting ignorance while continuing to seek answers is far more productive than pretending to know. It's also honest.

GO DEEPER . . .

• Gardner, Martin. *Did Adam and Eve Have Navels: Debunking Pseudoscience*. New York: W. W. Norton, 2001.

- Gardner, Martin. *The New Age: Notes of a Fringe-Watcher.* Amherst, NY: Prometheus Books, 1991.
- Nickell, Joe. *Looking for a Miracle: Weeping Icons, Relics, Stigmata, Visions & Healing Cures.* Amherst, NY: Prometheus Books, 1999.
- Piattelli-Palmarini, Massimo. *Inevitable Illusions: How Mistakes of Reason Rule Our Minds.* New York: Wiley, 1996.

OUT THERE

Chapter 10

"NASA FAKED THE MOON LANDINGS."

> **I agree with Neil Armstrong, who gave one of the best responses to the ridiculous Moon-hoax claims when he said, "It would have been harder to fake it than to do it."**
> —Andrew Chaikin

Have you ever looked up at the full Moon on a clear night and wondered how people ever managed to walk on that thing? At first glance it almost seems impossible, doesn't it? Well, maybe it was impossible back in the 1960s and 1970s, and nobody has walked on it. That's what some people believe. They say it was all a big hoax designed to show up the Soviet Union during the Cold War. America's political and military face-off with the Russians was high stakes, with the fate of the world possibly hanging in the balance. Therefore, it can be argued, it's not completely crazy to suggest that NASA and the US government might have faked the Apollo success in order to win a public relations battle and take the "Red menace" down a peg or two in the eyes of the world.

A journey to the Moon and back with twentieth-century technology was no easy feat by any stretch of the imagination, and no one should ever underestimate the ability of politicians and generals to lie and bamboozle the public. However, charges of a Moon-landing-hoax conspiracy are preposterous. In my experience, most of those who were personally involved or well educated about NASA's effort to reach the Moon find it difficult to take Moon-landing-hoax believers seriously. But while the believers may come across as comical in their denial of what was one of the most thoroughly documented events of all time, laughing at them or dismissing their claims without comment may not be the best reaction. They have reasons for not believing, and those reasons can and should be dealt with. So let's investigate this amazing claim that history's greatest technological achievement never happened.

THE WORLD'S GREATEST LIARS?

Neil Armstrong may have better name recognition, but my pick for the all-time greatest spaceman is John Young. His résumé reads like the history of human space exploration: US Navy fighter pilot, test pilot, two Project Gemini missions in Earth orbit, two Apollo missions to the Moon (commander of the *Apollo 16* mission), two space shuttle missions (commander of the first-ever space shuttle flight), and decades of service to NASA down on the ground, where he has worked on everything from the challenges of living in space long-term to the threat supervolcanoes pose to life on Earth. His most spectacular achievement came when he commanded the 1972 *Apollo 16* voyage to the Moon. Young spent approximately three days on the lunar surface, exploring, collecting rocks, and conducting experiments.

Or did he?

Was the Apollo program just another case of The Man sticking it to the naïve peasants and making fools out of all of us? It is not difficult to imagine evil conspiracies lurking in the shadows during the late 1960s and early 1970s when the Moon landings took place. None other than "Tricky Dick" himself, Richard Nixon, was president. The American social scene during this period certainly was chaotic and disturbing. There was the Vietnam War, race riots, the Robert Kennedy and Martin Luther King assassinations, horrific pollution, the constant threat of nuclear war, disco music, and so on. But faked Moon landings? I don't think so. To go from Watergate to Moongate is a huge leap, one so enormous that it should not be taken seriously without very good evidence to back it up. And this, no surprise, is the key problem hoax believers have. There is no evidence to support the claim. All they have are baseless accusations and a series of questions that intrigue people who don't know much about the Moon landings.

THEY LOOKED ME IN THE EYE

I have been fortunate in my career as a writer to have met and written about many key people from NASA's glory days when it successfully pulled off the greatest adventure in all of human history. Here's is a partial list of my close encounters:

Scott Carpenter (Mercury Seven astronaut)
Tom Stafford (*Apollo 10* commander)
Frank Borman (*Apollo 8* commander)
Rusty Sweickart (*Apollo 9*)
Jim McDivitt (*Apollo 9*)
Buzz Aldrin (*Apollo 11*)
Alan Bean (*Apollo 12*)
Gene Cernan (*Apollo 10, Apollo 17* commander)
Dave Scott (*Apollo 9, Apollo 15*)
John Young (*Apollo 10, Apollo 16* commander)
Charlie Duke (*Apollo 16*)
Gene Kranz (Mission Control Center flight director)
James O'Kane (Apollo spacesuit engineer)
Jack Cherne (lunar module engineer)
Walter Jacobi (engineer, member of Wernher von Braun's rocket team)

After many hours of both formal interviews and casual conversations with these men, I never once felt a hint of suspicion that any one of them was lying to me. Keep in mind that I'm the sort of guy who clings to skepticism like it's a life raft at sea. My baloney detector is always switched on. If I picked up on the slightest hint that even one of these men was pushing a made-up story, I would jump all over it. But their detailed memories of the space program and, specifically, the recollections of those who walked on the Moon were absolutely convincing to me. I have looked into their eyes and asked them about that "magnificent desolation" that orbits 240,000 miles away from Earth.

Their responses were every bit as credible as they were fascinating. Isn't it unlikely that so many men could tell such a massive lie in such great detail and with flawless delivery?

Another pertinent question is why many of the Moonwalkers would add personal and emotionally charged layers to a hoax. *Apollo 16* astronaut Charlie Duke lived on the Moon for three days in 1972. He told me how he left behind a photograph of his wife and children as a sort of eternal monument to his love for them. *Apollo 12* astronaut Alan Bean explained to me that his memories of walking on the Moon inspired him to become an artist. Today he is a commercially successful painter of space and lunar scenes. *Apollo 17* commander Gene Cernan told me how he felt humbled and spiritual as he stared up at Earth—his tiny, faraway home hanging all alone in the cold darkness of space. He also recounted for me how he wrote "T D C," his daughter's initials, into the lunar dust. "I guess it will be there forever," he said, "however long 'forever' is."[1] If these guys are all in on a big, fat hoax, then they sure are being excessive with the corroborating lies.

Cernan described to me how meaningful it was for him to have commanded the 1972 *Apollo 17* mission that put the first scientist on the Moon (geologist Harrison Schmitt). "The Moon was my Camelot for three days. The science was exciting," Cernan said. "We saw things that no human had ever seen before."

After spending more time on the Moon than any astronaut had before, Cernan says he thought to himself, "Pinch yourself one more time and make sure this is real. You're going to be out of here soon, so appreciate it."

"When you look back at the Earth," Cernan added, "it is so overwhelming, so powerful and beautiful. You see no borders, no language differences, no color differences. You don't see terrorism. I just wish I could take every human being up there and tell them to take a look."[2]

It is not only the detailed and personal stories that stand out. Obvious passion for space exploration—past, present, and future—is evident in all these men decades later. There is enthusiasm and pride in their words and facial expressions. Yes, they really did go to the Moon. It's either that or NASA had to have assembled a team of actors far better at the craft than anything Hollywood's best casting agents could have managed.

In addition to the astronauts, thousands of people worked on the Apollo program. In my view, there simply are too many people describing the same story of a grand vision, hard work, and spectac-

Figure 3. *Apollo 16* commander Gene Cernan on the Moon. The six Apollo Moon landings that took place in the 1960s and 1970s were among the most thoroughly documented events in all of history. Photo by NASA.

ular success. This is a critical problem with the Moon-landing-hoax claim: too many people and too many layers of detail. Literally hundreds of thousands of people worked on the project. Landing on the Moon was an extremely complex achievement, and if any one phase of it—the *Saturn V* design, the fuel, the command module, the lunar module, the computers, the navigation, the suits, life support—failed, then the landings could not have happened. Surely somebody—at least one honest person—would have spoken up and presented evidence by now if his or her little corner of the vast project were a sham.

The astronauts, Mission Control Center personnel, and engineers I have spoken with over the years certainly did not strike me as puppets in some Nixononian-NASA conspiracy. None of them seemed to me like the sort of personalities who would dedicate their entire lives to upholding a colossal lie. My sense is that these were honorable and proud people. If they took part in a hoax, isn't it likely that at least one of them would have come forward and confessed now that they are approaching their twilight years, if only to ease their conscience before dying?

We can't forget the Soviet Union in all of this. It is very unlikely that America's technologically advanced Cold War rival would have been fooled by such a daring and elaborate hoax. And, if the Soviets did know, it is difficult to imagine why they would have played along and not exposed the lie in order to claim a propaganda victory over the corrupt capitalists. Finally, if the United States had faked the Moon landings out of some perceived need during the Cold War, don't you think it likely that the collapse of the Soviet Union and the end of the Cold War in the 1990s would have loosened up a few tongues? Surely someone would have come clean by now.

Another curious problem with the hoax claim is that NASA did not go to the Moon just once. They made the trip nine times! Apollo missions 8, 10, and 13 orbited the Moon without landing. The *Apollo 13* mission was supposed to be a landing but was aborted due to a malfunction that nearly killed the crew. If the lunar landings were faked, why would NASA go to all the trouble of faking a failed landing too?

Apollo missions 11, 12, 14, 15, 16, and 17 landed on the Moon. The last two missions spent approximately three days each on the lunar surface. Why? If this was a NASA con job, why would they have repeated it so many times, increasing the risk of discovery with each mission? One faked landing followed by a big parade and an abrupt end to the program would have made more sense. Why would NASA stretch the fraud to include ten more Moonwalkers and five more command module pilots after the *Apollo 11* mission? Why make so many more men responsible for pushing the lie for the rest of their lives when limiting it to one three-man crew would have been safer?

INTRIGUING CHALLENGES, SIMPLE ANSWERS

I have written many newspaper and magazine articles relating to the Apollo program, including a few about Moon-landing-hoax claims. These have sparked many discussions (and a few arguments) with people who are convinced that men such as Armstrong and Young are among the greatest liars of all time. Over the years, I have learned that there are a few particular points that are most popular with hoax believers and always seem to come up. Each of them is easy to explain.

No Stars!

Look at photographs of Apollo astronauts on the surface of the Moon and you won't see any stars. There is only blackness behind and above them. I have several beautiful autographed photographs of astronauts on the Moon, and none of them have even one visible star. Hoax enthusiasts have seized on this as evidence that the photos were taken inside a movie set somewhere on Earth. This is a stunningly bad argument. Those who insist on believing that the landings were faked should not base their case on this one. There are better wrong arguments available.

Although writing has always been my primary gig, I'm an accomplished photographer as well. I'm no Ansel Adams, but I certainly have seen more of this world through a camera lens than most. Literally thousands of my photos have been published in newspapers and magazines. I once placed third in the British Broadcasting Union's Commonwealth Photographic Awards, an international competition open to professional photographers in more than fifty nations. I won the Canada and Caribbean competition section of that contest. I have taken photos of people, places, and animals in more than twenty-five countries across six continents. I have taken photos in just about every imaginable situation and environment. As a result, I find this specific hoax evidence particularly irritating because it is incredibly silly.

As a former sports photographer, I have taken countless photographs of athletes at night in lighted stadiums. Guess what? In every one I've taken of football players, soccer players, track athletes, and the like, no stars can be seen in the background of the photos. How can this be? Did I fake the sports photos? Did I really take them in a movie studio with a backdrop that was painted black? Where did the stars go?

The explanation is simple. To take a good photograph of a runner on a stadium track at night—or an astronaut on the Moon wearing a white suit—the camera's exposure must be set correctly according to the amount of light reflected by the primary subject. That means you have to make sure the camera will let in just the right amount of light for just the right amount of time. But here's the catch: If the exposure is correct for the person you are trying to photograph, it won't be correct for the faint light of stars in the sky behind them. Under routine photography conditions, it's one or the other. You can't get both the person and the stars in the same shot without making a special effort to do so. It works the same way on the Moon.

The Case of the Missing Crater

Another popular issue with hoax believers is the missing blast crater that they say should be under the landed lunar module but isn't. Pictures show no noticeable craters under the spacecraft that landed on the Moon. How, they ask, could a rocket ship with a powerful engine (ten thousand pounds of thrust) land on the dusty surface of the Moon and leave no visible crater? Again, this is an easy balloon to pop—embarrassingly easy. No conspiracy theorist worth his or her decoder ring should go anywhere near it.

The lunar module is one of the greatest technological achievements of all time. I believe it ranks up there with the Oldowan hand axe and the computer. It was built and used for space flight exclusively. That's why it looks so weird. It didn't have to be aerodynamic, so it wasn't. It's a testament to engineering efficiency and creativity. On six occasions, a lunar module dropped down from orbit like some invading metallic arthropod and safely delivered two astronauts to the Moon's surface. It was a superb machine, one we should all be proud of. So why didn't it create a giant crater when it landed?

The lunar module's descent stage engine didn't blast a hole on landing because its ten-thousand-pound-thrust engine was throttled back, way back, on descent. It had to be, right? Otherwise it couldn't land! Look at it this way, why aren't there hideous skid marks all over your driveway? After all, your car is probably capable of going in excess of one hundred miles per hour. So when you pull into your driveway at one hundred miles per hour and stop suddenly, you must make a noisy, messy, and chaotic stop, right? But wait, you don't do that. Just like the Apollo astronauts who flew the lunar module, you ease off the power well in advance of the stopping point so that you can control your arrival onto the driveway and make a gentle stop. No skid marks in your driveway, no crater under the lunar module.

The Flag Was Waving!

One of the lowlights of *Conspiracy Theory: Did We Land on the Moon?* a terrible pseudodocumentary that aired on Fox network in 2001, was video of astronauts standing around what the narrator claims is an American flag waving in the wind. There is no wind on the Moon, of course, so the charge was that the flag's movement was caused by a gust of wind from a door left open on the film set or perhaps an untimely blast from some air conditioner vent. Could this be it? Is this the smoking gun that proves the Apollo Moonwalks were all faked here on Earth? Of course not. The flag does not wave in the wind.

First of all, there is a support arm that extends out from the flag pole at a ninety-degree angle. It's there so the flag can be spread out and displayed in an environment with no wind. Second, the flag is not moving as a result of wind. It moves because of the astronauts' *contact with the flagpole*. As they drove it into the ground, vibrations and twisting caused the flag to swing back and forth. It moves for a bit after they let go of the flagpole because of momentum.

Who Believes NASA Lied?

Fortunately, only 6 percent of Americans polled admitted to being Moon-conspiracy believers in a 1999 Gallup survey. Unfortunately, 6 percent of the US population equates to several million people. A Gallup spokesperson suggested that the Moon hoax is likely not a very popular belief because 6 percent of the population will say yes to just about anything that a pollster asks.[3] I'm not convinced that this is a minor problem, however. I have encountered many Americans who may not qualify as committed hoax believers but are dedicated doubters nonetheless. Additionally, Moon-hoax belief rates are likely to be significantly higher in countries that have low rates of science and history literacy and/or a common mistrust of virtually anything the US government claims.

What does the future hold for the Moon-hoax belief? Will it fade away in time like most crackpot theories usually do? Or will it only get worse as the landings become "ancient history" in the minds of new generations? As the Apollo program recedes further into history, it may be easier for young people to fall for the hoax claim. A chilling warning of this may be a 2004–2005 survey that found 27 percent of

Americans aged 18–25 had at least "some doubt" that NASA went to the Moon, while 10 percent indicated that it was "highly unlikely" that anyone had ever landed there.[4] America is by no means the only place where this bizarre belief exists. A 2009 survey commissioned by an engineering and technology magazine found that a quarter of British people think the landings were faked.[5] Based on personal experience, I would estimate that the percentage of hoax believers in the Caribbean overall is closer to 50 percent at least. In Jamaica and Cuba specifically, random conversations I had with many local people led me to believe it could be beyond 75 percent there. It might be even higher than that in some Middle Eastern countries, where mistrust of anything to do with the United States is common.

The reasons for the strange idea that we never went to the Moon probably has a lot to do with the appallingly low levels of science literacy and historical knowledge that are common in the United States and throughout the world. For example, 47 percent of Americans don't even know how long it takes Earth to revolve around the Sun (one year).[6] Worse, 18 percent of Americans don't even know that Earth revolves around the Sun! They think it's the other way around.[7] It is not uncommon to hear about large numbers of high school students not knowing basic information such as whether the Civil War came before or after World War I or who the United States fought against in World War II. So is it really all that surprising that some people don't know that twelve men walked on the Moon more than forty years ago? As for the hardcore Moon-hoax believers, I suggest giving up on wind-blown flags, starless photos, and missing craters. Those are dead-end trails. While one certainly could argue that a large government conspiracy of this sort could have been attempted, it's difficult to imagine how it could have succeeded and never been exposed all these years later. Without evidence, this claim goes nowhere. It's nothing more than a wild story, no better than the Roswell spaceship crash.

I always try my best to look for the bright side in things, and I managed to find something positive in this Moon-landing-hoax belief. I see it as understandable, maybe even appropriate, that some people refuse to believe that human beings have visited the Moon. Apollo was such a fantastic achievement in the 1960s and 1970s that it *should* be difficult to comprehend, at least at first glance. Everyone won't readily absorb something as profound as our first journey to somewhere beyond Earth. Sometimes I look up at the full Moon on a clear night and still shake my head in amazement. "Wow, twelve guys walked around up there!"

To have made it to the Moon less than one hundred years after the Wright brothers flew at Kitty Hawk was a stunning feat, apparently one too spectacular for some people to accept as real. That's how special the lunar landings were. Hoax believers are just one more indication of Apollo's greatness.

GO DEEPER . . .

Books

- Bean, Alan. *Apollo*. Shelton, CT: Greenich Workshop Press, 1998.
- Cernan, Gene. *The Last Man on the Moon*. New York: St. Martin's Griffin, 2000.
- Chaikin, Andrew. *Man on the Moon: The Voyages of the Apollo Astronauts*. New York: Penguin Books, 2009.
- Collins, Michael. *Carrying the Fire: An Astronaut's Journeys*. New York: Farrar, Straus, and Giroux, 2009.
- Kranz, Gene. *Failure Is Not an Option: Mission Control from* Mercury *to* Apollo 13 *and Beyond*. New York: Simon and Schuster, 2009.
- Plait, Phil. *Bad Astronomy: Misconceptions and Misuses Revealed, from Astrology to the Moon Landing "Hoax."* New York: Wiley, 2002.
- Ottaviani, Jim, Zander Cannon, and Kevin Cannon. *T-Minus: The Race to the Moon*. New York: Aladdin, 2009.

Other Sources

- *For All Mankind* (DVD), Criterion.
- *From the Earth to the Moon* (DVDs), HBO Films and Tom Hanks.
- *In the Shadow of the Moon* (DVD), Velocity/Thinkfilm.
- *Magnificent Desolation: Walking on the Moon* (DVD), HBO and Playtone.
- *To the Moon* (DVD), NOVA.
- *When We Left Earth: The NASA Missions* (DVDs), the Discovery Channel.

Chapter 11

"ANCIENT ASTRONAUTS WERE HERE."

What can be asserted without proof can be dismissed without proof. . . .

—Christopher Hitchens

I owe a huge debt of gratitude to *Chariots of the Gods?* author Erich von Däniken. His 1968 book was a bestseller and later spawned a documentary in the 1970s that faithfully presented his case that advanced space travelers visited Earth thousands of years ago and interacted with ancient humans. At some point along the way in my childhood, I saw the documentary on TV and was amazed. I may have only been nine or ten years old at the time, but I was bright enough to recognize that this was a big deal. Aliens had influenced, maybe even started, human civilization! I was shocked and so impressed by how all the evidence converged on the same obvious conclusion: extraterrestrials were here! And the evidence is all over the place. Clearly, extraterrestrials helped build massive structures all around the world and many cultures remember them in folklore and art.

Yes, like millions of others, Von Däniken fooled me. He reeled me in, hook, line, and sinker, and had me believing that extraterrestrials built the Egyptian pyramids, placed those giant statues on Easter Island, had a spaceport with runways in Peru, and maybe even mated with us. The documentary inspired me to find Von Däniken's book at the library. An interesting side note is that the original 1968 publication of Von Däniken's book was titled, *Chariots of the Gods?* Today that question mark has been dropped, however, making the title a declaration rather than a question. But this is a complete reversal of reality. As more people have looked into the ancient astronaut idea over the years, Von Däniken's claims have only become less convincing and his evidence exposed as obviously out of line with the facts. I found the book at the library but don't recall if I actually read it. I do remember seeing the photos inside of it and feeling convinced that there must be something to this story. I was particularly impressed by a photo of ancient artwork that seemed to show an astronaut at the controls of a spaceship. Looking back, I can forgive myself for being a naïve child, but why exactly did I fall for Von Däniken's fairy tale?

The reason is obvious: I was a bright and curious kid who knew *what to think* when it was test time in school but didn't yet know *how to think* in everyday life. No one at school or at home ever explained to me that it's necessary to challenge weird ideas, to react to unusual claims with skepticism, to ask for evidence, and to make the effort to separate science from pseudoscience. Like most kids, I was left on my own to stumble around in a world teeming with errors and lies and hope for the best. I was a defenseless, trusting child, and Von Däniken was targeting me with precision-guided nonsense designed to herd me straight to the conclusion he wanted. It wasn't a fair fight.

Fortunately, something happened on my way to a lifetime of playing the victim to con artists, quacks, and fools. NOVA, the excellent PBS science program, produced an episode that analyzed the claims made in *Chariots of the Gods*. I happened to see *The Case of the Ancient Astronauts* sometime in the late 1970s, and it changed everything for me. The NOVA program did a brilliant job of showing how Von Däniken's claims were hollow and unscientific. Maybe extraterrestrials did land here thousands of years ago, but NOVA made it clear that *Chariots of the Gods* fails to prove it. What struck me is how misleading and loose with the facts Von Däniken had been. It was like a light went off in my head, "Oh, now I get it, you can't trust everything you read and see on TV, even if it looks and sounds like real science." The NOVA show put real experts and real facts up against Von Däniken's story, and it all collapsed like a house of cards. Credible archaeologists said that ancient Egypt didn't spring up from nowhere, as Von Däniken claimed. They also explained that ancient Egyptians were intelligent and perfectly capable of building the pyramids without alien assistance. I heard reasonable interpretations of the artwork Von Däniken highlighted that didn't require aliens at all. The experience of being hoodwinked and then enlightened changed me forever. Never again would I believe something important without pausing to question it. Being in a book or a documentary doesn't mean something is true. Post–*Chariots of the Gods*, I was different. Now I would think for myself and use skepticism and critical thinking to defend myself against people who would have me believe unproven things. Thank you, Erich von Däniken. If you hadn't suckered me with that ridiculous book of yours and set me up to learn an invaluable life lesson in childhood, I'm scared to imagine where my mind would be today.

WHO NEEDS ALIENS?

Let's explore some of the problems with the claim that ancient astronauts once visited Earth and left ample evidence. First of all, I have no problem admitting that it could have happened. Earth is approximately 4.5 billion years old and anatomically modern humans have been around for more than one hundred thousand years. The universe is about 13.7 billion years old, plenty of time for intelligent life to have evolved on many other planets. Other civilizations could be out there. And somebody from somewhere could have dropped by for a visit in the past, assuming they were patient travelers or figured out how to move around our very large universe at higher speeds than we think are possible. But an unusual and spectacular claim like this needs something more than "it's possible" to be worth believing. Good evidence and compelling arguments are not to be found in *Chariots of the Gods*, however. For example, Von Däniken makes a big deal about old artwork depicting what he says look like astronauts wearing spacesuits and helmets. But that's his interpretation. Another, more likely, interpretation is that these are nothing more than depictions of tribal people wearing clothes and headdresses. This explanation is simple and doesn't require going to extraordinary lengths to accept. But Von Däniken's claim is gargantuan and needs to be backed up by a mountain of supporting evidence, which he does not have.

Von Däniken and others who work so hard to push the ancient astronaut story on unsuspecting minds are at their most offensive when they claim that ancient people could not have built the things they built. In their view, humans who lived a few thousand years or so ago were too dumb to have engineered and constructed large, complex projects. Don't believe it. I have been inside a burial chamber deep in the Great Pyramid of Khufu at Giza. I also have been in several tombs in the Valley of the Kings near Luxor. While I admit to being thoroughly impressed by the scale, age, and effort behind these works, not once did I see anything that made me suspect it was beyond the capabilities of human beings—even those who lived a very long time ago. Yes, it was a mighty feat to work with an estimated two million stone blocks weighing more than a couple tons each. It's true that scholars do not know with absolute certainty exactly how it was done. Most likely it was accomplished with the clever use of cranes, ramps, and a huge labor force. However, what can be said for sure is that the reasonable idea that humans built the pyramids is much more likely to be accurate than the radical idea that aliens did. The former has evidence, logic, and common sense going for it, while the latter has none of those things.

University of South Florida archaeologist Nancy White's feelings about Von Däniken's claims are typical of most archaeologists who remain unconvinced. "I only read it once long ago, but the biggest problem is that it's not science," White said. "He was, by profession, a motel operator."[1]

White, the author of *Archaeology for Dummies*, strongly rejects the idea that ancient people were incapable of designing and building the impressive remains attributed to them.

"The ancient human mind of any prehistoric time period was just as complex as the modern mind, possibly more so before people had writing to help them remember things. Real archaeology should be exciting enough on its own."

The late Carl Sagan felt the same way:

> Fundamentally, what Von Däniken has done is to sell our ancestors short, to assume that people who lived a few thousand years ago or even a few hundred years ago were simply too stupid to figure anything out, certainly to work together for a long period of time to construct something of monumental dimensions. And yet people of a few hundred or a few thousand years ago where no less intelligent, no less capable, than we are. Perhaps in some ways they were better able to work together.[2]

It is shocking to me how many people believe that the pyramids in Egypt were built by extraterrestrials. I've encountered this belief all over the world—except in Egypt, interestingly. The Egyptians I met were all quite certain that their ancestors deserved the credit. I once encountered this weird belief while talking about one of my books, *Race and Reality*, on a radio talk show. A caller wanted to know why

I hadn't included information about "who really built the pyramids" in my book. There I was, all geared up to talk about the concept of race and human diversity, but somehow the silly and unrelated idea of pyramid-building aliens took center stage. I politely explained to the caller that the world's leading archaeologists have no doubts about who built these structures, as well as every other ancient structure we know of. I added that human brains back then were virtually the same as ours, so there is no reason to think that ancient Egyptians were not capable of building them. He scoffed and insisted that it's impossible. Even today we couldn't build pyramids like that if we tried, he declared. Straight out of *Chariots of the Gods*, I thought to myself. After all these years, Von Däniken still haunts me.

Donald Redford, professor of classics and ancient Mediterranean studies at Penn State, says there is no big mystery about the Egyptian pyramids. He estimates that twenty to thirty thousand workers probably built the Great Pyramid at Giza in less than twenty-three years. No magic or extraterrestrials required. It was primarily just ropes, lubricated surfaces and a lot of pulling. Whenever someone challenges him about this, he says he simply shows "a picture of twenty of my workers at an archaeological dig site pulling up a two-and-a-half-ton granite block. I know it's possible because I was on the ropes too."[3] There are still unanswered questions about how various aspects of the construction, but so what? The absence of an answer to something is not a reason to plug in miracles, magic, or aliens. Besides, the ancient Egyptians weren't the first people to do some heavy lifting. A site in Turkey called Gobekli Tepe may be the world's oldest religious temple. Dated at 11,600 years old, it includes massive stone pillars that are eighteen feet tall and weigh sixteen tons. Ancient people were smart enough to cut these stones and move them into place without the use of wheels or large domesticated animals. And they did it some seven thousand years *before* the Egyptians built their pyramids at Giza.

I am baffled by Von Däniken and others who claim that the pyramids at Giza suddenly materialized in human history and that such large and challenging construction projects could not have sprung up without alien involvement. I visited Saqqara in Egypt, where I saw the large stepped pyramid of Djoser. It's older than the more famous pyramids at Giza and—guess what?—it looks just like the less sophisticated earlier version of those pyramids one would expect. I also saw even older tombs, called mastabas, which also seemed to me like clear

Figure 4. The Pyramid of Djoser at Saqqara in Egypt is a problem for Erich von Däniken and others who claim that the more famous pyramids at Giza appeared suddenly in Egyptian culture, indicating that the Egyptians would have to have had extraterrestrial assistance in building them. This pyramid is older and less technically sophisticated than the Giza pyramids. It is an example of precisely the sort of architectural, engineering, and construction progression that one would expect to have led to the Giza pyramids. Photo by the author.

points along a learning curve or engineering development for ancient Egyptian builders.

Von Däniken's claims were wildly popular in the 1970s and still resonate with millions of people today. This is unfortunate because it leads people away from a connection they could be feeling for the many fascinating and brilliant people who came before us. Here is a small sampling of Von Däniken's condescending—and thoroughly unjustified—attitude toward ancient people as well as his odd underestimation of contemporary construction capabilities:

> If the Stone Age cavemen were primate and savage, they could not have produced the astounding paintings on the cave walls.[4]

> The Great Pyramid is visible testimony of a technique that has never been understood. Today, in the twentieth century, no architect could build a copy of the pyramid of Cheops, even if the technical resources of every continent were at his disposal.[5]

> Where did the narrators of *The Thousand and One Nights* get their staggering wealth of ideas? How did anyone come to describe a lamp from which a magician spoke when its owner wished?[6]

> It is an embarrassing story; in advanced cultures of the past we find buildings that we cannot copy today with the most modern technical means.[7]

All of this is incorrect. Even if extraterrestrials actually did visit Earth thousands of years ago, Von Däniken's arguments for it are wrong. Ancient people were obviously capable of brilliant and creative achievements. We know this because we have plenty of good evidence that proves it—including Gobekli Tepe, the pyramids in Egypt, stone statues on Easter Island, the Roman Colosseum, the Parthenon, and cave art left by prehistoric people. We also know that the human brain has been powerful, capable, and creative for many thousands of years. Finally, it is difficult to imagine any ancient structure anywhere being beyond the comprehension of today's engineers and builders. Von Däniken claims this repeatedly, but that doesn't make it true.

Sadly, the baseless claim of ancient space travelers visiting Earth has not faded away. If anything, it may be stronger today as more and more people are taken in by it. There are numerous websites and copycat books out there, and the History Channel seems to be on a mission to create as many new believers as possible with its steady stream of pseudodocumentaries about ancient alien astronauts. As I write this chapter, Von Däniken's book *Chariots of the Gods* is ranked number seven on Amazon in the "Astronomy and Space Science" category, right up there with Stephen Hawking's latest work. In the "Ancient History" category, *Chariots of the Gods* is ranked number nine.

Sigh.

GO DEEPER . . .

Books

- Colavito, Jason. *The Cult of Alien Gods: H. P. Lovecraft and Extraterrestrial Pop Culture*. Amherst, NY: Prometheus Books, 2005.
- Feder, Kenneth L. *Frauds, Myths, and Mysteries: Science and Pseudoscience in Archaeology*. New York: McGraw-Hill, 2010.
- Stiebing, William. *Ancient Astronauts, Cosmic Collisions, and Other Popular Theories about Man's Past*. Amherst, NY: Prometheus Books, 1984.
- Story, Ronald. *The Space Gods Revealed*. New York: Harper-Collins, 1986.
- Wenke, Robert J. *Patterns in Prehistory: Humankind's First Three Million Years*. New York: Oxford University Press, 2006.
- White, Nancy. *Archaeology for Dummies*. New York: For Dummies, 2008.

Other Sources

- *Birth of Civilization* (DVD), National Geographic.

Chapter 12

"UFOs ARE VISITORS FROM OTHER WORLDS."

If the evidence were good enough, my colleagues and I would abandon our antennas and begin crawling the countryside. It would be easier and cheaper. It would also offer the tantalizing possibility of communication that was up close and personal. But after more than sixty years of UFO sightings, we still seem unable to come up with the good stuff. Physical evidence—a taillight or knob from an alien craft—is in short supply.

—Seth Shostak, senior astronomer at the SETI Institute

What we see is only what our brain tells us we see, and it's not 100 percent accurate.

—John Medina, *Brain Rules*

Believing is seeing.

—John Slader

Nobody wants this one more than me. I even have the credentials. A heavy dose of *The Twilight Zone* and original *Star Trek* episodes in childhood primed me to think outside the sphere. While other kids played marbles and obsessed over baseball cards, I was busy contemplating ways to defend Earth against extraterrestrial invaders. In adulthood I once had a dream that involved the Drake equation.[1] That's not normal. I've probably watched the films *Contact* and *The Day the Earth Stood Still* (1951 original) more times than any adult not currently living in his parents' basement. I can't recall many song lyrics, but I have no problem reciting Charlton Heston's classic line from *Planet of the Apes*: "I can't help thinking that somewhere in the universe there has to be something better than man." It's also very likely that I've read more books and watched more documentaries about astrobiology and space exploration than most astronauts.

A few years ago, I attended a lecture by SETI senior astronomer Seth Shostak with the barely restrained enthusiasm of a twelve-year-old girl at a Justin Bieber concert. My pulse quickened when he announced that if any extraterrestrial signals are out there, fast-improving technology gives us a very good chance of detecting them within the next twenty or thirty years. I was far too dignified to ask Shostak for his autograph after the talk—but there's no denying that I wanted it. I don't pretend to know that intelligent extraterrestrials exist. However, the idea that they probably do excites me. I may not be an ET *believer*, but I am definitely an ET *hoper*. If they ever really do land, the UFO enthusiasts are going to have to get in line behind me. Some years ago I literally announced to my family that if a spacecraft ever lands anywhere near our zip code, I'm going in to say hello, offer the first handshake, give the Vulcan salute, or whatever. My precocious children were quick to warn me about vaporization, enslavement, uncomfortable probing, radiation, and alien pathogens. But I don't care about their nagging over details. It would be a moment too big, too exciting, and too important to shrink from. First contact is mine; I'm going in.

I have always preferred to give UFO claims a fair hearing rather than dismiss them automatically. Maybe it's because I have the heart of a UFO believer, if not the mind. Or perhaps it is because I have seen weird things in the sky too—twice. The first time I was ten years old, cruising down some dreary Florida highway in the car with my father when I spotted several lights streaking across the sky at high speeds. They did this repeatedly and flew together in loose formation. I can't remember much detail, only that I was amazed by what I was seeing and pretty sure they couldn't be airplanes based on how sharply they turned. If someone had pressed me afterward to explain what I saw, I assume I would have said that what I saw was definitely something strange—probably alien spaceships. Today, however, if I saw the same thing I would stop well short of suggesting that I saw alien spaceships. I'm not ten years old anymore, and I now understand how easily the human mind can be misled, particularly when the eyes feed it unusual and unexpected images. Looking back, a more likely explanation for what I saw might be that it was nothing more than the reflected lights of passing cars on the window of the car I was in. Then again, maybe it *was* an alien armada passing through the solar system. I'll never know for sure, and I have to live with that.

A few years later, I was outside playing in my neighborhood and saw several tiny lights in the sky moving very fast. Again, I didn't think they moved like jets. Maybe it was a meteor shower. Or maybe it was squad of alien anthropologists observing the play habits of Earth boys from high altitude. What were they? What did I see on those two occasions? The only honest answer I can give for both events is "I don't know." Not everyone is willing to accept that kind of uncertainty, however.

People all around the world often see weird things in the sky that they cannot readily identify. But some are unwilling to admit that these things they see are unidentified and leave it at that. So they jump to the extraordinary conclusion that the objects must therefore be extraterrestrial spaceships. The point that needs to be emphasized and repeated often when talking about UFOs is that seeing something that is an *unidentified* object is not the same as seeing an *identified* alien spaceship. The jump from one to the other is not reasonable or logically defensible. The unidentified nature of such sightings means they could be labeled just about anything. If, for example, fairy belief was widespread today, then all these hovering and streaking lights in the night sky would likley be identified as fairies rather than spacecraft from another planet. No matter how strange a light in the sky may seem, there can be no justification for pretending to know that it's a spaceship unless one can clearly see the hull of a star cruiser with nonhuman beings waving from the portholes.

Since not everyone is aware of this, it's always worth stating the fact that no one has ever produced any scientifically confirmed evidence of a spaceship having visited us. This absence of evidence is crucial to the issue and should be the focal point of every UFO discussion. Nothing, not one piece of a flying saucer has ever been found and passed on to the world's scientists and skeptics so they could examine and verify its extraterrestrial origins. Nothing. We do not have a single piece of exotic metal that could not possibly have been crafted by humans, no devices that baffle the best engineers, and—despite what you may have heard—not even one alien body. Sure, it's *possible* that such evidence exists in Area 51 or some top-secret government vault in a hollowed-out mountain somewhere. But until somebody releases it, leaks it, or steals it and shows us, it's silly to pretend that we know it's there. All we have are eyewitness accounts of weird things in the sky and stories of people being abducted or interacting with aliens. Based on what we know about the reliability of eyewitness accounts, however, this is just not good enough for something so unusual and important. It is known beyond a shadow of doubt that sane, intelligent, honest human beings are capable of misinterpreting what they see and misremembering what they saw. Therefore, before coming to the conclusion that extraterrestrials are buzzing around our planet, we must have hard evidence that scientists can analyze, test, and confirm. Short of an alien spacecraft landing in Times Square, nothing else will do for a claim this extraordinary and important. Tens of millions of people have decided not to wait for evidence, however.

According to a Gallup poll, 24 percent of Americans, 21 percent of Canadians, and 19 percent of Britons believe that "extraterrestrial beings have visited Earth at some time in the past."[2] A study by the National Council on Science and Technology and the National Institute of Statistics and Geography calculated that as much as one-third of the US population and 38 percent of Mexicans think that alien spaceships visit us.[3]

Although UFO belief is embraced by many millions of people worldwide, this does not necessarily mean that there is something bizarre going on above us that involves intelligent beings from another world. One often hears the "where there's smoke, there's fire" reasoning applied to the UFO issue. But we know that millions of people can believe in things that are 100 percent wrong. For example, millions of American adults—nearly 20 percent of the US population—think that the Sun revolves around the Earth![4] It wouldn't matter if 100 percent of Americans believed that; the Earth would continue to orbit the Sun. Never allow yourself to be too impressed by a belief's popularity. Reality is not determined by a majority vote. If some people can be wrong about a claim, then it's possible everybody could be wrong about it. Even if everyone on Earth believed in alien visits, that alone would not prove it. Only good evidence can do that. And good evidence is the one thing missing in the UFO phenomenon. After more than five decades of enthusiastic interest and investigation, we have nothing more than stories. I am not suggesting that anyone should belittle believers or dismiss all claims without a second thought. But if we strip away the hype and unjustified assumptions, it turns out that the vast majority of UFO reports describe precisely what that abbreviation was supposed to mean in the first place: *unidentified* flying objects. The failure to identify something as a plane, a satellite, a cloud, a flare, a balloon, Venus, or a remote control toy blimp simply does not justify jumping to the fantastic conclusion that it therefore must be an alien spacecraft. I think of most UFO belief as the intelligent design of the sky. Creationists/intelligent design believers point to the inability of today's biologists to answer every question about life and then draw extraordinary and wholly unwarranted conclusions. Many UFO believers duplicate the same error when they point to the fact that astronomers, the Air Force, the FAA, police, and so on cannot or will not identify every object in the sky to their satisfaction. The absence of readily available explanations is seen as evidence that an intelligent being created life or that unidentified objects must be spaceships. But this is flawed thinking; not knowing means *not knowing*. It may be distasteful because everybody loves closure. But sometimes we just have to be grown-ups and swallow a bit of frustration. There will always be unanswered questions, and inventing fictional solutions is not the best way to react to them. Don't forget, I'm a guy who would sacrifice body parts in order to make first contact and hitch a ride with extraterrestrials for a sightseeing tour through the M16 Eagle Nebula and beyond. But I understand how important it is to keep my hopes and fantasies properly corralled so that they don't compromise my skepticism and critical thinking skills. I think fantasy is wonderful, but I still want to live in the real world. The urge to believe should never overwhelm the need to think. And the lack of satisfying answers to everything should never excuse us to swap reason for made-up answers.

Most people underestimate how easy it is to misinterpret "things" we see in the sky. For example, an airplane approaching you from a distance can appear to hover even though it may be traveling at a high rate of speed. This might lead to the incorrect conclusion that "it can't possibly be a plane." Furthermore, some large planes *can* hover and then fly away. The unique twin-engine V-22 Osprey, for example, is capable of flying at speeds of more than three hundred miles per hour as well as hovering in one spot. An Osprey flying at night might move in ways that confuse an unaware ground observer into thinking that it couldn't be a plane or a helicopter. In addition, many military aircraft can eject flares that float down slowly on small parachutes to illuminate the ground below. Many planes and helicopters also use flares as defensive countermeasures to confuse or redirect heat-seeking missiles that may be targeting them. I have seen video of countermeasures on fixed-wing aircraft that rapidly spiral away from the aircraft on independent trajectories. Several flares launched rapidly from a jet at night at high altitude could be very difficult to correctly identify from the ground, especially if the supersonic jet that deployed them is already long gone. Many military planes and helicopters can also release chaff—numerous tiny bits of metal, plastic, or glass—as a countermeasure to defend against missiles that home in on targets using radar. Under some conditions, a cloud of these tiny objects might reflect light in ways that could mislead someone on the ground into thinking it is something it is not.

I proposed the possibility of flares from military aircraft being responsible for some UFO reports to Mark "Gunner" Lewis, a former US Marine avionics technician who spent many years in and around combat aircraft. He said he could easily imagine people who are already inclined toward UFO belief confusing countermeasures for starships. "[Countermeasure flares] make a distinct pattern," he explained.

> From a distance, far enough for the engines and propellers not to be heard, they may appear as slightly elongated lines or ovals. But I am guessing at that. They can be fired in a more staggered pattern as well.
>
> Although not a countermeasure, a field illumination flare might confuse civilians, and noninfantry related military, for that matter. This type of flare, when deployed, puts out enough candle power to rival an entire football stadium of lights. It is designed to illuminate a battlefield for several minutes and cause the enemy to lose their

night vision. There are different colors for different applications. The unusual thing about these flares is that they are suspended by mini-parachutes. The heat rising from the flare lifts the chute like a hot-air balloon, keeping it aloft. This intense light could act in many odd ways. It could appear to hover in one place and then disappear as it extinguishes. It could spiral and or bounce around at altitude. Being carried by thermal drafts or shear winds can make it look as if it were moving under power. I have seen some "UFO" video and recognized them as these types of flares.

Do I believe that people can be misled or, if predisposed, confuse countermeasures for alien spacecraft? My answer would be yes. With the intent to hoax and access to field flares, I think I could do a great job of making a "UFO sighting." I think anything from fireflies to the orbiting space station can be an alien craft for those who look for it.[5]

IT'S RAINING UFOs

Here's something else to think about: Every day some *four thousand tons* of rock and dust from space burn up in our atmosphere.[6] Sometimes this rain of debris creates brief displays of light that might be misinterpreted by whoever happens to be looking up at the time. And don't forget the way our minds work. When we see multiple points of light in the dark, it can be a natural and subconscious response to connect those dots and create a larger, solid object where there is none. Then we have all the celestial bodies that account for many UFO sightings. Venus, for example, can be very bright in the evening and has fooled people into thinking it's a close spaceship rather than a faraway planet. When you don't know what you are looking at in the sky, size and distance are difficult, if not impossible, to determine. The reason we can accurately estimate the size of birds and planes when we see them is only because we already know the size of birds and planes. The next time someone tells you he or she saw a strange gigantic green and orange light in the sky, explain this to them.

I'm slightly embarrassed to admit it, but I was fooled by the Moon. You may have noticed that a full Moon in the early evening can sometimes look extraordinarily large and dramatic when it's low and close to the horizon. Later, however, it seems to be back to normal full-Moon size when it's high in the night sky. Several years ago I actually chased this gigantic Moon with my camera and tripod in the hopes of capturing a spectacular image. Speeding around in my car, looking for the perfect spot to shoot from, I had high hopes for capturing the mother of all Moon photos. But I ended up frustrated, disappointed, and bewildered by pictures that showed nothing but a small Moon. I later learned that I had fallen victim to a well-documented Moon illusion. It's just a trick of the mind and eye, one that cameras don't fall for. If I couldn't even get the size of the full Moon straight in my head—even though I knew exactly what I was looking at—then I know I have no business making confident estimates about the size of some unidentified object I might see in the sky. Finally, add to all this the fact that every now and then some mischievous person launches a balloon with lights or flares attached just to see how many people fall for it and think it's a spaceship. Be on guard when you look up; it's an optical minefield up there.

DON'T FORGET HOW MEMORY WORKS

We all would be better off if we understood more about how our brains can mislead us when it comes to seeing and remembering. Perhaps, then, visual miscues and memory mistakes wouldn't trip us up so often. But this necessary knowledge can be harsh. Sometimes after learning what scientists have discovered about the workings of the human brain, I want to curl up in the fetal position and cry. Much of what I had assumed about the way in which we perceive the world around us has turned out to be far off the mark. For example, we don't actually "see" anything. What really happens is that the mind *processes patterns* received from the optic nerves and then the mind *creates pictures* of what it decides the scene being stared at is—or should be.[7] And memories? Forget about it.

A 2011 study found that more than half of the public, 63 percent, think that memory works like a video camera.[8] However, as mentioned in chapter 1, our brains do not record what we see, hear, and experience; and memory is not a simply playback of the recording. That's not even close to what happens when we remember things. There is not some sort of organic high-tech audio-visual record and playback system in your head. The way in which human memory actually works is more like having a little old man who lives inside your skull and *tells you stories* about what you saw or experienced in the past. But every time he tells you one of these stories about your past, he changes it by leaving out some parts and adding other parts that never happened. He might even decide to change the order of events, which means that the storyline in your head won't match reality. The old man in your head does this because he decides it's best for you to remember it that way. He alters your memories in ways he believes make them more sensible and more useful to you. Usually he's right, most of our memories do need to be edited and condensed to be functional. But sometimes this process causes problems, like in courtrooms, for example, where many innocent people have been sent to prison based on the inaccurate memories of witnesses. This is one more reason to be skeptical when considering eyewitness accounts of UFOs, ghosts, Bigfoot, angels, and so on, no matter how credible and sincere the person telling the story may be.

WHAT GORILLA?

Ever heard of inattentional blindness? If not, you better get up to speed on it because things are happening in front of your eyes all the time that you aren't seeing. Wizards, magicians, shamans, and pickpockets figured this out centuries ago. Because of the way the brain functions, we can be shockingly bad at noticing things. For details on inattentional blindness and other important brain phenomena that impact how we perceive reality, I recommend the books *Sleights of Mind* (by neuroscientists Stephen L. Macknic and Susana Martinez-Conde) and *Invisible Gorilla* (by cognitive psychologists Christopher Chabris and Daniel Simons). The authors of *Invisible Gorilla* are the researchers who conducted the now-famous experiment in which they asked people to count the number of times players on one team passed a basketball in a short film. (If you want to test yourself, don't read any further until after viewing the video.[9]) The test is challenging, as players circle while passing the ball while another team is doing the same thing. In the middle of the film, a woman in a gorilla suit walks into view, thumps her chest, lingers for a moment in the middle of the scene, and then walks off. The gorilla is on screen for about nine seconds. This experiment has been conducted numerous times with consistent results. About half of test subjects fail to notice the gorilla. It seems impossible, but it's true, *half* never see a gorilla right in front of their alert mind and open eyes! And it's not a question of intelligence or education. Harvard students fared no better than nonstudents. This is the "illusion of attention," scary stuff if you think about. It means we can stare directly at an object and miss something very unusual or important that is literally right next to or even passing in front of the target of our attention. We may think we see everything we need to see when something is right front of our open eyes, but often this is not the case. This is how human brains work, and it has obvious implications for many things—including UFO sightings.

IT CAN HAPPEN TO ANYONE

Virtually anyone can be fooled by an unexpected object or light in the sky. Philip Plait, a popular astronomer and skeptic, once had his own encounter with UFOs. He was waiting to watch a shuttle night-launch in Florida when he noticed a dozen or so unusual lights hovering in the distance. Even with binoculars he couldn't tell what they were. He eliminated likely objects such as planes or satellites and became increasingly excited until he was finally able to see that they were ducks. Yes, a world-class astronomer who lectures and writes with great passion about the virtues of skepticism—especially when it comes to UFO sightings—was perilously close to thinking a flock of waterfowl was the advance strike force of an alien invasion fleet. For the record, Plait says that, even in the heat of the moment, he never allowed himself to believe that the lights were extraterrestrials, but he does admit to having an "odd feeling" in the pit of his stomach and then feeling vaguely disappointed when the quacking former UFOs flew by. By the way, the birds glowed because of reflected light from bright lights at the launch pad.[10] If something like that can (almost) happen to an elite sky watcher like Plait, is anyone safe from being fooled? And just imagine if the birds had not flown his way but changed course and faded out of sight before he was able to identify them. Or imagine if it wasn't a skeptical astronomer observing them but instead was someone with a prior belief in UFOs/spaceships. Most likely it would be very difficult to convince that person that he or she probably only saw something as simple as flares, escaped birthday party balloons, planes, meteors, or a flock of birds.

I have waded through many books on this subject. One of the more intriguing UFOs-are-spaceships books I have read is *UFOs: Generals, Pilots, and Government Officials Go on Record*, by investigative journalist Leslie Kean. I think it promotes unjustified belief in UFOs and would have been much better if the author had balanced it with skeptical critiques of UFO claims. It's still an interesting book though, with accounts of some of the more credible UFO sightings. Kean complains that UFO claims are too often suppressed by governments and automatically dismissed by skeptics. I can't speak for how governments handle UFO sightings, but I certainly disagree that this is the case with most skeptics. No UFO skeptic I know, myself included, would reject a claim of an unusual sighting without first listening and thinking it over. We also don't feel the need to explain away every sighting because we understand that the absence of an explanation is not in itself proof of anything. I believe I'm in sync with the vast majority of UFO skeptics when I say we are open-minded about the possibility of a UFO being an extraterrestrial spacecraft but are not impressed by claims without evidence. Most UFO skeptics tend to be fans of science and would love to discover that we are not alone in the universe. Show us compelling evidence and you will have our undivided attention. Show us a spaceship and our skepticism about visiting extraterrestrials will vanish in an instant.

The most important point to be made about Kean's book is that it presents an impressive collection of highly credible eyewitnesses, people in positions of authority with professional aviation experience. However, after an initial flash of excitement while reading their stories, it's a quick return to Earth. One only has to remember that fighter pilots, airline pilots, generals, and aviation officials are human too. And as such they are vulnerable to the same natural mistakes of perception, interpretation, and memory that the rest of us are. An example of this would be police officers. They are trained to look for details that typical citizens would miss, but they still can and do make significant mistakes in seeing and recalling important aspects of events they witness.[11] If training, skill, and experience formed an impenetrable fire wall against misperception, illusion, hallucination, and faulty memories, then UFO stories from military and professional pilots would be all the solid evidence we need. But we know that this is not the case.

An astounding forty million Americans say they have seen a UFO.[12] Assuming respondents defined *UFO* as spaceship, that's a lot of people who are seeing what they think are alien visitors. But if extraterrestrials really are constantly buzzing around above all these witnesses, then where are all the high-quality photos and HD video of them? Think about the age we now live in. Who doesn't have a cell phone with a camera in it these days? Have you been to Disney World or a kid's birthday party lately? Virtually *everyone* is armed with a camera of some kind. We can add to these hundreds of millions of potential UFO photographers and videographers the hundreds of satellites and UAVs (unmanned drones) that are constantly monitoring and photographing the Earth. Shouldn't they be able to detect or capture images of all these low-flying spaceships that are supposed to be here? Finally, astronomer Andrew Fraknoi asks why the *tens of thousands* of amateur astronomers around the world who look up every night are not the ones who are responsible for the majority of UFO claims. He suspects the reason for this is because they almost always know what they are looking at. They, for example, are unlikely to mistake Venus or a meteor shower for spaceships.[13]

Seth Shostak of the SETI Institute points out that not only is the case for even one extraterrestrial visit unproven, but there is also nothing to show for all these decades of research and attention.

"Despite the fact that about one-third of the populace believes that aliens are visiting our planet, we've really learned nothing from that," he explained. "If the aliens are really here—which I strongly doubt there's been precious little effect on us. But space exploration has revealed countless new, fabulously interesting facts. As a simple example: Until the 1970s, the moons of Jupiter were just bright points of light in our telescopes. Now we know them in detail and have reason to think that some of them could be habitable. The same could be said of the Saturnian system, and of course Mars. Space exploration does lots more than merely fuel conspiracy theories and provide us with tales of strange lights in the sky."[14]

Since UFO sightings seem largely culture driven—why aren't unidentified flying objects assumed to be high-flying harpies or dragons?—I'll make a prediction. For the last five or six decades, virtually all UFO sightings have suggested large vehicles with roughly human-sized aliens aboard. Currently, unmanned aerial vehicles (UAVs) are being utilized increasingly by the US military and CIA around the world. Other countries such as Great Britain, Israel, and Russia are investing in UAVs too. Now the US military is adding insect-sized UAVs to conduct surveillance, and who knows what else, to their robotic fleet. Some are pure machines while others are living insects turned into obedient cyborgs. As the popular culture eventually becomes aware of these new technologies in the coming years, I predict we will begin to see new waves of UFO reports that describe extremely small extraterrestrial vehicles with tiny aliens aboard. Wait and see.

GO DEEPER . . .

- Bennett, Jeffrey. *Beyond UFOs: The Search for Extraterrestrial Life and Its Astonishing Implications for Our Future*. Princeton, NJ: Princeton University Press, 2008.
- Chabris, Christopher, and Daniel Simons. *The Invisible Gorilla and Other Ways Our Intuitions Deceive Us*. New York: Crown, 2010.
- Darling, David. *Life Everywhere: The Maverick Science of Astrobiology*. New York: Basic Books, 2002.
- Macknik, Stephen, and Susana Martinez-Conde. *Sleights of Mind: What the Neuroscience of Magic Reveals about Our Everyday Deceptions*. New York: Henry Holt, 2011.
- Roach, Mary. *Packing for Mars: The Curious Science of Life in the Void*. New York: W. W. Norton, 2011.
- Sagan, Carl. *Pale Blue Dot: A Vision of the Human Future in Space*. New York: Ballantine Books, 1997.
- Sheaffer, Robert. *UFO Sightings: The Evidence*. Amherst, NY: Prometheus Books, 1998.
- Shostak, Seth. *Confessions of an Alien Hunter: A Scientist's Search for Extraterrestrial Intelligence*. Washington, DC: National Geographic, 2009.
- Webb, Stephen. *If the Universe Is Teeming with Aliens . . . Where Is Everybody? Fifty Solutions to the Fermi Paradox and the Problem of Extraterrestrial Life*. New York: Springer, 2010.

Other Sources

- Bad UFO (blog), http://badufos.blogspot.com/.

Chapter 13

"A FLYING SAUCER CRASHED NEAR ROSWELL, NEW MEXICO, IN 1947 AND THE GOVERNMENT KNOWS ALL ABOUT IT."

Roswell is the world's most famous, most exhaustively investigated, and most thoroughly debunked UFO claim. It's far past time for UFOlogists to admit it and move on.

—B. D. Gildenberg, Project Mogul participant

It is one of the most important events in all of history. Technologically advanced extraterrestrials traveled some vast unknown distance to reach our planet in the summer of 1947. Tragically, however, the spacecraft crashed, killing the entire crew. The cause of the disaster remains a mystery. Perhaps a collision with a flock of birds doomed them, or maybe the alien crew made the mistake of opening a window and then lost control after being infected by Earth germs, a la H. G. Wells's *War of the Worlds*. Whatever the reason, their journey ended tragically in a lonely field near the small town of Roswell, New Mexico.

Some witnesses say there were multiple crash sites, indicating that more than one spaceship went down. A rancher discovered strange metallic debris, providing hard evidence of at least one downed space vehicle. The US military was quick to recover the wreckage and a few alien bodies as well. Initially, the army said it had "captured a flying disc," and this was reported by the local newspaper.[1] Afraid that an unhinged public would panic and riot in the streets if people learned what had happened in New Mexico, the government made the decision to execute a cover-up and initiate a strict policy of denial that continues to this day.

OK, SO WHAT REALLY HAPPENED?

The Roswell crash is a great story. Much as I would love to believe it, however, I can't. The story is just not good enough because there are very sensible, credible, and more believable alternative explanations for what happened. When one learns the real story of the 1947 incident and the way in which the Roswell myth was rehashed years later and then nurtured by the media and a town that likes tourism dollars, it becomes clear that if there's one place we can be sure aliens *did not* crash in 1947, it would be Roswell, New Mexico. After all, no other location on Earth has been as scrutinized and analyzed—without uncovering any evidence—as Roswell has been by UFO believers, UFO skeptics, military investigators, and journalists.

Something did happen near Roswell in the summer of 1947. A strange object really did fall from the sky, and the US military really did lie about what it was. Unfortunately for space enthusiasts like me, flying saucers and aliens had nothing to do with it. In the late 1940s, the US military was concerned about the Soviet Union becoming the second nation to have nuclear weapons. It was just two years after World War II and the Cold War was already under way. In the days before spy satellites, the Army Air Force (the US Air Force did not yet exist as a separate branch) established Project Mogul, a top-secret program to develop ways of monitoring Soviet nuclear bomb tests. Project Mogul used high-altitude balloons to carry electronic listening devices that were designed to detect the sound of distant explosions. An aboveground nuclear blast is so loud that researchers believed they could pick up the sound waves at high altitudes even halfway around the world.

According to Project Mogul participant B. D. Gildenberg, the work was extremely sensitive. It was so secretive, in fact, that many of the people involved didn't even know the name of project until many years later.[2] It was not declassified until 1972. The concern was that if the Soviets learned about what was going on, they would move their testing underground and make detection even more difficult.

The official explanation Project Mogul researchers gave to anyone who asked what they were up to was simply, "weather balloon research." But these were much more than basic weather balloons. For some flights, several very large high-altitude balloons were joined together with cords to form a "flight train" that could be as long as six hundred feet. And then there was the equipment hanging beneath the balloons, including spiked silver-foil-covered reflectors that enabled the balloons to be tracked by radar. Gildenberg says these elaborate balloon trains were the cause of many UFO sightings in the region during the project's run.[3]

Project Mogul was active in multiple locations. One place where balloon trains were launched was Alamogordo Army Air Base in New Mexico—just one hundred miles west-southwest of Roswell. Gildenberg is certain that the famed Roswell wreckage was nothing more than the remains of a Project Mogul radar reflector. When rancher Mack Brazel found the debris scattered across the ground on June 14, 1947, he initially ignored it. Several days later, however, the modern UFO craze took flight when Kenneth Arnold, a private pilot flying over Oregon and Washington State, reported seeing unusual objects that came to be known as "flying saucers." Soon after Arnold's story was published in newspapers nationwide, UFO sightings began pouring in across America. Gildenberg believes Brazel heard about these sightings after driving into town on July 5, nearly three weeks after he had found the wreckage. He then reevaluated the debris he had initially thought was unimportant and told the Roswell sheriff about it. The sheriff reported it to Roswell Army Air Field, a base that had nothing to do with Project Mogul and knew nothing about it.[4] Then somebody at Roswell Air Base gave the gift that keeps on giving to UFO believers everywhere. An overly enthusiastic press officer at the base issued a press release stating that recent rumors of "flying discs" had become reality and the Roswell Army Airfield had recovered one. The next edition of the *Roswell Daily Record* carried a front-page story with the headline: RAAF CAPTURES FLYING SAUCER IN ROSWELL

REGION. At this point, it's worth pointing out that "flying saucer" did not automatically mean "extraterrestrial spaceship" to everyone in 1947. Furthermore, Kenneth Arnold, the private pilot who made the famous "flying saucer" sighting several days before the Roswell "incident" never said he saw a flying saucer or disc. He said he saw flying objects that were shaped like large boomerangs. The press incorrectly reported his description, however, so what should have been the beginning of the "flying boomerang" craze became instead the "flying saucer" craze.

In the meantime, Major Jesse Marcel flew the material recovered from Brazel's ranch to Fort Worth Army Air Base. Once there, it was immediately identified by people who knew what it was. Photographs were taken of the material, and the press was told it was from a weather balloon. The media reported this to everyone's satisfaction. End of story—or it should have been.

It has to be emphasized that there was nothing exotic or mysterious about the debris. It consisted of balsa wood, thin aluminum foil-like material, and rubber—hardly the stuff of interstellar flight. If it really was wreckage from a spaceship, it means that the aliens are smarter than we could ever imagine because they would have traveled across the galaxy in something similar to a child's kite.

"The Roswell debris was simply and obviously a radar reflector from a balloon," states Gildenberg. "Once available, this official explanation was accepted as self-evident. All one had to do was look at the photo to be convinced."[5] And that was it; the Roswell crash story was dead. By the way, notice that there was no mention by anyone at the time of alien bodies being carried away from the crash site, no alien autopsies, no multiple crashes, nothing said about strange metal, and so on. All those claims came later, much later.

In fairness, conspiracy theorists are technically correct about the military covering up the truth and lying when they said the wreckage was from a "weather balloon." Clearly it was not. It was wreckage from a "spy balloon." Most people probably will agree, however, that this was not an evil or significant lie, certainly understandable during the early days of the Cold War.

A MYTH IS BORN

The big "flying saucer crash" of 1947 was exciting for about twenty-four hours. Then Americans moved on and forgot about it. But sometimes you just can't keep a good story down. The Roswell legend roared back with a vengeance thirty years later and doesn't seem to be going away anytime soon. Energized by decades of UFO sightings, science fiction books, TV shows and films, and some very questionable journalism, the story has become deeply entrenched in pop culture. Today Roswell, New Mexico, has a museum, bus tours to the "crash site," and even an annual festival dedicated to the 1947 non-event. The "crash" is often mentioned on TV and in films.

How did a spaceship crash that never happened become part of America's unofficial history? How did this happen? It certainly wasn't due to new and compelling evidence that emerged after 1947, that's for sure. The Roswell story returned from the dead because a few people made the decision to "reopen the case" and start asking people what they remembered. In 1978, UFO believer Stanton Friedman interviewed the major who recovered the material, Jesse Marcel. The *National Enquirer* also interviewed Marcel, who now added new information to his story, claiming that the debris was unusual and couldn't be burned, for example. Charles Berlitz, the same guy who wrote books about the Bermuda Triangle and Atlantis, coauthored a popular book about the Roswell crash.[6] This time the story was too hot, and too profitable, to flame out. The timing was convenient as well. In the late 1970s, the festering wounds of Watergate and the Vietnam War left many Americans well primed to believe that their government was lying to them about hoarding the wreckage in a secret facility somewhere. All this new attention led to more witnesses coming forward with increasingly astonishing claims.

In the 1980s, the Roswell story grew to include the recovery and autopsy of alien bodies. This is particularly interesting since nothing had been said about this by anyone back in 1947. Why now, after so many years, did people remember alien bodies? It would seem that witnesses were either lying or really did see military personnel recover dead aliens from the crash site. But there is a third possibility, one that makes a lot of sense.

In the 1950s, the US Air Force conducted more unusual balloon projects in the region. Some of these involved dropping equipment, test dummies in silver suits, and even a real live human from high altitudes. The test dummies in particular—falling from the sky and then being picked up and carried away by military personnel—likely played a role in feeding the Roswell myth.

MY INTERVIEW WITH A ROSWELL ALIEN

In 2001, I interviewed Joe Kittinger, one of the great aviation pioneers of the last century. He had a remarkable career as a test pilot, fighter pilot, and balloonist. Kittinger was first to solo across the Atlantic in a hot-air balloon, and he performed a successful parachute jump from the upper edge of the atmosphere in 1960. That spectacular leap from a balloon at 102,800 feet (nineteen miles) still stands as the highest parachute jump ever. Kittinger broke six hundred miles per hour during a free fall that lasted more than four minutes. In total, it took nearly fourteen minutes for him to reach the ground. With so much adventure and history to talk about, I was reluctant to even bring up Roswell during the interview. I eventually did, however, because I had heard that he may have inadvertently contributed to the myth by being mistaken for an alien. I sensed in his voice a bit of frustration over the subject but, to his credit, he seemed eager to clear the air.

"It never happened," Kittinger said. "There was a very top secret army project that was designed to detect when the Russians detonated a nuclear weapon. They sent a balloon aloft with a very long antenna array, almost five hundred feet long. It had very exotic-looking equipment on it. The balloon landed on a ranch near Roswell. The so-called alien spaceship was that balloon. It's turned into a cottage industry, and it put Roswell on the map. A lot of people want to believe it was aliens, and they want to believe there was a big cover-up. But I'll tell you, it never happened."[7]

Kittinger, it turns out, was the closest thing to a real alien back then, having plunged to Earth from the edge of space himself. He is certain that high-altitude balloons and those test dummies inspired the Roswell myth.

"Absolutely they did. These dummies that we dropped from balloons were dressed in pressure suits, so they looked unusual. One time we dropped one and it fell way up in the mountains. These dummies weighed more than two hundred fifty pounds. So how do you carry one out of the mountains? We put it on a stretcher and carried it to the back of an ambulance to take away. Now if somebody is back in the weeds watching this they are going to say, 'Wow, look at that alien they have there.' We think that a lot of the alien sightings were actually us doing our work with the test dummies."[8]

Project Mogul veteran Gildenberg also believes the dummies were behind the eyewitness accounts that came out decades after 1947: "Many aliens were described wearing flight suits identical in color and detailing to suits used on our dummies."[9] Additionally, the crash of a large KC-97 Stratotanker airplane in 1956 might have contributed to stories of alien bodies showing up at the Roswell Army Hospital. That accident killed eleven men. Their badly burned and disfigured bodies were recovered and taken to the hospital, where they may have been seen by future "Roswell witnesses." This possible explanation for some of the alien body sightings is detailed in the US Air Force's official study on it, "The Roswell Report: Case Closed," published in the 1990s.[10] Kittinger strongly endorses the report as the final answer to this myth. "Anyone who has any doubts about what happened at Roswell should read it," he said of the official Air Force report. "When you get to the end of it, you won't have any doubts. Anyone who is interested in the truth and the real facts should read that report."[11]

I know what you are thinking. The dates don't add up. There is an obvious problem with the timeline when the supposed crash happened a number of years before the military was using test dummies in that area and the KC-97 crashed. Assuming the witnesses are being honest, how could they remember seeing things in 1947 that actually took place in the 1950s or 1960s? By being human, that's how. Don't forget how memory works! The human mind doesn't file away archival footage of everything we see and hear, in correct order, and then wait for us to request a perfect playback. Our memories are *constructed*. This means they are edited, embellished, and shuffled around. And because memories are *associative* as well, details that were not part of the original event as it really happened often get tossed in. Connections our brains "think" make sense are made in an effort to give us coherent and useful memories. It may seem like we are being constantly lied to by our own brains, but they don't do all this to fool or harm us. They do it because we don't need to remember every detail about everything. It would be inefficient to spend time and energy trying to recall everything, so the brain does its best to give us what it thinks we need. This is why it would not be so unusual or unexpected for someone to blend a 1947 memory with the memory of something they saw, read, or imagined they saw, years later.

Some readers may find all this a bit hard to believe. I have no such doubts, however, because I once caught my brain feeding me a memory it thought made sense but didn't match reality. In the summer of 2011, I was driving in my car listening to Colin Cowherd's ESPN radio talk show. The topic was baseball and Cowherd mentioned Pete Rose's run at Joe DiMaggio's revered fifty-six-game hitting streak. Immediately I remembered the night Rose came up short after getting a hit in forty-four games. The scene was crystal clear in my mind. It was during my college days and I was in a friend's dorm room while that game was on his TV. I even remember pausing our conversation when Rose came up to bat so I could watch. Rose struck out and DiMaggio's record survived. But then Cowherd mentioned the date for that game: 1978. "Ha," I thought to myself. "He blew it. He's way off. That happened in the mid or late 1980s." The year 1978 couldn't possibly be right, because I saw Rose's streak snapped on TV in a friend's dorm room when I was in college in the 1980s. In 1978 I was a sophomore in high school. I assumed the commentator just made a simple mistake with the year. Later, however, I was curious enough about the exact year to check, and what I found shocked me. Pete Rose's hit streak was stopped in a game that took place in 1978! Even then, even after seeing the date verified by a credible source, I still remembered it incorrectly. I could still "see" myself, in college, in my friend's dorm room, watching Pete Rose strike out. But it couldn't have happened that way. I wasn't in college then and didn't even know that particular friend back in 1978. I was just a kid then, and that game was probably on past my bedtime. What happened? The likely answer is that my constructed memory of Pete Rose striking out was edited, shuffled, and put together for recall by my brain in a way that totally violated the integrity of the actual time line. Maybe I was in that dorm room with my friend in the 1980s and glimpsed a program that included video of Pete Rose's career highlights, including that specific strikeout. My subconscious mind then combined the real 1978 event with the viewing of a replay of it years later to create one memory that seemed to make sense—except that it was not accurate. The gap of at least six or eight years was compressed and eliminated—without my conscious permission, by the way. After experiencing this phenomenon firsthand, I have no problem understanding how someone around during the Roswell incident in 1947 might end up with a memory of it that also includes memories of events that occurred years later.

It is also possible for a person to be prompted into a false memory just by exposure to a compelling presentation. This was shown by a 2011 study that revealed how disturbingly easy it is to fool human memory. Researchers found that simply showing people a commercial would lead many of them, just a week later, to remember trying the product advertised even though they never could have because the product was fictional. They call it the "false experience effect."[12] When I read about this, I immediately thought of the Roswell claim and other strange things people remember seeing or experiencing. Given the popularity of attention-grabbing alien stories that spread through popular culture like wildfire in the 1960s, 1970s, and 1980s, how many Roswell witness accounts might be attributable to the "false experience effect"?

It's not difficult to manufacture convincing memories of things people don't really remember at all. There is a cute little story about my youngest daughter that I occasionally share with anyone who will listen. When she was around one year old, Marissa had a fever and I carried her around the house while singing some silly made-up song in the hope of distracting her from the discomfort. At one point I paused, held her up, face-to-face at eye level, and thought about how wonderful and perfect she appeared. Even while sick she was impossibly beautiful. Her little face was the cutest thing in all the universe—and then she showered me with vomit. My little angel unleashed a relentless stream of foul demonic fluid, so much of it that I would've sworn it exceeded her body weight. And then she smiled and giggled. I stood there for a long time, soaked with milk and half-digested baby food, unsure what to do.

The interesting thing is that Marissa, now ten years old, remembers that incident very well. But after talking with her about it, I'm convinced that her memory of drenching daddy is not of the actual event but is a constructed memory created from hearing me tell the story. A story heard today can shape tomorrow's memories. What if I had added into each retelling of that story that it was raining, there were seven puppies in our living room, and I was dressed up like an Elvis impersonator at that time? Would she "remember" those details if she heard the story several times over the years? Probably. It's easy to imagine how Roswell witnesses might have heard or read stories about aliens and then subconsciously constructed convincing memories that mix in elements of those stories with their real experiences at Roswell decades ago.

WHAT WE HAVE HERE IS A LACK OF TRUST

The absence of good evidence and the availability of reasonable alternative explanations have not stopped millions of people from believing that a spacecraft crashed in Roswell more than sixty years ago. A staggering 75 percent of Americans reject the military's official explanation of the Roswell story.[13]

This figure of 75 percent is odd because only 24 percent of Americans polled by Gallup believe that "extraterrestrial beings have visited Earth at some time in the past."[14] It seems, therefore, that a significant number of people hold the strange position of not believing aliens crashed at Roswell while also not believing the government when it says that aliens didn't crash at Roswell. This suggests that much of the Roswell story's popularity is owed to a general mistrust of government. If everything had happened in roughly the same way, minus government involvement, the incident probably would have been forgotten long ago.

Do governments lie? Of course they do. Sometimes it's for good reasons and sometimes it's because politicians, military leaders, and civil servants want to get away with things the rest of us would not approve of. This does not mean, however, that governments lie all the time about everything. Sometimes when a government says it did not recover an alien spaceship and hide extraterrestrial bodies, it just might be telling the truth.

GO DEEPER . . .

Books

- Frazier, Kendrick, Barry Karr, and Joe Nickell, eds. *The UFO Invasion: The Roswell Incident, Alien Abductions, and Government Cover-ups.* Amherst, NY: Prometheus Books, 1997.

- Kaufman, Marc. *First Contact: Scientific Breakthroughs in the Hunt for Life beyond Earth*. New York: Simon and Schuster, 2011.
- Kittinger, Joe, and Craig Ryan. *Come Up and Get Me: An Autobiography of Colonel Joseph Kittinger*. Albuquerque: University of New Mexico Press, 2011.
- Klass, Philip J. *The Real Roswell Crashed-Saucer Cover-up*. Amherst, NY: Prometheus Books, 1997.
- McAndrew, James. *Roswell Report: Case Closed*. Grand Prairie, TX: Books Express Publishing, 2011.
- Ryan, Craig. *Pre-Astronauts: Manned Ballooning on the Threshold of Space*. Annapolis, MD: US Naval Institute Press, 2003.
- Saler, Benson, Charles A. Ziegler, and Charles B. Moore. *UFO Crash at Roswell: The Genesis of a Modern Myth*. Washington, DC: Smithsonian Books, 2010.

Other Sources

- *Skeptic* 10, no. 1 (2003).

Chapter 14

"ALIENS HAVE VISITED EARTH AND ABDUCTED MANY PEOPLE."

We had demons from ancient Greece, gods who came down and mated with humans, incubi and succubi in the Middle Ages who sexually abused people while they were sleeping. We had fairies. And now we have aliens. To me, it all seems very familiar.

—Carl Sagan

The extraordinary claim that aliens have not only made contact with people but also abducted or restrained some of them in order to conduct strange and horrifying procedures is one of the more important paranormal beliefs, in my opinion. Often mocked and dismissed without a second thought, these beliefs are so extreme that they deserve close scrutiny. It is one thing to imagine that the position of stars and planets influence daily human life or that some fuzzy light in the sky is an alien spaceship, but it is something else entirely to believe that extraterrestrials came into your bedroom and experimented on you. To "remember" a visitor from space having sex with you, extracting semen from you, or placing an electronic device up a nostril or in your anus takes things to an entirely new level. Perhaps these remarkable stories, made by many people at least since the 1960s, can tell us something important about paranormal claims in general. Maybe there is a lesson here about the power of cultural influence and the remarkable ability of a human mind to create its own "reality."

First of all, let's be fair and address whether or not alien abduction incidents could have happened. The only sensible answer to that is that it is possible. There are trillions of stars in the universe, and the discovery of planets orbiting stars has become routine in recent years. The universe is about 13.7 billion years old and it's very large. We might be alone, but it seems unlikely given all the chances for life out there. There likely are trillions of worlds within the three hundred billion or so galaxies spread across the observable universe. Of course there could be extraterrestrial life, but we might be the only intelligent life capable of space travel. Maybe we are the first intelligent life so far. Maybe we are the last. But if there is life in the universe and it's smart enough to figure out an efficient way to travel here, then maybe we have been visited. And maybe some humans were selected for study. This is an extraordinary claim and, as we shall see, there is nothing remotely approaching proof, but it is possible. For example, on a scale of "what has a better chance of being real," I certainly would place alien visitations ahead of astrology, ghosts, and people who have conversations with dead people. At least the idea of extraterrestrials existing doesn't conflict with the laws of nature. For this reason I tend to be a bit defensive on behalf of people who believe in alien abductions. Why are these stories a big joke to so many people when other more unlikely claims enjoy considerable respect? Shouldn't abduction believers get at least as much respect as, say, the 75 percent of Americans who believe in angels?[1]

While there is no evidence at this time indicating that life, intelligent or otherwise, exists anywhere else but Earth, the only sensible position is to maintain an open mind on that issue. Having an open mind, however, does not mean one should let every wild belief creep in. Since childhood I have had a strong attraction to both astronomy and science fiction. The possibility of alien life thrills me. I couldn't ignore an alien abduction story if my life depended on it. But after I have heard the story and there is no evidence to back it up, then I know it's *just a story* and nothing more. Anecdotes are not evidence. Never forget that stories alone do not prove anything.

The problem with the idea that aliens have been abducting people is that after all these years, after all these claims, there is nothing to show for it. To date, there is not one verifiable case ever of an abductee who produced hard evidence or shared important information that could only have come from a technologically advanced extraterrestrial species. There have been claims about people having alien devices implanted in them. Great! Send one of these gadgets to the *New York Times* science desk and we will finally know the truth. There also have been many stories about extraterrestrials giving abductees vital messages to share with all of humanity. Fine, but don't just tell us about how the aliens want us to live in peace and harmony. Anyone could make up that stuff. Let's hear something only an advanced alien species would know, like how to cure cancer, how to travel faster than light, or an explanation for dark matter.

Anyone who cares about evidence, logic, truth, and reality has no choice but to conclude that alien abduction claims are probably not true. This does not mean, however, that all those who make the claim are lying or are not worth listening to. I believe researchers and the general public should pay more attention to people who say they were abducted, not because I think their stories are accurate, but because they offer us an opportunity to learn more about delusions, sleep, false memories, and the influence of culture on beliefs. These cases are excellent examples of how easy it is to believe in things that are not real. If a person can somehow end up with vivid and authentic-feeling memories of a home invasion by a gang of big-eyed aliens with ray guns, then there should be no underestimating the brain's ability to betray us when it comes to thinking about much more mundane things like psychic readings, alternative medicine, lights in the sky, ghosts, the Bible code, and so on.

Dr. Susan A. Clancy, author of *Abducted: How People Come to Believe They Were Kidnapped by Aliens* spent five years of her life studying hundreds of people who believe they were abducted by aliens. After listening to some of the most bizarre and disturbing stories ever told, Clancy settled on very down-to-earth explanations for why these people believe what they believe. She is not convinced that her interview subjects or anyone else have been abducted by aliens. She also does not think the vast majority of abductees are mentally ill, lying, or unintelligent. "Yes, they held some strange beliefs without any strong evidence to support those notions," she writes in her book, "But don't many of us do the same thing? They weren't much weirder than the people I see at family reunions. . . . The truth is that almost all of us can believe things without much evidence. The only thing unique about the alien abductees I have met is their particular belief."[2]

According to Clancy, what is going on in many of these cases is likely an incident of sleep paralysis mixed with false memories. Something real happened—a nightmare on steroids—and then that event was "explained and confirmed" by a hypnotherapy session. A person could walk away from such treatment with the powerful memory of an alien abduction that no lecture on the virtues of critical thinking could easily undo.

She also discovered that belief in abduction is not the starting point. In most cases people were disturbed by weird things they couldn't explain like mysterious bruises on their bodies, specific events such as waking up and finding their pajamas on the floor, or general feelings of being an outsider in society. Clancy says that often the belief is part of an "attribution process," an attempt to answer questions. "Alien abduction belief," she explains, "reflect attempts to explain odd, unusual, and perplexing experiences."[3]

So how do people end up with the false memory of an alien abduction in their mind? It can happen far more easily than you probably think. Psychologist and memory researcher Elizabeth Loftus conducted experiments in the 1990s that showed how easy it is not only to modify a person's memory but to give them an entirely false memory as well.[4]

A standard component of the typical abduction scenario is that the victim's memory of the event is "wiped clean" by the aliens—although they never make a complete job of it. A nagging suspicion that something terrible has happened haunts the person who then seeks out someone to help put the pieces together, maybe a hypnotherapist who specializes in retrieving memories of alien abductions. But there is a problem with the popular perception of hypnosis and memory recovery: it's not supported by good science. "A wealth of solid research, conducted over four decades, has shown that hypnosis is a bad way to refresh your memories," argues Clancy. "Not only is it generally unhelpful when you're trying to retrieve memories of actual events, but it renders you susceptible to creating false memories—memories of things that never happened, things that were suggested to you or that you merely imagined. If you or your therapist have preexisting beliefs or expectations about 'what might come up,' you're liable to recall experiences that fit with those beliefs, rather than events that actually happened. Worse, neither you nor your therapist will realize this, because the memories you do retrieve seem very, very real."[5]

After decades of work on false memories, Loftus concludes this: "Just because it's vivid, detailed, expressed with confidence and emotion, doesn't mean it's true."[6]

The smart way to proceed when confronted by an extraordinary claim or event is to look for the easy answers first. If I walk out in my driveway tomorrow morning and find that my car is missing, my first instinct will be to suspect that it was stolen, it was borrowed, or somebody is playing a trick on me. I would have to eliminate all those possibilities, and many more, before I arrive at the possibility that it has been taken by aliens.

If the extraordinary event happens to be waking up in your bed, finding yourself surrounded by strange beings and unable to move, then it seems to me that sleep paralysis with hallucinations would be an easier explanation than bringing extraterrestrials into it. Have no doubts, sleep paralysis is a real phenomenon, and it's not as rare as you may think. Some 20 percent of people are believed to have experienced at least one sleep paralysis episode with hallucinations. It happens when the natural transition between deep sleep and waking up is somehow derailed. The brain can still be in a sleep state with motor output from the brain blocked, as is normal during sleep so that body movement is restricted, but the person "wakes up" and feels paralyzed. Add to this the possibility of a dream in progress, and one could be in for a very scary ride. In an awakened state, or something close to it, a dream might be impossible to separate from reality.[7]

Andrea, a thirty-something Canadian schoolteacher, has had so many sleep paralysis episodes that she has learned to simply relax and ride them out. She says that it is easy for her to understand how people can be terrified and misled by a sleep paralysis experience. "It's happened so often now—about a dozen times in my life—that it's lost a bit of its edge," she explained. "Also, I'm quite sure I'd heard of it before it happened to me, so I didn't suffer long with the, 'Oh-shit-what's-happening-to-me?' feeling. I can't remember one stand-out episode, really, but I do know you still definitely hear and smell things. So, if you didn't know what was happening, you'd be able to hear noises in the house, the TV, voices, and things like that. It would be *so* terrifying if you didn't know what was going on, and being conscious but unable to move does make you feel like you're being crushed somehow. You so badly want to open your eyes, but can't. You try to imagine what's going on in the room, and it would be pretty easy to think of something terrible, since this terrible thing is happening to you. My only concern is, How long will this last? I'm pretty good at calming myself and getting back to sleep, thankfully."

Trevor, a very bright university graduate and former soccer star now in his midtwenties, says he has experienced four or five sleep paralysis episodes in his life. "Usually I feel like someone is in the room with me or lying down next to me," he said. "Sometimes I can see a face and sometimes I can only see a shadow-type figure. I always feel like I'm conscious but can't move or make any sounds. A lot of the time I'm trying to talk or yell but nothing comes out. When I really concentrate hard I can move a finger, and once I make that small movement then I wake up. The experience is damn scary."

Based on clinical descriptions and personal experiences like Andrea's and Trevor's, an episode of sleep paralysis certainly would terrify most people. Imagine going to sleep and then "waking up" to find that you can't move your arms and legs no matter how hard you try. You can't speak or cry for help. Now imagine scary figures congregating around your bed. Maybe one of them touches you, hurts you, or even rapes you. During all of this you might feel like you are fully awake but immobilized and helpless. Maybe you can see, hear, feel, and even smell everything around you. Then you wake up in the morning feeling terrible. You are confused, tired, and maybe even haunted for weeks and months by a sense that something is very wrong. But you can't quite figure out what really happened that night. Perhaps you might become obsessed with "the problem" and begin grasping for answers anywhere you can find them. What was done to you? Who did it? Why did they do it? A few millennia ago, a mischievous god would have been a credible answer. Back then, gods came down from Mount Olympus or wherever to assist, torment, challenge, and have sex with humans all the time. A few centuries ago, the suggestions that it was a witch or ghosts likely would have satisfied your curiosity. During those times, virtually any unidentified noise in the night was deemed to be a demon, a ghost, or some such supernatural creature. Today, however, we have the mythical "grays," little aliens with big brains and large, creepy dark eyes. They are the night creepers of our time. It's the ideal upgrade for the high-tech space age period we now live in. Everybody knows about the agreed-upon look of aliens today. They are in books, films, and television shows, and they star in numerous contact and abductions stories. I have a rubber one posing on a bookshelf in my house right now. The standard alien abduction and experimentation storyline is near universal as well. It has spread around the world to most countries. Even most young children know how aliens are supposed to look and how the script is supposed to go when they come calling: first abduction, then experimentation, followed by memory erasure and release back into the wild. Yes, cultural saturation appears complete. This means, of course, that the stage is set for little extraterrestrials to visit many more people in their sleep.

Sleep paralysis and false memories may account for the alleged victims' belief, but what about the millions of people who do not claim to have had such an encounter but still believe these incidents occur to others? Why in the world would they choose such an improbable explanation when a far more reasonable one is available? I have been curious about this for many years and have asked every abduction believer I've ever encountered. The responses are almost always the same. First they retell a story, such as the famous Betty and Barney Hill abduction that supposedly occurred in 1964, and ask why anyone would make up such a thing. I counter that some people might do it in order to get attention or possibly profit from it. Or maybe sleep paralysis, hallucinations, and manufactured dreams from hypnotherapy make the event seem real to them even though it never happened. Maybe they are not lying and really do believe it happened, but this alone does not mean it did happen. I explain that people can have hallucinations, sleep paralysis is a known phenomenon, and psychologists have proven that it's not difficult for someone to change real memories and create false memories in another person's mind. It's almost always at this point that believers try to switch lanes and pull me into a debate over the possibility of extraterrestrial life. After I make it clear that I'm a big fan of astrobiology and think there is a very good chance that the universe is teeming with life, maybe even a few million intelligent species, they accuse me of contradicting myself. How, they ask, can I be so closed-minded about alien abductions when I'm open to the possibility of alien life? My final answer is that I'm not closed-minded. The idea of advanced extraterrestrials coming here, even if they are too aggressive and sexually perverted, is exciting and I would want to know all about it. But just because the possibility of contact thrills me doesn't mean I'm willing to pretend that it has happened until there is a sensible reason to believe it has.

GO DEEPER . . .

- Clancy, Susan. *Abducted: How People Come to Believe They Were Kidnapped by Aliens*. Cambridge, MA: Harvard University Press, 2005.
- Shostak, Seth. *Sharing the Universe: Perspectives on Extraterrestrial Life*. Berkeley, CA: Berkeley Hills Books, 1998.

Chapter 15

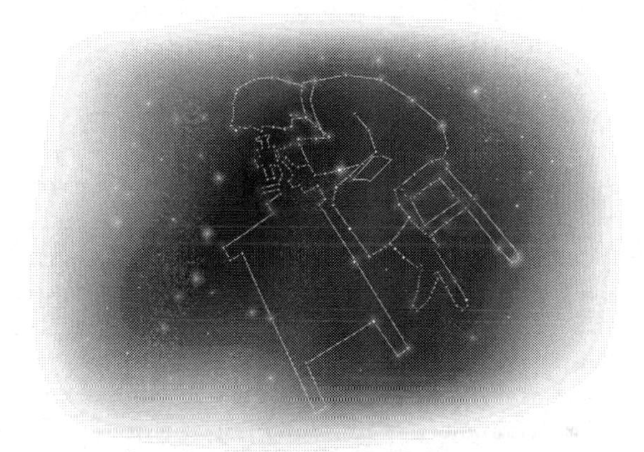

"ASTROLOGY IS SCIENTIFIC."

Throw away the tarot deck and ignore the astrology column. They are products offered you by charlatans who think you are not the marvelous, capable, independent being you are.

—James Randi, *Flim-Flam!*

There is harm, real harm, in astrology. It weakens further people's ability to rationally look at the world, an ability we need now more than ever.

—Phil Plait, BadAstronomy.com

I can remember being mildly impressed with my sign, Libra, many years ago because it seemed to offer amazing insight into my personality. According to astrology, I was intelligent, fair, resourceful, kind, funny, and charming. Yep, that's me! But just when I might have been in danger of being seduced by the power of the stars and planets, skepticism and critical thinking came to the rescue. Astrology never had a chance.

It's one thing to believe in something for no other reason than you want to or it feels good, but it amazes me how many people say astrology is scientific and supported by evidence. They don't "believe" in astrology, they "know" it's true because it's so logical. You would not believe how many people I have encountered who think astrology is synonymous with astronomy. I found this to be common in the Caribbean, where astrology and astronomy books are more often than not shelved together as the same subject in bookstores. Many Caribbean people speak of astrology like it is legitimate science. I found the same to be true throughout Asia. During a visit to Chicago's fine Adler Planetarium, I was shocked to see an astrology book in the gift shop, shelved alongside books by Stephen Hawking and Carl Sagan. Thousands of newspapers, including the one I currently write a column for, print horoscopes on their pages every issue.

Given the extraordinary nature of astrology's claim—that the positions of stars and planets at the time of one's birth reveal or determine both personality and future events that can be foretold—it is remarkable that so many people believe it. But believe in it they do. In the United States, for example, 25 percent of the population thinks astrology's claims are valid.[1] Despite having no theory to explain it, and no evidence or scientific basis to prove it, astrology has been touted, taught, sold, bought, respected, and "used" for thousands of years with no signs of going away anytime soon. Many of history's most powerful and important people have taken this pseudoscience seriously—and not just in ancient times. For example, many of President Ronald Reagan's daily meetings and movements were set by his wife based on his horoscope.[2] Many other prominent leaders in recent years have been rumored to consult the stars before making decisions. That should scare everyone who understands that astrology is worthless as a source of useful information.

To be crystal clear, astrology is not scientific—not even close. It was originally rooted in magical thinking when it gained popularity some four thousand years ago and, as far as I can tell, it has remained loyal to those roots to this day. Astrologers love to point out that modern astronomy grew out of ancient astrology. Sure, ancient astrologers engaged in real astronomy too, but that overused claim to fame is smothered under the weight of all the pseudoscientific baggage that astrology carries.

One only has to consider the source of astrology's alleged ability to know a person's personality and destiny to recognize that there is nothing to this. It may surprise some believers, but there is no complex mathematical formula, no ancient equation derived from the collected wisdom of previous generations, that reveals personality traits according to birth date. No, the traits associated with a particular sign are primarily based on how ancient people viewed the creature or object that they constructed with stars after playing connect the dots with constellations. I'm a Libra, for example, which is represented by scales that somebody centuries ago imagined they saw in the sky. For this reason I'm supposed to be a fair person and very concerned about justice. I'm not joking; that's really what it is based on. The ancient Babylonians never launched a scientific quest to find a genuine link between human behavior and the locations of astronomical objects. No, they just looked up at the night sky and then made it all up. So why do people still believe in it all these years later?

Astrology endures because most people don't appreciate the need for skepticism and critical thinking when confronted with such claims. It also obviously responds to a natural human desire to know ourselves better. Who wouldn't want to know if tomorrow will be a good day or a bad, if love is on the way, or if there is danger lurking? It also feels good to have our egos stroked, something professional astrologers are well aware of and make sure to deliver to clients. This is why you will never see a horoscope like this:

> You are not attractive and you are probably never going to be very successful. Your dreams are unrealistic and will not come true. Maybe you should just give up now and stop wasting everybody's time. Face it, we all can't be winners. Somebody has to lose, and that somebody is you. It is also time for you to accept that nobody really likes you. You are annoying. Do your friends and family a favor and move to another country.

Now that's a horoscope I might respect. But we never seem to see one like that in the newspapers, do we?

I have a short list of key points that I like to share with astrology believers. I don't attempt to bully them into a corner over the unlikely effect of Venus's gravity on my pet dog's happiness or specifics like that. Not only is that tactic usually a waste of time but, as with other paranormal or pseudoscientific beliefs, I figure it's best for the believer to do his own mental work and navigate his own way out of the dark and into the light. Remember that one or two well-crafted questions that are delivered in a friendly manner can do far more damage to an irrational belief than a hundred thundering facts declared by a smug skeptic. I tend to offer the following friendly advice and leave it at that:

- Believers should ask professional astrologers, or anyone who actively promotes astrology, to explain how they come to their conclusions. Don't read astrology books, articles, and horoscopes uncritically. Pay attention and notice that no sensible explanation is ever given for how we can know people born on June 4, for example, act one way while people born on September 4 act another way. Always ask the key questions. What *causes* the differences in personalities? Where does the information that horoscopes are based on come from? What specific force is exerting all this influence on us? Is it gravity? If so, how does the extremely weak gravitational effect from distant stars and planets do it? Is it altering our brains or genes? How? Once one realizes that there are no sensible and consistent answers to any of this, that it's all based on ancient superstition, it becomes much more difficult to believe in astrology.
- Don't stop with your horoscope. Most people who regularly check their horoscope in the newspaper or on a website probably never think to read the other horoscopes. Pick another sign and read the horoscope for it as if it's yours. Do this every day for a month and you are bound to be amazed at how "accurate" that one is too. Horoscopes are more convincing when you only read one of them because they are intentionally written to be one-size-fits-all. Once it's revealed that one horoscope works just as well as another, it's easy to recognize what's going on.
- Watch out for the trap of remembering a few predictions that came true while forgetting the hundreds and thousands that didn't come true. Astrologers, just like psychics, know that if you make enough guesses and predictions, at least some of them are bound to score.

Astronomy is like kryptonite to astrology. Learn some of it and you will see that the real science of stars and planets makes astrology wither and die out in the cold darkness of space. Good riddance.

GO DEEPER . . .

- Culver, Roger B. *Gemini Syndrome! A Scientific Evaluation of Astrology.* Amherst, NY: Prometheus Books, 1988.
- Culver, Roger B., and Philip A. Ianna. *Astrology: True or False? A Scientific Evaluation.* Amherst, NY: Prometheus Books, 1993.
- Goldberg, Dave, and Jeff Blomquist. *A User's Guide to the Universe.* New York: Wiley, 2010.
- Hart-Davis, Adam. *The Cosmos: A Beginner's Guide.* London: BBC Books, 2007.
- Stewart, Joseph V. *Astrology: What's Really in the Stars.* Amherst, NY: Prometheus Books, 1996.

SCIENCE AND REASON

Chapter 16

"ALL SCIENTISTS ARE GENIUSES AND SCIENCE IS ALWAYS RIGHT."

There is a widely used notion that does plenty of damage: the notion of "scientifically proven." Nearly an oxymoron. The very foundation of science is to keep the door open to doubt.

—Carlo Rovelli, "The Uselessness of Certainty"

I'm smart enough to know that I'm dumb.

—Richard Feynman

It's a major problem that science is overlooked, underappreciated, discouraged, denied, obstructed, and outright rejected by many people around the world, but what about those who give science *too much* credit and respect? While there is no data I know of on how many people have an excessive faith in science and believe it can do no wrong, I suspect it is more common and more of a problem than most would think. I have traveled extensively on six continents, not as a pampered tourist in air-conditioned buses with guides paid to smile at me, but mostly as a lonely writer and photographer roaming the back alleys and hillsides of the real world. Even in lands where superstition reigns supreme, I encountered many people who possess a weird reverence for science. For them it is just another form of magic or religion, and those who do science are only more sorcerers and holy men. They believe science is always right and every scientist is a genius. They view technology, the products of science, as secular miracles and proof of some imagined infallibility among scientists. This is far from reality, of course. It is also dangerous when people place too much stock in scientists and the things they say. One of the common paths to irrational beliefs is placing too much trust in authority figures. It is not only a problem in politics and religion. Trusting the words of scientists uncritically can be just as risky.

Look hard enough and it's possible to find doctorate-holding scientists somewhere in the world who will tell you that cigarettes are not all that bad for you, UFOs are definitely alien spaceships, minds can bend spoons, and the Earth is less than ten thousand years old. Isaac Newton, arguably the greatest scientist ever, was not content to invent calculus and figure out how planets move. He also spent a lot of his time trying to turn lead into gold and calculating the date on which the world would come to a supernatural end. Leading scientists in Europe and the United States were once convinced that the shape of a person's head was a reliable measure of intelligence and morality. Racism and sexism were once sanctioned by mainstream science. How can any of this be true? Scientists are supposed to come to conclusions based on evidence and sound reasoning. They are smart and spend a lot of time in school. They understand the scientific method and know the difference between science and pseudoscience. They are supposed to be the people who always get it right and know everything about everything, right?

Not even close.

For all its greatness, science has one flaw that will forever keep it humble and stop it well short of perfection: *Humans* are involved in it. The bias, greed, arrogance, fear, overconfidence, susceptibility to illusions, tendency to see patterns where none exist, excessive trust in tradition, and many more traits that define us guarantee that mistakes will be made. Even with the contaminating effect of human involvement, however, science still helps us to understand ourselves and the universe better than anything else we have. If you have a serious injury or illness, your chances of survival are much better in the hands of someone trained in medical science rather than magic, prayer, or alternative medicine. If you want to know something about space, it's best to consult an astronomer rather than an astrologer. Magic spells won't take you very far, but a rocket or airplane—products of science—will. Most people neglect to give credit where credit is due. But let's be honest, it was science that revealed to us all so much of the important stuff that we now tend to take for granted. Prophets, soothsayers, and fortune tellers did not come up with cell theory, germ theory, special relativity, plate tectonics, atomic theory, the theory of gravity, the theory of evolution, and so on. Without science, we would still be blaming demons for illnesses and reading bird entrails to set our daily agendas. Without science, we would know little about the galaxies, nebulae, and planets beyond our solar system, nor would we know about the genes in our cells or the microbes all around us and inside of us. Only science offers the ladder by which we may one day climb to faraway planets. Magic will never take us to Mars. Prayer will never unlock the remaining secrets of our genome. Science is the one and only path to continuing enlightenment about the universe and ourselves. As a tool of discovery, it is nothing less than wonderful and absolutely necessary for modern civilization. However, we must not let science's ability to produce so many wonderful answers and discoveries hide the messy and error-prone way in which it works.

Science is imperfect and based largely on failure. It advances by riding on the backs of mistakes. Scientists make errors and then try to learn from those failures. The first thing that happens when a new discovery is made public or a new theory proposed is that scientists around the world try to figure out how they can tear it down. Unfortunately, this process of error correction might take decades, even centuries. Therefore, we have to be on guard and always give only conditional loyalty. Science is a brutal and unforgiving gauntlet that answers have to run into and come out of alive before gaining acceptance. Actually, nothing ever escapes that gauntlet because nothing is ever really proven "true" in the final sense. In science, everything is left open to correction forever. If it's found to be wrong tomorrow or one thousand years from now, then the answer is amended. Nothing is sacred or untouchable. No scientist is infallible or beyond challenge. At least that is how it's supposed to work.

Science produces the goods over and over for a number of reasons. First of all, real science is not based on majority vote or popularity. The influence of politics, ego, and money may exist, but in the end evidence and experiments determine what is real. And there is no trusting anyone when it comes to important matters of science. Evidence must be available for scrutiny by other scientists, and experiments must be reproduced by others if the idea is to stick. Another great feature of science is that anyone can do it. While it may be impossible to land a prestigious professorship at Harvard without the right credentials and connections, it's not difficult at all to demolish an incorrect theory and establish a correct one, no matter what credentials you have. If you really do have the goods and can prove it, then the world's scientists will listen and accept the new reality staring them in the face. This is where many of those who push the creationism/intelligent design agenda show themselves to be blatantly dishonest. If it were real science, then it would be established in the scientific community first with discoveries, evidence, and experiments—openly documented in respected journals—for all to analyze. Only then would it filter down to high school classrooms. Proponents of creationism/intelligent design apparently know that it will never happen this way because it's not real science. So they focus on marketing, political campaigning, and legal cases. Their actions speak louder than words. Real science—for all its faults, mistakes, frauds, and failures—works. It confirms reality while weeding out false claims and bad ideas, which is probably why so many paranormal and pseudoscience peddlers fear to go anywhere near the scientific method.

While professional scientists do tend to be very smart people with high levels of education, of course, they certainly should not be viewed as some sort of inerrant demigods among us. In the same way that many highly intelligent people can believe in unlikely things such as psychics and ghosts, scientists can have downright kooky ideas bouncing around in their skulls as well. For example:[1]

- Aristotle thought the brain's primary purpose was to cool the blood.
- Astronomer Percival Lowell believed Martians built irrigation canals on Mars.
- Nuclear physicist Edward Teller thought we should use nuclear bombs to excavate a new harbor in Alaska and to crack open the Moon in order to study its interior.

- Nobel laureate Linus Pauling was convinced that megadoses of vitamin C could cure cancer.

Not only can scientists be wrong, they also can be bad, very bad. The infamous "Tuskegee experiment" was a forty-year study of nearly four hundred Alabama men who had syphilis. Even when treatment became available for these men, researchers made the decision to continue the study without telling them that penicillin could help them. This decision was not only devastating for the men but also for their wives and children.

In 2010 it was revealed that scientists in the 1940s infected Guatemalan prisoners, prostitutes, and others with venereal diseases in order to study them. The Pan American Health Organization knew about the experiments, according to Guatemalan president Alvaro Colom.[2]

I suspect that I will be haunted forever by the interview I conducted with Eva Mozes Kor. As a young girl imprisoned at Auschwitz, she had to endure a series of painful scientific experiments conducted by Dr. Josef Mengele. Several years ago I interviewed "the Father of the H-bomb," Edward Teller. I recall feeling uncomfortable when he defended his prominent role in developing weapons many times more powerful than the bombs dropped on Hiroshima and Nagasaki during World War II. He declared that I would be speaking Russian if not for American nuclear weapons. This may have been a fair point, but I was disappointed that he did not seem to recognize the great danger that some scientists have imposed on all of us by creating weapons capable of destroying civilization.

It is important to understand and remember that science is a tool that can be used for good or bad. As much as I love science and admire many scientists, I never forget that there is nothing inherently good or safe about science. Yes, it is the source of lifesaving vaccines and it brings us those stirring images of distant galaxies, but it is also where napalm and weaponized anthrax come from, too. Apart from the rules and standards we decide to impose on it, science cannot be counted on to be a force for good only. This is why I stop well short of idolizing all scientists and adopting science as my ersatz religion. I've been a science lecturer and I continue to do my best to popularize and explain science to others through my writing. But I never suggest that it's perfect, safe, or a risk-free path to utopia. Science may give us that wonderful *Star Trek* future, or it may deliver doomsday.

I love that good scientists openly admit that they can give us only a tentative version of reality and truth. They don't keep it a secret that everything is up for revision, correction, or rejection in science. Unlike politicians and religious leaders, scientists change their minds and rewrite the textbooks all the time. They do this because they believe in science not as a permanent, fixed set of laws written into stone, but rather as an ongoing process of discovery. I have no doubts that a number of scientific facts that I think are correct today will turn out to be incorrect in the future. This is why I have always made sure to include disclaimers when giving science lectures to students: "According to the best current evidence" punctuates many of my statements. I want people, especially younger students, to know that science is nowhere near finished, that there is plenty of work left to do. No one should think of scientists as another priesthood that possesses some ultimate truth. A good scientist is one who wallows in failure and is never quite 100 percent sure of anything.

Science has been able to give us so much because it relies on evidence and experiments to make discoveries and answer questions. Scientists are able to do great work because they accept failures and learn from them. Do not allow yourself to be fooled; there is no perfection here, no purity, and no safety from evil. But in science there is always another wonderful and important discovery waiting just around the corner. It's also the best way we have of determing what is real and what is probably not real. It's the best thing we have to help us navigate through our mysterious, complex, and often-dangerous existence. Never forget that what is most important about science is not who made the big discoveries, but how they did it.

GO DEEPER . . .

- Brockman, John. *What Have You Changed Your Mind About? Today's Leading Thinkers Rethink Everything*. New York: Harper Perennial, 2009.
- Brooks, Michael. *Free Radicals: The Secret Anarchy of Science*. London: Profile Books, 2011.
- Brooks, Michael. *13 Things That Don't Make Sense: The Most Baffling Scientific Mysteries of Our Time*. New York: Vintage, 2009.

- Hanlon, Michael. *10 Questions Science Can't Answer (Yet): A Guide to the Scientific Wilderness*. Hampshire, UK: Palgrave Macmillan, 2010.
- Marks, Jonathan. *Why I Am Not a Scientist*. Berkley: University of California Press, 2009.

Chapter 17

"THE HOLOCAUST NEVER HAPPENED."

Those sent to a camp associated with an industrial plant run by the SS economic branch were usually worked to a state of enfeeblement before being sent to the gas chambers, though the old, the weak, and the young might be gassed immediately. Auschwitz, the large camp in Poland, served both purposes. Those sent to the extermination camps, like Treblinka and Sobibor, were gassed on arrival. In this way, by the end of 1943, about 40 percent of the world's Jewish population, some six million people, had been put to death.

—John Keegan, *The Second World War*

If five to six million Jews were not killed, where did all those people go?

—Michael Shermer

I have always been deeply troubled by the Holocaust. Any mass killing is a horrible event, of course, but the manner in which Hitler's "Final Solution" was undertaken seems to damn us all. The efficiency and creative use of technology made it more than just another one of history's bloodbaths. It was the industrialization of murder, so disturbingly modern in its planning and execution. I suspect I may process the Holocaust differently than most. I'm not a Jew but I feel pain and loss; I am not a Nazi but I feel shame and guilt. The world calls it genocide. But I see it as suicide. Take a peek over our false walls, the ones built of manufactured divisions called race, nations, and religions, and you will see too that it was people killing people, one more episode of a dysfunctional species cutting its own throat.

During a visit to Jerusalem, I spent an afternoon at the Yad Veshem Holocaust History Museum. It was a miserably hot day. Appropriate, I figured. Comfort would be rude in that place. I remember draining the last drops from my water bottle and thinking, it was good that the Holocaust will never be forgotten. I only wished that the murdered Native Americans of the Americas and the Caribbean had a memorial somewhere as impressive as this one. The Africans who were shipped to the New World as cargo should be remembered in this way too. Every shameful chapter of history, every bloodbath, ought to have a prominent monument and museum somewhere in the world. Yes, it would take up a lot of real estate, but maybe the sight of so many tangible reminders of our worst failures would shock us into confronting our depraved past and finally resolving to mature and do better.

The architecture, landscaping, photos, artifacts, statues, and somber silence blend seamlessly to great effect at Yad Veshem. Together they annihilate ignorance and indifference. I was imprisoned for several minutes by the black-and-white photograph of an adorable little child who never got to grow up because he was born on the wrong side of Hitler. Looking back on my visit, it's difficult to imagine anyone not being moved by the weight of the place. But, of course, there are millions of people around the world today who would equate Yad Veshem to Disney World. For them it's a fantasyland for propaganda and profit. These people are known as Holocaust deniers, and they accuse mainstream historians of stupidity at best, complicity with fraud at worst. It's all a hoax, they say, perpetrated by Jews upon the world for political and economic gain. This is a staggering claim. Given the evidence for the Holocaust and all the pain attached to it, this makes the Moon-landing-hoax claim seem almost reasonable. The Holocaust never happened? Hitler wasn't trying to eliminate Jews from Europe? Six million Jews never died? All those dead bodies never piled up in pits and gas chambers? Nazi prison camps were not murder factories? Are the deniers serious?

While no one person can bear witness to an event that spanned several years and multiple countries, a collection of voices is difficult to deny. I have personally spoken to several people who lived through key events in World War II. They certainly have no doubts about the reality of the Holocaust. Carwood Lipton saw more combat than most as a sergeant in Easy Company of the 101st Airborne Division. He enjoyed a bit of fame as a prominent character in the Tom Hanks/HBO miniseries *Band of Brothers*. Lipton told me that during the war he arrived at Landsberg prison camp in Germany, shortly after another group of US soldiers had liberated it. What he saw horrified him, and the experience never completely released its grip on him. I asked him what he thinks about people who deny that the Holocaust happened.

"Oh . . . [long pause] it was absolutely terrible. They [Holocaust deniers] should have been there to see it. The smell was terrible. I can still remember the smell."[1]

Lipton and other Easy Company veterans attended special private screenings of *Band of Brothers* episodes in 2001. He chose not to watch episode nine, however. That was the one about the horrors discovered at Landsberg. "I just didn't want to bring back those memories," he explained.

In 2002, I interviewed Armin Lehmann, a member of the Hitler Youth in Germany who won the Iron Cross for bravery in combat against the Russians. At age sixteen he served as Hitler's personal courier in the Berlin bunker during the final days of the war in Europe. Lehmann described to me a disturbing childhood in which young schoolchildren had their heads measured to confirm the Nazi ideology of racial superiority. He and his classmates were taught propaganda designed to make them fear and despise Jews. "Jews were presented as evil people who were out to destroy the world," he said.[2]

Lehmann told me that he feels the percentage of Germans who knew about the Holocaust while it was happening is "debatable."

"Most say they did not know," he said. "But in retrospect, all of the signs and signals were there. I think more should have known than admit it."

Barbara LeDermann was a childhood friend of Anne Frank. In 1943, the Germans began rounding up all Jews in Amsterdam for transport to "labor camps." Barbara had heard, however, that these were in reality places where Jews go to be killed, so she made the bold decision to go into hiding. But she didn't do it the way Anne Frank did. Barbara decided to hide in plain sight by changing her last name and pretending to be a non-Jewish German girl. It worked. She was able to live with non-Jewish friends and avoid the fate of her parents and younger sister—all of whom died in prison camps. Years later, Barbara was recognized as a hero by many for her role with the Jewish underground. During the war, she had risked her life to deliver food and newspapers to Jewish families in hiding.

"I never knew about gas chambers back then," Barbara said. "If people told me about things like that I wouldn't believe it. It is beyond comprehension. Who could believe that there were people who could do that to innocent people? After the war I used to go to the railway station, hoping to see my family. But they were never there. It took me a long time to accept that they were never coming back."[3]

Finally, there is Eva Mozes Kor. I don't usually cry during interviews, but I teared up as this little woman described how the infamous Nazi doctor Josef Mengele injected her with toxins and germs as part of cruel experiments he conducted on twins at Auschwitz. Eva was ten years old.

"The first night I was there [Auschwitz prison camp] I went to the latrine and found three dead children on the floor. . . . Mengele was the god of Auschwitz. We always knew that when he came in we would have to be very still and do whatever he needed us to do. He would come in, morning after morning, to count us and see how many guinea pigs he had."[4]

Eva said she was measured and examined for hours at a time. Sometimes the nurses would take so much blood from her that she would faint. She says she learned later this was part of a study to determine how much blood wounded soldiers could lose before dying.

"They would inject me with a minimum of five injections, three times a week. Those were the deadly ones. The majority of twins died in these experiments. Once a twin was injected with a germ, the other twin would be kept nearby under surveillance. When the twin that had been injected with the germ died, the other twin would be killed so that Mengele could do comparative autopsies."

Eva and her sister survived the Holocaust. The rest of their family did not.

It is important to understand that Holocaust deniers are not necessarily screaming neo-Nazi skinheads who show up at public rallies wearing SS uniforms. I have attended neo-Nazi and Ku Klux Klan meetings in Florida, and many of the people I saw who were cheering on the memory of Hitler and scoffing at every mention of the Holocaust did not look anything like the stereotypical foaming-at-the-mouth racist. Many of them could have passed me on the street and I never would have imagined that they were passionate Holocaust deniers. It turns out that most of the people who drive this movement by writing the books and speaking at conferences tend to present themselves in public as very polite and sophisticated people.

Science historian and skeptic Michael Shermer researched the movement for a book and says the deniers are "relatively pleasant" in public.[5] Who knows what is going on in the privacy of their minds, but most of them talk primarily about honoring the truth, checking facts, and doing proper history. While analyzing their process of reaching and defending their positions, Shermer recognized tactics nearly identical to those used by many creationists and intelligent design proponents in their battles against modern biology:[6]

- They concentrate on their opponents' weak points rather than strengthening their own position and focusing on it.
- They exploit errors by mainstream scholars and suggest that if some things are wrong, everything must be wrong.
- They take quotations out of context to bolster their position.
- They claim that debate among mainstream scholars on specific points suggests disagreement about the validity of the entire subject.
- They focus on the unknown and ignore what is known. They point to data that fit their claims and ignore data that do not fit.

Shermer argues that, just as it is with the theory of evolution, there is no one piece of evidence that proves the Holocaust happened. We know it happened through a "convergence of evidence."

"Deniers seem to think that if they can just find one tiny crack in the Holocaust structure, the entire edifice will come tumbling down," Shermer explains. "This is the fundamental flaw in their reasoning. The Holocaust was not a single event. The Holocaust was thousands of events in tens of thousands of places, and is proven by millions of bits of data that converge on one conclusion. The Holocaust cannot be disproved by minor errors or inconsistencies here and there, for the simple reason that it was never proved by these lone bits of data in the first place."[7]

It makes sense to side with the evidence, especially when there is so much of it. In this case, it's overwhelming: There are blueprints of crematoria and gas chambers; damning documents; numerous quotes from Nazi leaders about exterminating Jews; thousands of photographs; and, most important of all, we have thousands of chilling testimonies from survivors and those who lost family members. We also have testimonies from some of the killers themselves. The Holocaust happened.

"My wife is Jewish, so I know from personal experience the loss of family members who simply disappeared and cannot be found," said historian Nick Wynne. "It has a profound effect on surviving family members. Dwight Eisenhower knew that the world would have a hard time believing the Holocaust happened, so he ordered that the death camps be filmed for posterity and to disprove the deniers. Historians should be prepared to deal with the deniers by accumulating as much information as possible. Should they be silenced? No, everyone has the right to believe and say what they want—even if they're wrong! That's the essence of freedom of speech."[8]

As repugnant and just plain wrong as Holocaust deniers may be, Wynne is correct. Attempting to legally muzzle them is not the solution. People should have the right to believe and say incorrect and offensive things. The best way to respond to Holocaust deniers is not with subpoenas and indictments but with education and evidence-based rebuttals. Dragging deniers into courtrooms and making public spectacles of them—as was done in Canada with Holocaust denier Ernst Zundel—rewards their provocations by giving them undeserved relevance and an aura of importance. It is the claims themselves, not the people who make them, that need to be defeated. This can only be done by slaying the lies and distortions with superior evidence and personal testimonies—two things the Holocaust has plenty of.

GO DEEPER . . .

- Gilbert, Martin. *The Holocaust: A History of the Jews of Europe during the Second World War.* New York: Holt Paperbacks, 1987.
- Keegan, John. *The Second World War.* New York: Penguin, 2005.
- Lipstadt, Deborah. *Denying the Holocaust: The Growing Assault on Truth and Memory.* New York: Plume, 1994.
- Posner, Gerald. *Mengele: The Complete Story.* Lanham, MD: Cooper Square Press, 2000.
- Shermer, Michael, and Alex Grobman. *Denying History: Who Says the Holocaust Never Happened and Why Do They Say It?* Berkley: University of California Press, 2009.

Chapter 18

"GLOBAL WARMING IS A POLITICAL ISSUE AND NOTHING MORE."

> **I believe the biggest problem to solving global warming is the role of money in politics, the undue sway of special interests. . . . Policy decisions are being deliberated every day by those without full knowledge of the science, and often with intentional misinformation spawned by special interests.**
>
> —James Hansen, *Storms of My Grandchildren*

I am convinced that global warming is real and that our industrialized civilization is the cause or at least a key contributing factor. I don't believe it will bring about our extinction or the total collapse of civilization, but I do think it will be devastating for some and costly for all one way or another. I could be wrong, of course, but currently this seems to be the most sensible position one can hold, based on what I have read and heard over the last thirty years from scientists who are directly involved in researching this issue. But the reality of global warming is not what this chapter is about. For the moment, let's not concern ourselves with whether or not it's a real phenomenon. Instead, let's deal with how illogically and irresponsibly politicians, the news media, and the general public have dealt with the global warming issue to date.

Being a liberal or a conservative, Democrat or Republican, should not have anything to do with how one thinks about whether or not global warming is real. This is a real-world issue with potentially dire consequences for billions of people. Inexcusably, however, global warming has been something that most people assess first and foremost in terms of political tribe affiliation. This is profoundly irresponsible. It makes no sense to draw conclusions about scientific evidence and ideas based on whether it is liberal or conservative politicians who first side with the scientists who bring it to light. Shouldn't reality and responsibility to the world be the priority? Liberals should be skeptical of global warming and analyze it honestly and intelligently like anyone else might. And conservatives should have sense enough to get their science from scientists rather than from politicians. If liberals are convinced of global warming's legitimacy, then they should be most concerned about promoting science and defending scientists against unreasonable attacks on their credibility, not toeing the party line in a childish political skirmish. Conservatives should be smart enough to know that politicians with law degrees and science illiterates who happen to have their own radio and TV shows should not be anyone's go-to source on serious science matters. Earth's climate potentially impacts all life everywhere—including conservatives and liberals alike. Clearly a rapidly changing global climate is more important than picking candidate A or candidate B in the next election. This is about figuring out what is happening with our climate and what is coming in our collective future. Don't you think this is one issue that should rise above childish playground politics? After all, climate and greenhouse gas emissions are not political. They are not liberal or conservative. Our reactions to them through new legislation and changed behavior can be political, of course, but the climate and gases themselves are not.

Polls have confirmed the obvious repeatedly: For most Americans, political slant strongly influences how they think about global warming. A recent survey found that 73 percent of Democrats believe greenhouse gas emissions cause global warming, but only 28 percent of Republicans believe it.[1] Again, this chapter is not debating whether global warming is real and human-caused. It's about people deciding how they feel about this important science issue based, not on science, but on shallow politics. Even if we assume here that global warming is real, as I believe, Democrats who accept it only because their political leaders do are not much better than Republicans who reject it because of their leaders. It sounds obvious, but millions just don't get it so it bears repeating: people should rely on science when assessing the validity of scientific issues, not politics.

Sadly, people in general, regardless of political affiliation, are losing trust in the scientific consensus on global warming. It seems that the illogical and stubborn nature of this relentless debate seems to have eroded public trust in science overall to the point that the level of understanding and acceptance of what most scientists say about global warming has fallen among Democrats, Republicans, and independent voters across the board in recent years.[2] It appears that it's not just the Republican or conservative perspective that is compelling to people, the "don't bother listening to scientists" mantra has traction everywhere in America. Right or wrong, it would not be so depressing if people were rejecting global warming after thinking about it independently, objectively, and rationally. But that's not what is happening. They are being bludgeoned into either conformity or apathy by the whirlwind of political shouting and antiscience rhetoric.

There are many factors one could blame for the tragic politicization and intellectual corruption of an important scientific issue, but I point to just one—Al Gore. The former vice president was the wrong man at the wrong time. Despite whatever good intentions he may have had, Gore as the point man for global warming crippled the issue from the start due to his prominent status over on one half of the playpen of American politics. He was a politician, first and foremost, so he was divisive and suspicious to millions before he even said one word on the issue. Truth, reality, and science would never be allowed to overshadow the view that he was a Democrat and therefore any agenda he pushed was seen by millions as Democratic and not Republican. So, of course, the global and borderless challenge of global warming immediately morphed into a partisan shouting match based on political allegiance rather than science. It's probably not fair to fault Gore for this. It's unlikely that he could have foreseen his negative impact on this issue. One could also argue that he at least raised awareness, a good thing. But to the degree that the awareness he raised has been largely drowned out by the most immature political babbling and squabbling imaginable, I don't see his efforts as a net gain. If the global-warming issue had been pushed early on by a more neutral crusader, I believe we would be in a better place right now.

In the minds of many Americans, Al Gore is a running joke. To them he is the inventor of both the Internet and global warming, a profit- and ego-obsessed man who has promoted history's greatest hoax. It's clear how many people think: If you are a conservative and don't like Gore, then you can't possibly believe in *his issue* of global warming. Because of Gore's role as poster boy, climate change became just another political piñata for the children who run America to take turns whacking. It's not about scientists trying to warn us, some believe. It's about a Democrat trying to sell us something bogus for his own gain. Meanwhile, of course, the world keeps warming.

The truth, of course, is that Gore did not invent or discover global warming. He didn't even contribute any key research. He is a nonscientist who seized an issue that he apparently felt was important and worth pushing. Perhaps he also believed that it would benefit him politically to do this. He is a politician, after all. Given the way this has played out over the last couple decades, it seems clear to me that it would have been better if Gore had never become the public face of global warming. That role should have gone to a key scientist or key scientists who were immersed in the problem. After all, they are the people who understand it better than anyone and they are not politicians first and foremost. This is not to say that scientists didn't try to tell the world. I can recall reading about global warming as far back as the 1980s. But most people don't follow science news. It took Gore's involvement to put it on the radar and get the yelling started, but that meant little listening and less thinking would follow. Perhaps if the general public was less concerned with celebrity affairs and sports results and more attentive when it comes to important science news, people would have been aware of the global-warming issue long before it was attached to a political figure that half of America would never trust.

Here's a scenario to consider: What if a polarizing politician who happened to be a conservative, say George W. Bush or Sarah Palin, had made global warming their pet issue? What if that conservative politician produced and starred in an Oscar-winning documentary and also pushed hard for related legislation and lifestyle changes? Given the nature of current American politics, is there any doubt that the poll figures would likely be reversed? A majority of Republicans would probably believe in global warming and most Democrats might deny it. Only the stubborn refusal to respect science likely would be the same.

To be clear, the problem here is not the arguing itself. Debate is healthy. Being skeptical and asking questions about something as important as climate change is a good thing. It should be challenged and questioned—*everything* should be challenged and questioned. But it's not the science that is being debated in this case. It's been a case of confirmation bias runing amok. Most people dismiss everything about global warming that contradicts their *political* position and accept everything that supports it. Rush Limbaugh, Sean Hannity, and Glenn Beck are not climate science experts—not even close—so it makes no sense that they should be trusted as leading minds on this issue by millions of Americans. Regardless of political party, everyone should be paying attention to what credible scientists are saying. It's really not that difficult: expert opinion should be sought from experts. Relying on politicians who are shackled one way or the other by their tribe is a foolhardy way to go about figuring out a science issue. Their goal on most days is to defend and promote their position—whether it's right or wrong. It's even worse to rely on professional rabble rousers on radio and TV for the final word on important scientific matters. Have no doubts, their primary concern is advertising revenue, certainly not scientific accuracy or even the state of our planet. They succeed through controversy, mistrust, and division. They are not scientists. They do not do science. They do not know science. Scientists know science.

GO DEEPER . . .

Books

- Hansen, James. *Storms of My Grandchildren: The Truth about the Coming Climate Catastrophe and Our Last Chance to Save Humanity*. New York: Bloomsbury, 2010.
- Mooney, Chris, and Sheril Kirshenbaum. *Unscientific America:*

How Scientific Illiteracy Threatens Our Future. New York: Basic Books, 2009.

- Pierce, Charles P. *Idiot America: How Stupidity Became a Virtue in the Land of the Free*. New York: Anchor, 2010.
- Schmidt, Gavin, and Joshua Wolfe. *Climate Change: Picturing the Science*. New York: W. W. Norton, 2009.

Other Sources

- Climate Central, www.climatecentral.org.

Chapter 19

"TELEVISION NEWS GIVES ME AN ACCURATE VIEW OF THE WORLD."

For every five hours of cable news, one minute is devoted to science.

—Chris Mooney and Sheril Kirshenbaum, *Unscientific America: How Scientific Illiteracy Threatens Our Future*

The one function that TV news performs very well is that when there is no news, we give it to you with the same emphasis as if there were.

—David Brinkley

Whom the gods would destroy, they first give TV.

—Arthur C. Clarke, *Voices from the Sky*

I am deeply conflicted when it comes to today's television news media. Ask me what I think of it and I'm likely to answer that it's wonderful—but also terrible. It's invaluable—yet mostly a waste of time. TV news is necessary for keeping people informed—and a primary reason so many people are astonishingly ignorant. It's crucial to a healthy democracy—but also the cause of so many voters being seduced by idiot candidates.

I preach to my children that it is important to be aware of current events and follow the news every day. Read newspapers and watch TV news, I say. The more news, the better, I tell them. But, like all parents, I'm a hypocrite. *Do as I say, not as I do.* Over the years, my mind has migrated away from television news. It's obviously more about entertainment, excitement, and fear than news and information. The goal of cable news executives is not to make me an informed citizen of Earth. Their mission is to tickle the dark reptilian depths of my brain and hook me so that they can then barter with my soul for advertising revenue. I suppose I could have spent the summer of 2011 keeping up with the Casey Anthony murder trial that seemed to captivate most of America. But I chose instead to use that time hanging out with my kids, reading, writing, and doing other things I felt were a more valuable use of my time. Isn't it fitting, by the way, that in the same month ABC announced the cancellation of the fictional soap operas *One Life to Live* and *All My Children*, news departments were serving up the nonfictional soap opera of a troubled mother charged with killing her child? Why pay actors and screenwriters when you can just point cameras at human train wrecks and rake in the money? It's also curious that so much attention was given to the Anthony case—centering on the death of one child—considering the fact that during the forty-two-day trial more than one million children under the age of five died in the developing world from malnutrition and preventable diseases. How much coverage of those child deaths did you see during the summer of 2011?

In the month of July 2011 alone, cable news and major network news covered the US government turmoil over raising the debt ceiling. While it was an important story, I would estimate that at least 90 percent of what aired was meaningless back-and-forth babbling between party loyalists that contributed nothing to understanding the issues or following the progress of the story. I wonder how many American TV news viewers were aware that in that same month of July nearly thirty thousand Somali children died in the worst famine East Africa had seen in twenty years.

These days I spend the bulk of my daily allotted "news download time" reading science magazines and visiting science news websites. I care about the world and sincerely want to know what is going on, but not so much that I'm willing to sit through news roundups that are mostly violent crimes, gossipy nonsense about celebrities, and political soundbites crafted for sixth-grade-level comprehension. There was a time when I thought that in addition to reading a daily newspaper it was my obligation as a thoughtful person to watch a lot of television news. I still like the idea that TV news is there and I do have it on most days as background noise when I'm shaving or whatever, but I can't stomach very much of it in single sittings anymore.

Somewhere along the way I realized that my life was not being enhanced by watching metrosexual androids and former beauty queens fake concern while reporting on the deranged giraffe that bit off the nose of some tourist at the zoo earlier today. *"Exclusive analysis from our Beverly Hills plastic surgeon correspondent after the break!"* The reporting on environment, health, and science topics is almost always too shallow and designed to scare rather than educate. *"Which is more dangerous, the flu or the flu vaccine? A concerned soccer mom weighs in on tonight's 'Health Zone' report."* And many of the news packages are just plain pointless by any reasonable standard. *"Should pets have their own Facebook accounts? Viewer tweets shed light on the controversy, next!"* Nor am I necessarily better informed on important political issues after watching an endless parade of political hacks defend their respective parties at all costs, with little or no thought given to truth, reason, or what might be best for society. *"Tune in at eleven! A former Republican campaign strategist and a former Democratic campaign strategist will tell us which party they think is doing the best job for America."* I also don't have time for all those professional pollsters who specialize in telling me what I am likely to think about things I don't care about. *"In this segment of 'News Watch' we'll look at the new survey that has everyone talking. It turns out that 37 percent of Americans think members of Congress are satisfied with the public's perception of Congress."* Politics in general is treated as a shallow sports competition by television news, at the expense of sensible priorities and truth. For example, both sides of an issue are usually given equal amounts of time and respect—no matter if one side is completely illogical, untrue and in opposition to scientific fact. *"This morning on* Wake Up AM *we'll hear from two senators as they square off on an important and contentious issue: Does the Earth revolve around the Sun or vice versa?"*

Finally, it's a minor point, but what is it with hurricanes? Why do reporters keep standing out in them? It might have been cool when Dan Rather did it in 1961, but at some point in the 1970s or 1980s that routine became a terrible cliché and an insult to higher forms of life. *"Tonight we say farewell and thank you to Hank, our beloved meteorologist, who was struck in the head by a flying toilet seat while bravely reporting on Hurricane Hematoma in Louisiana. We'll miss you, buddy."*

THEY AREN'T SHOWING YOU THE REAL WORLD

The primary problem with most television news today is that it's just nowhere near the reflection of reality that most viewers probably assume it is. Much is said about conservative and liberal biases in the news, but political favoritism is a trivial concern compared to the irrational fears and warped perspectives that TV generates. Political leanings are not the biggest problems with Fox News and MSNBC. The primary problems are that they illogically prioritize news coverage, cover politics like sports events, present tremendous amounts of nonsense as important news, fail on competent science reporting, and stoke fears unnecessarily. Anyone who doesn't know how to assess television news for what it is and recognize the nonsense is likely to end up with a wildly inaccurate view of the world. Longtime skeptic investigator and journalist Benjamin Radford details many of the key problems with television news in his excellent book, *Media Mythmakers: How Journalists, Activists, and Advertisers Mislead Us*. He warns that sensationalism, predefining news, and selective news coverage have resulted in a "news bias" that leads viewers far astray from reality. Radford writes:

> Television, by its very nature, distorts the reality it claims to reflect and report on. Events are compressed, highlighted, sped up. Thus a person who occasionally watches sports highlights on TV will likely see more home runs and touchdowns than a person who attends local games regularly; television viewers are likely to see more murders than a police detective, more serious car crashes than a tow truck driver, and more plane crashes than a crash investigator.[1]

Radford also points out that TV news viewers are given "wildly disproportionate" coverage of crime compared to the amount of crime actually occurring. He cites studies that found crime devoured nearly a third of air time while topics such as education and race relations were given 2 percent or less.[2]

I've seen the problem of reality distortion from both sides of the television screen. For more than fifteen years I held a variety of jobs in journalism. Most of those years were in print journalism, but for a (mercifully) brief time I also worked in television news. Doing local TV reporting was fun, and I felt like it had some value to the community. But I was also one of those plastic-haired automatons who reads "news" from a teleprompter and pretends to be personally wounded while going through the laundry list of disasters and mayhem around the world. Of course the truth is that I was far too distracted by the challenge of correctly pronouncing the names of exotic countries and vowel-heavy dictators to think much about the true horror of the events. I also had to be ready to gear up for that cheery toss to the sports anchor. It was all incredibly shallow, but so was I at the time, so I suppose it was a nice fit. The good news is that this part of my journalism career was short and I moved on before the makeup soaked in and I became addicted to quasi-celebrity status.

After the TV experience, I worked in various positions as a reporter, features writer, travel writer, sports editor, world news editor, and photographer for newspapers and magazines. It's been a fun ride, one that is still not over as I continue to write a newspaper column that focuses on human rights, science, and skepticism. Journalism is a great way to learn new things and meet interesting people you otherwise never would cross paths with. And it really is a vital source of important information that our world needs. So I stop well short of condemning it, of course, but it really does need improving because the news media as it exists now—especially television news—breeds far too much ignorance and confusion about the real world.

My experience working in the news taught me how important it is as a news consumer to keep all those images, commentaries, and soundbites in proper perspective. I know better than to react too strongly to everything I see on a television news report because I know that I'm being presented with images selected specifically to get attention or deliver the greatest shock value. What we see on television is almost never a fair representation of what is really going on. I'm not suggesting that all reporters and videographers/photographers are being dishonest and intentionally fooling viewers. They are just doing their best to capture and present words and images that will catch the public's eye and make an impression. Nothing wrong with that—it's their job. We certainly can't expect journalists and photographers to seek out the most mundane quotes and to aim their cameras at average images.

Journalists look for interesting people that stand out and say interesting things. Good photographers instinctively rush to the most eye-popping and powerful scene. Average or typical scenes repel them. I did it all the time. It's automatic; it's the job. In 1992, I roamed around Homestead, Florida, after category 5 Hurricane Andrew blew through. My cameraman and I sought out the most traumatized and emotional victims to interview because it made better TV. From everything they said on camera, we selected their most dramatic quotes to use in the final reports. For my on-camera stand-up bits, I blabbed away while posing in front of the most spectacular scenes of destruction we could find. I recall doing one stand-up in front of a gigantic tree that had been completely uprooted by the winds. It was great; it looked as though I could have been reporting on day three of the apocalypse or maybe a nuclear war. No attempt was made to present a balanced presentation of Homestead, Florida. We were not trying to be dishonest, but the fact is what we did was misleading to viewers who don't understand how television news works. As bad as Homestead was, I suspect many of my viewers were left thinking it was even worse. We weren't doing an academic research paper or a scientifically balanced assessment of a Florida town's destruction. We wanted to show the absolute worst for dramatic effect. We definitely were not concerned with the median experience of storm survivors. We felt no obligation to show a fair or random sampling of the broad spectrum of destruction. Some houses survived the storm very well. As I recall, however, we didn't shoot a single one of them. We made a conscious effort to show the most severe destruction we could find in order to impress our audience. Clearly our reporting was slanted to show the absolute worst of the event we covered. It was not reality.

This news media habit of presenting the extreme at every opportunity would not be so much of a problem if viewers understood it and processed what they see accordingly. But too many people think television news mirrors the real world in some vaguely accurate way when the truth is that it does not. For example, when people watch a cable news report about some disgruntled shoe salesman who dyed his pet ferret's fur green, ate his iPod, and then set his wife's hair on fire, they tremble in fear and then head to Walmart to buy more ammunition. They fail to consider the fact that about seven billion other people made it through the day without doing those things.

Any perception of the world based on TV news is bound to be horribly distorted. For example, try asking Americans who regularly watch TV news to name five of the most violent places in the world today. I have done this on a few occasions and found that their answers invariably match TV news coverage as opposed to reality. I hear Iraq and Afghanistan, of course. The "Middle East" always comes up. But then the answers become inconsistent and delivered without much confidence. Colombia sometimes comes up because of drug violence and, thanks to George Clooney, Darfur might get a mention. One guy said Detroit.

What is interesting is that every time I have asked people to list the world's most violent places, I never once got the Democratic Republic of Congo (DRC) for an answer. But how can this be when this central African country has been suffering death and destruction far beyond anything seen in Iraq, Afghanistan, Israel, Palestine, Colombia, or even Detroit? Precise statistics are impossible, but war in the DRC has claimed anywhere from three million to seven million lives over the last several years. It is likely the most deadly conflict since World War II. But few Americans know anything about it. They think Jerusalem is a more violent place. Why is this? Why does the DRC war draw a blank in the minds of so many Americans? I suspect the primary reason is because TV news doesn't cover it anywhere near as intensely and consistently as they do other hot spots. If a few terrorist attacks in a particularly bad month kill fifty or a hundred people in Israel, for example, it will be all over CNN, Fox News, MSNBC, ABC, NBC, and CBS news. They will devote hours and hours of coverage, day after day, to it. But if fifty thousand people, half of them children, are killed in the DRC that same month, it probably wouldn't get much more than thirty seconds. The problem is that there are not scores of major network journalists and camera crews roaming central Africa looking for stories. No video, no TV time. News media people defend this by saying it's too expensive to send reporters everywhere. But shouldn't combat correspondents be sent to where the death and destruction are greatest? Assuming that human lives are equally valid, shouldn't an ongoing conflict that already has claimed a death toll in the millions top every news organization's list of priorities?

BE AFRAID, BE VERY AFRAID

There is no denying that television news breeds irrational fear. Consider the example of terrorism. Americans have been on high alert since the 9/11 attacks in 2001, mostly because of the way TV has thoughtlessly catered to politicians who exploited the danger in order to consolidate power and win votes. Repeatedly watching images of buildings blowing up and falling down, the constant stream of soundbites about terrorism threats turned America into a nation of very scared people. But was the level of fear justified by the reality? The RAND-MIPT terrorism database shows 14,790 deaths due to terrorism worldwide from 1968 through April 2007. That equates to an average annual death toll of 379.[3] Not good, of course, but compare this to the more than 40,000 people killed in automobile accidents each year in America. Heart disease is a far greater threat to American lives than al Qaeda. In his book, *The Science of Fear*, Daniel Gardner compares that figure of 379 terrorism deaths to other death rates. In 2003, in the United States alone, 497 people accidently suffocated to death in bed; 396 were accidently electrocuted; 515 drowned in swimming pools; and 16,503 Americans were killed by criminals who were not terrorists. Gardner also points out that the 397 terrorism deaths per year figure *overstates* the risk for Americans because most deaths from international terrorism occur in places like Kashmir and Pakistan, not in the United States. Between 1968 and 2007 in North America, 3,765 people have been killed by international terrorism. And that number *includes* the 9/11 attacks that took 2,977 lives. This death toll over nearly forty years is only slightly more than the number of Americans killed riding motor cycles in one single year. Another interesting revelation in Gardner's book is that if you remove the Middle East and South Asia from the equation, there has been a worldwide *decline* in terrorism since the early 1990s.[4] Based on television news coverage of the terrorism threat, however, one might think al Qaeda is everywhere, killing Americans and everyone else by the thousands. The TV news media sells fear. The good news is that you don't have to buy it. If you want to, you really can choose to keep your mind in the real world.

PRESENTING BOTH SIDES OF A ONE-SIDED STORY

Another major problem with news is that most journalists try to be fair. This is not necessarily because they are exceptionally fair-minded people. In most cases it's probably for no other reason than they don't want their bosses or the public accusing them of being biased—supposedly a mortal sin in journalism. Being fair and presenting both sides of an issue may sound like a no-brainer, but it's not that simple.

Fairness, defined as an even split of words or air time, is overrated and can be the worst possible objective in many instances. Attempts by journalists to be fair often leaves them with a warped view of reality just like their customers. For example, the controversy between modern biology and creationism/intelligent design in the United States is usually reported on as though it is a clash between two rival, but equally valid, scientific theories. But nothing could be further from the truth. Evolution and intelligent design are not equally valid scientific theories, and that should be made clear in the reporting. As any credible biologist can explain, creationism and intelligent design are not based on real science, have no compelling evidence, and offer no competent theories. I understand the social controversy and the need for balance if the story is about that particular aspect of the issue. But otherwise there is no place for the "other side of the story." Look at it this way: Should journalists give equal time for comments from Holocaust deniers every time they report on the Holocaust? Should medical journalists give psychics and tarot card readers equal time with neuroscientists when reporting on brain issues? Should reporters check in with astrologers every time they cover a NASA launch? Some topics do not warrant a "fair" hearing from multiple interests simply because some sides have not earned their way onto the playing field. Good journalists should be educated enough, honest enough, and brave enough to know this and report accordingly.

IS MISLEADING NEWS BETTER THAN NO NEWS?

Despite my problems with it, I'm not calling for the end of television journalism. I understand that it does play a valuable role and I appreciate the many brilliant and brave people who report on important and dangerous events that are happening near and far. If not for the light they help shine, we would all be worse off. But is this really the best the industry can do?

For another view I sought out a veteran television news reporter and anchor whom I have known for years and respect a great deal. I asked her about what I see as TV news' misrepresentation of the world. She readily agreed that TV news is not a perfect reflection of reality. "I think that news by its very nature cannot provide an accurate view of the world," she explained. "It can provide snapshots of reality, but in and of itself, it can't be your guide. The parts of life that are least newsworthy make up the majority of our days, and that is the way it is supposed to be. News is supposed to be about the aberrations, the extremes, the unusual."

Fair enough, but the cable news stories that fill the hours these days are usually not *important* aberrations or events. Lindsay Lohan's troubles with the police may qualify as aberrations or extreme events, but are they more important than the malaria crisis in the developing world or the James Webb telescope? I would say no, yet millions of TV news viewers might be led to think otherwise, based on the way air time is allocated by news producers. It's reasonable to describe TV news as "snapshots of reality" and therefore incapable of comprehensibly showing the world as it really is. But there is still the problem of too many snapshots of the wrong subjects as well as too much gossip, entertainment, and fearmongering packaged as important news.

For all its faults, television news is still immensely valuable. I'm honest enough to admit that a sensationalist, entertainment-based, and income-driven news media is better than no news media. Imperfect as it is, I wouldn't want it to go away. Imagine how much worse the world could be if those in power never had to worry about how their evil deeds would play on CNN and the BBC. I simply wish more people would recognize television news for the exaggerated, distorted, and misleading view of the world that it is and not be so easily pulled down paths of fear and distortion. I also wish that the people who run major news broadcasting corporations could figure out a way to make their profits while also producing news that is intelligent and socially valuable.

So what is the solution for those who would like to avoid ending up with an abused mind that has been dimmed and misled by cable news? I don't recommend a total news blackout because then you become one of those scary people who stumble around not knowing anything about anything. I do, however, advise cutting way back on daily TV news consumption because too much of it is simply mindless fodder that exists primarily to fill time until the next commercial. I still make sure to skim newspaper headlines and browse a few reputable news sites online every day. But I spend as little time as possible doing this. If something is obviously important, then I read the article, otherwise my eyes keep moving. I avoid giving significant chunks of my days to TV news, opting to check it only sporadically to make sure World War III didn't kick off while I was cleaning the garage. I only watch TV news for extended periods when there is a major breaking news event or I feel the need to check in on the industry that almost stole my soul. What do I do with the time I save? I mostly read articles in science magazines. News and current events are more than the talking points of political parties and which celebrities are divorcing or languishing in rehab. News and current events also include the discovery of a new species or exoplanet. I love wading into new issues of *Scientific American*, *National Geographic*, *New Scientist* and *Discover*. In my view, the information I find in those publications is no less important, relevant, and entertaining as anything I might find in mainstream news. For example, I could be wrong, but I feel I'm much better off reading about a new breakthrough in primate studies than I am watching CNN's in-depth coverage of congressional blustering and presidential soundbite deliveries.

I also read books. I wish more people appreciated the unique power of a book. Not because I happen to write books, but because I sincerely believe that books (most of them) are good for the world. One could make a case for the book being the most important and powerful invention of all time. Given the impact of the book on human history, it certainly has to make top ten, no doubt. Books are amazing little worlds that contain ideas and stories that never die, even when the authors do. Sadly, reading is not as valued or as common as some might assume. Did you know that once a third of American high school students graduate, they never read another book for the rest of their lives? Even worse, 42 percent of college graduates never read another book—ever. Perhaps most disheartening is the fact that more than three-fourths of American families did not buy or read a single book in all of 2007.[5]

One of my favorite books is *The Demon-Haunted World*, by Carl Sagan. I first read it more than ten years ago but still flip through it occasionally. Whenever I read the words in that book I can "hear" Sagan speaking to me. Just by opening that book, he pays me a visit and shares some of his ideas, even though he died in 1996. If one is short on time and has only thirty minutes or so per day for reading, I suggest spending five of it doing a high-speed headline-skim of a couple of the most reputable news sites or newspapers, then use the remaining twenty-five minutes turning the pages of a good mind-expanding book or science magazine. High-quality nonfiction books and science magazines offer two things television news is unwilling or unable to provide much of these days: information presented in depth and a realistic perspective on the world.

GO DEEPER . . .

Books

- Baym, Geoffrey. *From Cronkite to Colbert: The Evolution of Broadcast News*. New York: Paradigm, 2009.
- Boorstin, Daniel J. *Image: A Guide to Pseudo-Events in America*. New York: Vintage, 1992.
- Hedges, Chris. *Empire of Illusion: The End of Literacy and the Triumph of Spectacle*. New York: Nation Books, 2010.
- Postman, Neil, and Steve Powers. *How to Watch TV News*. New York: Penguin, 2008.
- Radford, Benjamin. *Media Mythmakers: How Journalists, Activists, and Advertisers Mislead Us*. Amherst, NY: Prometheus Books, 2003.
- Shabo, Magedah E. *Techniques of Propaganda and Persuasion*. New York: Prestwick, 2008.

Other Sources

- *Archaeology.*
- *Cosmos.*
- *Discover.*
- *Focus.*
- *National Geographic.*
- *Nature.*
- *New Scientist.*
- *Popular Science.*
- *Science.*
- *Science Illustrated.*
- *Scientific American.*
- *Smithsonian Air and Space.*

Chapter 20

"BIOLOGICAL RACES ARE REAL."

The reality of human races is another commonsense "truth" destined to follow the flat Earth into oblivion.

—Jared Diamond

The lay concept of race does not correspond to the variation that exists in nature.

—Joseph L. Graves Jr., *The Emperor's New Clothes*

Race is a human invention. . . . We made it, we can unmake it.

—Evelyn Hammonds

Most people today believe with absolute confidence that humankind comes prepackaged as a collection of biologically sorted subgroups called "races." These categories, according to the views of most people around the world, are natural, obvious, and undeniable. They also are widely believed to have profound implications for intelligence, physical ability, and even moral behavior. These beliefs are destructive, deadly, unscientific, and just plain wrong. There is no such thing as races, no naturally occurring biological categories of humans that match popular racial categories such as "black," "white," "Asian," and so on. Race belief is one of the most important targets for skepticism because of the havoc it has wreaked throughout history and the great harm it continues to cause today.

Few people ever think to doubt the concept of race because it's considered to be common sense. For example, anyone can look at a dark-skinned African, a light-skinned European, and a typical Chinese person and recognize immediately that they belong in different biological groups or "races," right? If we can see races with our own eyes, then why question it? No one is going to confuse the African for the Asian, or the European for the African. They are different "kinds" of humans that fit into different biological categories. How can that be wrong? It's a no-brainer, right? Wrong; as we shall see in this chapter, it's very easy to dismantle that three-person lineup example and show why biological races are illogical, inconsistent, and nonexistent—except in our minds.

Before we plunge in and explore some of the problems with race belief, however, it is important to be clear that this chapter is not about racism. Race belief does often lead to racism, of course, but this is not a lecture on morality, fairness, or how to make friends. Nothing in this chapter has anything to do with liberal versus conservative, evil versus good, or any form of political correctness. This chapter is concerned only with the reality of our biological diversity and the delusion of biological race categories. It also is important for readers to understand that rejecting the race concept is not a denial of the real biological diversity that exists, nor is it a rejection of the existence of "races" entirely. *Biological* races may not exist but *cultural* races certainly do. Yes, there really are "black people" and "white people" in America, for example. But these are cultural groups of our own creation. They are much more like clubs or organizations than subspecies based on origin and kinship. What we mistakenly see as nature's divisions are instead canyons of our own creation. This is good news, however, for if we invented biological races, then we can certainly decide to move on from them and begin to view ourselves in a more honest and realistic way.

Aren't we different? Isn't biological diversity real? Don't people look differently and don't they have different colored skin, different facial features and hair types? Can't we often tell if someone is more or less related to other people? Of course; all that is true. But the key is that our biological diversity does not sensibly translate into our species being spilt into meaningful biological territories with firm borders around them. The more than one billion humans commonly referred to as "black people," for example, are so genetically diverse that placing them all into one category, defined by kinship and distinct from other groups, is laughable in the light of scientific facts. Biological races do not exist because our species is too young, too closely related or blended—and the blending continues every moment of every day. This means we cannot intelligently divide ourselves up into anything like the traditional notions of racial groups such as "black," "white" and "Asian." It's like trying to draw lines in flowing water. It just doesn't work.

After giving lectures about race and even writing a book on the subject, I have found that the most effective way to enable people to see why biological races are not real is to present them with simple thought experiments. Here are some that seem to work well:

How many oceans are there? If you did well in middle school geography class, you probably answered "five": Pacific, Atlantic, Indian, Arctic, and Southern. But wait! That's not quite right. Five is the *cultural* answer, based on our twist of reality. The correct answer —*nature's answer*—is just one. Look at a globe if you don't believe me; we live on a one-ocean world. We *made up* five ocean names and attributed them to different regions of the same vast body of water. There may have been practical reasons for giving different names to different areas of water, but it doesn't give us an excuse to believe in the literal existence of five distinct oceans, as most people probably do. Any of this sound familiar?

Bad boxes. Imagine that you are some sort of a giant god, and you have decided to organize the little people scurrying about around your feet by tossing them into boxes. Let's say you have five boxes and you decide to sort the people by their skin color, or if you like, some other physical trait such as hair or nose width. You carefully place them into their respective boxes, but immediately you see a problem. The boxes aren't very good. They don't seem to have sides or tops and the people keep running back and forth to other boxes. To further complicate matters, many of the people you dropped into a specific box keep mating and having babies with people in different boxes. Soon you realize that your sorting is in vain. This is what biological race categories are like in the real world. They are boxes without sides or tops that have never done a good job of containing people and their genes.

That three-person lineup in your head. Remember the dark-skinned African, the Chinese person, and the light-skinned European I mentioned at the beginning of this chapter? In the minds of many people, a mental image of three such different-looking people standing side by side wins the case for race. But it shouldn't, and here's why.

The human species cannot be fairly represented by a three-person lineup. It's a false and misleading example that fools us into thinking that we all fit neatly into a few distinct and obvious natural categories when in fact we do not. Consider what would happen if I presented you with a three-person lineup made up of an indigenous Fijian, an Ethiopian, and a Navajo Indian—all dressed in neutral clothes so no cultural or geographical clues were apparent. Most people would struggle to assign them to races. Race identification seems easy when we conveniently omit the billions of people who fit "in between" the members of that original three-person lineup.

Consider how silly it would be to present a seven-foot-tall man next to a five-foot-tall man and declare: "Behold, proof that our species consists of a 'tall' species and a 'short' species." We would laugh at that because we know immediately that there are a variety of heights in between the seven-footer and the five-footer. We should react the same way when we imagine a dark-skinned African standing next to light-skinned European or an Asian.

A long, unbroken chain of humanity. If we somehow could see humankind, all seven billion of us, in one glance we would might recognize the reality of our borderless diversity immediately. We would see that there is no such thing as "obvious" or "commonsense" racial categories based on superficial traits such as skin color, hair texture, and facial features. Imagine if every human in the world were lined up, shoulder to shoulder, from darkest to lightest. You could walk that line for ten thousand years and you would never find a breakpoint, a natural border, where one race ends and another begins. If you were determined to create races, you would have to just arbitrarily create divisions were none exist, which is essentially how we ended up with races today. We drew borders around groups of people and then pretended that nature did it.

How many races? If races are real and plain to see, then shouldn't we all be able to name them? So what are they? How many races are there? See, for all the talk, paperwork, and obsessing over race, there is no agreement on this most basic question. Ask ten random people to list "the races of humankind" and you are certain to get ten different lists. Why is this? For something that is supposed to be obvious and commonsense, race is awfully difficult to pin down. The reason for this is clear: races are make-believe. We could have from three races to one million races depending on whom you listen to, what criteria you decide to emphasize, and what criteria you decide to ignore.

Race against time. As human-made categories and not natural biological categories, races are created and defined by the whim of cultures. The rules of race—written, rewritten, and often unwritten—have never been logical or consistent throughout history and never will be. They never can be because races are not based on solid science or lucid logic. And nothing is more flexible than fantasy.

Let's imagine that we have a time machine and send a "white" Irishman back to the United States in 1820. Guess what? When he emerges at that time, he will still be Irish but he won't be "white" anymore. How can that be? It's because race rules change. Believe it or not, there was a time in US history when many Irish immigrants were viewed as nonwhite.

Magic above the clouds. Did you know that a person's race can change today simply by flying? I know people in the Cayman Islands, for example, who do not identify themselves as "black" and their culture does not identify them as "black." However, if they fly to the United States they somehow turn into "black" people upon landing. How does this amazing transformation occur at thirty-five thousand feet? Does something physically happen to their skin color, facial features, hair texture, DNA, blood, and ancestry? No, their race changes because race is a made-up game and different societies play the game with different rules.

I saw this firsthand during my university days when I knew a light-skinned Haitian student who was "black" while attending school in the United States but then became "white" whenever she returned to her home country. This bizarre flip-flop of race could only happen because races are imaginary and not based on anything logical or scientific. Her race changed because Haiti reverses the "one-drop rule" as it is known in America. In the United States, some small observable African ancestry has traditionally meant a person is "black." In Haiti, some small observable European ancestry traditionally means a person is "white." Neither country can be said to be right and the other wrong in the way it determines race labels. It's just the way it's done, and it's a good example of how racial categories are constructed by culture rather than by nature.

Genes versus races. For those who may think that the science of genetics will somehow validate the concept of race, think again. It certainly has not done so yet, and there is no good reason to think it ever will. If anything, it will continue to make it clear that the traditional race concept doesn't work. In 2010, a paper was published in *Nature* about the sequencing of sub-Saharan African genomes. One remarkable discovery to come out of that research was that the genetic distance between two bushmen who had lived their entire lives within walking distance of one other was greater than that between any one of them and a typical "white" European or Japanese Asian. Think about this: One of the South Africans and a random "white" European or Japanese Asian are more closely related than the two South Africans are to each other. Now, imagine if we made a police-style lineup with those two South African men, a "white" European, and a Japanese Asian. Which two would most people place into a race together? But while the two South Africans may look very similar on the outside, the biological reality beneath the surface tells a very different story.

The truth before us is clear if we choose to recognize it. Our species simply does not accommodate naturally occurring race borders between vast groups of people. Cultures have created and artificially imposed them. The fact that so much death, cruelty, suffering, and social inefficiency has been caused or facilitated by this delusion demands that we finally accept the reality of who we are and abandon race belief.

GO DEEPER . . .

Books

- Diamond, Jared. *Guns, Germs, and Steel: The Fates of Human Societies*. New York: W. W. Norton, 2005.
- Gould, Stephen Jay. *The Mismeasure of Man*. New York: W. W. Norton, 1996.
- Graves, Joseph L. *The Race Myth: Why We Pretend Race Exists in America*. New York: Plume, 2005.
- Harrison, Guy P. *Race and Reality: What Everyone Should Know about Our Biological Diversity*. Amherst, NY: Prometheus Books, 2010.
- Olson, Steve. *Mapping Human History: Genes, Race, and Our Common Origins*. New York: Mariner Books, 2003.

Other Sources

- "Race—The Power of an Illusion," www.pbs.org/race.

Chapter 21

"BIOLOGICAL RACE DETERMINES SUCCESS IN SPORTS."

> **If you tell a Euro-American kid he can't play basketball because he cannot jump or an African American kid he can't swim because he can't float, you limit what they can become and further reinforce segregation. The message that genetically determined racial traits are responsible for athletic performances, as opposed to desire, coaching, and culture, reinforces racism. The effects of this racist ideology are felt far beyond the world of sports.**
>
> —Joseph L. Graves, *The Race Myth*

The games we play may seem like a trivial sideshow, but they are far from that in reality. Sports not only account for many billions of dollars in the world economy and consume countless waking hours, they also have a profound influence on how we perceive the biological diversity of our species. As we saw in the previous chapter, common racial groups such as "black," "white," "Hispanic," and "Asian" are culturally created categories and do not hold up to scientific scrutiny as biological subsets of our species, which is what most people think they are. There are multiple reasons why belief in biological races continues to thrive even as many scientists reject it, but sports certainly is a major contributor to this misreading of our biological diversity. Even people who are not fans can't help but notice the barrage of suggestive images sports serve up to the world every day: once again, all the finalists in the men's Olympic 100-meter race are black; most college basketball and football coaches are white; white men own NFL teams and sit up in luxury boxes watching while their mostly black employees run around sweating and smashing into each other. Billions of people worldwide undoubtedly see confirmation of their biological race beliefs played out before them on the courts and playing fields of sports. But what they see confirms nothing of the sort.

The intersection of race and sports is a tragic illusion that has worked to solidify the incorrect and disastrous belief that races are genetic prisons of destiny for us all. People have long tried to portray sports as a deep and accurate reflection of greater society. They see it as an honest manifestation of the way our species naturally sorts itself out. If one race produces the most champions in this sport or that, it is because success was written upon that race's DNA. It's a notion that comes in handy as reinforcement for traditional race belief. How can the existence of races be an illusion or a cultural creation when we see racial divisions and hierarchies rise to the surface on the sporting stage for all to see? The racial layering of sport results proves that the racial layering in society is natural and not the result of unjust history, immoral choices, evil actions, or unfair laws. No need for anyone to feel bad about anything or try to change anything. People of different races do some things better or worse than other races. All of this is wrong, of course, because the racial categories themselves make no sense. A black East African is likely to be more closely related to a white European than she is to a black South African. The rules of race—who gets assigned to which race—vary by culture today and they have varied over time. There is no logic or consistency to them. And, although race is believed to be all about kinship and ancestry, our popular race groups make a mockery of people's real kinship and ancestry. This alone makes it ridiculous to interpret the results of sports competitions through the lens of biological race.

There are many examples available to illustrate how the misinterpretation of sports and race misleads people and unjustly validates in their minds an invalid concept. However, we will look at just one example that should make it clear just how wrong it is to believe that sports prove the existence of biological races and reveal inherent abilities of people based on race.

Track and field is a sport I once competed in and have followed closely since I was thirteen years old. As the sports editor for a newspaper several years ago, I interviewed numerous Olympic track and field gold medalists. I have photographed and written about the sport at every level, from school races on grass fields in the Caribbean, all the way up to the Olympic Games. For me, track and field is special, the most beautiful sport of all. It's performance art that shows off the human form in all its glory. Unfortunately, it's also a deep well of irrational race belief. Its results often are misinterpreted as evidence of inherited race-based talents and glaring justification for believing in biological races.

Ask anyone which racial group is the best at running and you are sure to hear that it is the black race, of course. It's a no-brainer. Just look who collects all the medals at the Olympics in track and field, right? Not so fast. Don't forget how superficial traits such as skin color can mislead us about the realities of our biological diversity. Sure, people with darker skin seem to win races more often than people with lighter skin. But what does this really mean?

First of all, we need to consider who is actually winning which events. In recent decades, runners from East Africa have had remarkable success in long-distance events. Kenyans and Ethiopians routinely sweep the top spots in major competitions. In the sprints, North American, European, and Caribbean athletes with at least some recent African ancestry have dominated. So, if all these winning groups are members of the black race doesn't it follow, then, that the black race is superior at running and has a race-based genetic advantage? No, and here's why: The "black race" everyone speaks of is a culturally created category that is absurdly vast and makes no sense in biological or genetic terms. Because race rules are determined by societies rather than nature, millions of people can watch athletes such as Daley Thompson, Dan O'Brien, and Bryan Clay win gold medals in the Olympic decathlon and then walk away believing they have seen further proof of innate black athletic superiority, ignoring or never knowing that all three men had a non-black parent. We cannot look at an athlete and then know for certain his or her ancestry and make sweeping assumptions about race-based abilities. Superficial features such as hair, skin color, and facial dimensions easily mislead us.

It is also worth noting that the widespread belief of black racial superiority in sports is not only wrong logically but also unfair to black athletes. To claim that Michael Jordan's greatness was linked primarily to him being black is to overlook or deny the intense work ethic, dedication, and focus he poured into his career. Many of Jordan's peers commented on how his capacity for work and his obsession to win were far above everyone else's.[1] Biological race did not create Michael Jordan. He was born with individual talent, perhaps, but it seems clear that Jordan's success was owed mostly to his drive.

When he was nine or ten years old, Magic Johnson would practice basketball for an hour or two *before* his school bus picked him up on school mornings. If he was just a gifted black man born to play the game of basketball, then why did he have to work so hard for so many years in order to succeed? All those Kenyans you see winning marathons and track races don't just show up and win by default after showing their passports. They don't coast to victory on golden DNA. No, they ran more miles in childhood—much of it over mountainous terrain and at high altitudes—than most serious runners ever do in adulthood. As adults, the best Kenyan runners log training loads that give average runners nightmares. Writing off Kenyan success to their membership in something called "the black race" is not only inaccurate, it also robs them of respect due for the investment of all their years of toil.

None of this is to suggest that there may not be some genetic advantages beyond those of the individual that benefit East African runners. However, we have to be clear about what those group-genetic advantages are—if indeed they do exist. This may surprise many sports fans, but not only is it unjustified to say black people are great runners, it is not even justified to claim that Kenyans and Ethiopians make great runners! The fact is, virtually all of the elite runners in those two nations come from very small populations within the respective countries. I recall a telling moment during a trip to the Great Rift Valley, the small region within Kenya where most of the running stars come from. My driver, a Kikiyu man from Nairobi, commented on how amazed he was by the running abilities of the Kelenjin, a small tribe that has produced most of Kenya's Olympic champions. Here was a dark-skinned Kenyan who was as in awe as much as I was of Kenyan runners and seemed to feel every bit as genetically distant from them as I, a light-skinned American, did.

WHY CAN'T AFRICANS SPRINT?

Here's a fact that seems to escape the notice of most sports fans: there has never been a black African Olympic champion in the 100 meters, 200 meters or 400 meters. Isn't that interesting? According to the standard race-sports mythos, Africa is the source of the genes that fuel the fastest race of people on Earth. So why haven't Africans won anything in the Olympics shorter than 800 meters? It doesn't make sense. If black genes, *African* genes, are somehow the key to world-beating speed, then shouldn't the continent with the most African genes by far be the greatest source of sprinters and produce far more than the United States, the United Kingdom, Canada, and the Caribbean?

It seems clear that culture and possibly some smaller-group genetic advantages are behind whatever racial breakdown of sports we may see. If black men in America have the most success in basketball, it's most likely because so many black boys in America dream about making it to the NBA, believe they can do it, and then put in thousands of long, hard hours on the court trying to get there. This surely is the same reason we see mostly white men in professional ice hockey: the vast majority of kids dreaming about it and playing obsessively are white kids.

It is also important to be aware that it is foolhardy to look at something so fleeting as today's sports results and try to draw sweeping conclusions from them. So what if darker-skinned athletes have been winning most of the top-distance races over the last few decades? That's just not enough to conclude that the results are proof of some racial superiority at work. Just imagine if oil was discovered in the Great Rift Valley and the Kelenjin tribe was suddenly awash in cash. The Kelenjin children might decide they don't want to put in those ten-mile runs to school every day and prefer to have the family driver take them to school in the Range Rover™. They also might put down the porridge and begin feasting on American-style junk food while watching television and playing video games. Several years of that, and we might not see as many Kenyan champions coming up as we are used to. And what if the United States plummeted into a full-scale depression? What if little white kids all across America had to run to school and back home every day? What if they had to make do on a diet of fruits and vegetables? A few decades down the road *they* might be winning all the gold medals.

Imaginary scenarios aside, I would not be surprised if at some point during this century a few Native American populations, perhaps high in the Andes, begin producing distance-running champions. It's certainly possible. It would take a perfect storm of opportunity, a spark of success, good coaching, desire, and then snowballing confidence—just like what we saw in East Africa in the latter decades of the last century. Of course, if such a shift in running success happened, sports fans would no doubt declare: "Native Americans are the superior running race. It's obvious, just look who always wins in the Olympics!"

I focused on track and field as a single example of why it doesn't make sense to believe in biological races as a determining factor in sports success but I could have done the same with any sport that race believers see as confirmation for their delusion. Do "Asians" dominate table tennis and badminton because of a biological race advantage? No. Do white Southerners rule NASCAR racing because their biological race gives them a head start? No. Do Canadians dominate curling in the Winter Olympics because of some racial superiority or because nobody else cares about curling? Nothing we see in sports confirms the notion that popular race categories biologically lift some athletes to victory while holding others down. Virtually everything we see in sports can be explained by the immense power of culture and environment in our lives, with their ability to activate or leave inactivated genes. In addition to that, we must always consider the unique and complex path every individual athlete travels.

GO DEEPER . . .

- Harrison, Guy P. *Race and Reality: What Everyone Should Know about Our Biological Diversity*. Amherst, NY: Prometheus Books, 2010.
- Hoberman, John. *Darwin's Athletes: How Sport Has Damaged Black America and Preserved the Myth of Race*. New York: Mariner Books, 1997.

Chapter 22

"MOST CONSPIRACY THEORIES ARE TRUE."

> **Conspiracy theories result from a pattern-perception mechanism gone awry—they are cognitive versions of the Virgin Mary Grilled Cheese.**
>
> —Christopher Chabris and Daniel Simons, *The Invisible Gorilla*

Claims described as conspiracy theories can be a real challenge for skeptics because they tend to be built upon collections of minor and reasonable events, none of which is necessarily impossible or unlikely. It's the conclusion, the big evil plot, which overreaches and often requires a skeptical takedown. I usually give conspiracy theorists more time to make their cases because, unlike most of the other popular beliefs addressed in this book, many conspiracy theories don't require a violation of the laws physics or pseudoscientific delusions. There is no denying that people really do get together in secret to plan and execute weird and terrible things. For this reason, "conspiracy theories" really shouldn't be thought of as one distinct claim in the way Atlantis or ghosts are, for example. While someone may one day prove that ghosts are real, that can never happen with "conspiracy theories" because it's such a vague and loosely applied description. There is no doubt that while many are false, at least some are always going to turn out to be true. After all, there really was a conspiracy of Confederate loyalists behind the assassination of Abraham Lincoln. The Tuskegee experiment, in which medical researchers conspired to allow men with syphilis to suffer for decades, actually happened. There may not have been a conspiracy to cover up a crashed spaceship at Roswell, but there was a secret cover-up to hide the truth about the Cold War activities of Project Mogul. I don't think extraterrestrials are kept at Area 51, but I have no doubt that the US Air Force and CIA do keep many secrets there. The Watergate conspiracy was real. There really was a 9/11 conspiracy. The overwhelming evidence points to al Qaeda and not an inside job by the Bush administration, of course, but it was still a conspiracy. Many conspiracy stories are true—just not all of them, as some people seem inclined to believe. Fortunately, there is an easy way to sort through them and avoid falling for every crackpot claim that comes along. Remember, that same brain that tempts us to fall for every story and connect every random dot we encounter is also capable of saving us from falling into irrational beliefs. We just have to use it.

Whenever someone asks me if I believe in conspiracies I am obligated to answer: "Of course I do. Conspiracies are real; they happen all the time." I quickly add, however, that I would never accept any major conspiracy claim that is not backed up with sufficient evidence. It's easy to point to unanswered questions, coincidences, and possible connections between different events or people. But none of that is proof. As a rule, it's wise to be very skeptical about any conspiracy theory that makes extraordinary claims but fails to produce extraordinary evidence.

The thing that has long intrigued me about unproven conspiracy theories is how they are able to snare bright minds. I know very intelligent and highly educated people who are convinced that one tiny secret club or another is running the world, for example. One told me about vast underground "cities" from which the world is governed. Another told me that every major US politician—including the president, all governors, and all of Congress—are either homosexuals or pedophiles and controlled by sinister people as a result. I've lost count of how many people have told me that the Moon landings were faked, and sometimes I suspect that I'm the last American who thinks Lee Harvey Oswald acted alone. Discussions with the more passionate conspiracy believers can be fascinating. Their brains seem to work overtime at putting together the pieces of their imaginary puzzle. It's not that they are dumb or lazy. To the contrary, they work very hard to make sense of it all and then defend their conclusions at all costs. What I find to be a consistent problem is that smart conspiracy-theory believers behave just like most people do when clinging to any irrational belief: Their emotional investment is too great to give a fair hearing to contradictory ideas and evidence. And if anything does manage to creep under their defensive wall, the confirmation bias promptly dispatches it. This gets at the key difference between a hardcore conspiracy believer and a good skeptic. Maybe not all, but most devout conspiracy believers seem to find it very difficult, if not impossible, to change their minds no matter how strong the explanations and evidence are against their position. On the other hand, any sensible skeptic would jump on board with any conspiracy theory the moment someone provides proof for it. One mind-set is concerned with clinging to a specific belief above anything else while the other mind-set is concerned with the true story, whatever it may turn out to be.

Skeptics are often accused of having closed minds when it comes to conspiracy theories. That is nonsense. A mind that is unwilling to hear new ideas or change direction when the evidence demands it is the sort of mind I would never want to have. Thoughtful skeptics can and will believe in conspiracy theories—*after* they are shown proof. I, for example, don't believe that there was a vast and complex conspiracy to assassinate John F. Kennedy. It seems reasonable to me that Oswald did it and then the Warren Commission made mistakes in its report, which opened the door for lingering questions. However, I wasn't on the grassy knoll with binoculars that day, and it is possible that there was more to the JFK shooting than a lone assassin. My mind is open on the matter. If it ever turns out that there really was more to it, then I will accept it. Until then, however, I'm sticking with the Oswald explanation because that's where the best current evidence points.

It's no different with the 9/11 conspiracy theory. I've looked over the claims and remain unconvinced that the US government intentionally blew up the World Trade Center buildings with controlled detonations, killing thousands of its own citizens, in order to further some world domination or economic agenda. I suppose it could be true. But for now,

I recommend keeping the following points in mind when thinking about conspiracy theories:

- Keep an open mind. Be careful about dismissing conspiracy theories in blanket fashion. Discounting one or two versions of a popular conspiracy theory does not necessarily mean there was not a conspiracy. For example, just because Lee Harvey Oswald probably didn't get help from the Cubans, the communists, the mafia, the military industrial complex, aliens, or Lyndon Johnson doesn't mean he couldn't possibly have been aided by somebody else. Don't make the mistake of closing your mind to possible realities just because one popular version of a conspiracy seems thoroughly unlikely to be true.
- Sometimes big and important things happen for reasons that are unsatisfying or unknowable to us. When we are faced with these events it can be tempting to fill in the blanks prematurely in order to satisfy our curiosity or provide more meaning. But making up answers is a weak response that is unworthy of sensible and honest people. Resist the urge. Sometimes we have to be grown-ups and accept uncomfortable realities or the irritation of not knowing.
- Beware of confirmation bias. Most conspiracy theories depend on it to accumulate believers. Do not embrace every bit of information that supports a particular conspiracy claim while rejecting everything that contradicts it. Remember that we are naturally drawn to explanations that fit with our prior conclusions and beliefs. You have to fight against this tendency if you want to think clearly and make rational decisions about what to believe.
- The more unusual and complex a conspiracy claim is, the more good evidence you should expect to see before believing it.

it seems pretty clear that the only reasonable conclusion one can come to is that there was a 9/11 conspiracy but it involved Islamic terrorists, not George W. Bush and Dick Cheney. After all, Islamic terrorists had been saying for many years, loud and clear to anyone who would listen, that they were going to attack America. And it wasn't just talk. They had already detonated a truck bomb in the World Trade Center in 1993, killing six people. They blew up two US embassies in Africa in 1998 and attacked a US warship in 2000. There was an obvious series of events leading to 9/11, and the evidence links it to al Qaeda. I have interviewed one of the world's leading al Qaeda experts, Peter Bergen, twice and he certainly is convinced that Osama bin Laden and his followers were responsible for 9/11. It also seems reasonable to me—and more importantly to most engineering experts—that intense fires in the buildings *weakened* (not necessarily melted) the steel girders sufficiently to cause floors to pancake and the towers to collapse. But millions disagree. According to a 2006 Scripps Howard/Ohio University poll, more than one-third of Americans believe that the US government "assisted in the 9/11 terrorist attacks or took no action to stop them."[1]

GLORIFIED GOSSIP

I suspect that the popularity of conspiracy theories also has a lot to do with our obsession with gossip as well. If you haven't been paying attention, take a good look the next time you pass a newsstand. Celebrity magazines, essentially vehicles for gossip, are booming. Gossip is big business because it's an integral part of human interactions. All those who say they can resist listening to or passing on a juicy bit of gossip are lying—and we should call them liars behind their backs. To gossip is to be human. Anthropologically speaking, it probably helps us bond and helps us predict who we can trust and who we can't. Based on this, I think most conspiracy theories qualify as glorified gossip. The big ones are just like rumors about the neighbors or the new stranger in town, only multiplied and amplified a thousand times. If this is right, then conspiracy theories probably aren't going away anytime soon because some researchers think our brains are genetically or culturally predisposed to download and spread juicy information about others.[2] Listening excitedly to the dirt on someone and then passing it on to a friend the first chance we get is not abnormal behavior. It's who we are. It's what we do.

GO WHERE THE EVIDENCE LEADS

Not all, but many conspiracy believers seem to have lost perspective. The more passionate ones are locked into their conclusions more out of loyalty to a position than anything else. When they should be pursuing real evidence and true answers, no matter where the trail leads, they have instead dug in and refuse to budge no matter what evidence and counterarguments come along. This is not an intellectually respectable strategy, one I encourage them to reconsider. I certainly don't care about being lined up for or against any particular conspiracy theory. I just want to be aligned with the truth—no matter what that truth may be.

GO DEEPER . . .

Books

- Aaronovitch, David. *Voodoo Histories: The Role of the Conspiracy Theory in Shaping Modern History*. New York: Riverhead Books, 2010.
- Goldwag, Arthur. *Cults, Conspiracies, and Secret Societies: The Straight Scoop on Freemasons, the Illuminati, Skull and Bones, Black Helicopters, the New World Order, and Many, Many More*. New York: Vintage, 2009.
- *Popular Mechanics* editors. *Debunking 9/11 Myths: Why Conspiracy Theories Can't Stand Up to the Facts*. New York: Hearst, 2006.

Other Sources

- 911 myths, www.911myths.com.
- Polidoro, Mossimo. "Facts and Fiction in the Kennedy Assassination." *Skeptical Inquirer* 29, no. 1 (January/February 2005). www.csicop.org/si/show/facts_and_fiction_in_the_kennedy_assa ssination/.
- *Popular Mechanics* editors. "Debunking the 9/11 Myths: Special Report." *Popular Mechanics*, February 3, 2005. www.popular mechanics.com/technology/military/news/1227842.
- Radford, Benjamin. "Top Ten Conspiracy Theories." Live Science, May 19, 2008. www.livescience.com/11375-top-ten -conspiracy-theories-934.html.

STRANGE HEALINGS

Chapter 23

"ALTERNATIVE MEDICINE IS BETTER."

There is a fundamental flaw that runs through virtually all reasoning about alternative medicine. It is simply this: the body is likely to heal itself in time, regardless of what you do. This means whatever you are doing at the time of this natural healing will receive undue credit for the improvement.

—Hank Davis, *Caveman Logic*

Your worst enemy cannot harm you as much as your own thoughts unguarded.

—The Buddha

While any irrational belief may contain inherent risks, some are more dangerous than others. For example, believing in Bigfoot and the Loch Ness monster is relatively harmless compared to believing in a real monster called alternative medicine. No amusing sideshow, this belief drains trillions of dollars from customers and sometimes even kills them. Complementary and alternative medicine (CAM) is a strange mix of safe, dangerous, effective, ineffective, and just plain weird treatments. Many CAM treatments do great harm to people directly or by leading people away from science-based treatments that might have helped them. In a later chapter on homeopathic medicine I include the story of a baby who died an agonizing death from septic shock. The child was treated with homeopathic medicines (magic water) all the way to the bitter end by parents who rejected medical science and put their faith in alternative medicine. A 2008 Harvard study estimated that more than 365,000 people suffered premature deaths from AIDS unnecessarily in South Africa between 2000 and 2005 because of government policies that rejected science-based treatments in favor of alternative treatments.[1] According to the study, the South African government diverted attention away from antiretroviral drugs that were tested and proven to be effective and promoted unscientific remedies, such as lemon juice, beetroot, and garlic.[2] Have no doubt; CAM kills.

It is important to define complementary and alternative medicine because I have found that many people have no idea what the terms refer to. It's just medicine and healthcare, some assume, no different from medical science. It's probably best to understand CAM for what it is not rather than what it is.

Alternative medicine differs from science-based medicine in one very important way: *it has not been proven to work using scientific testing methods*. CAM is outside of the system that produced the drugs and treatments that have extended life spans and improved the quality of life dramatically for so many people last century. This does not necessarily mean that all alternative medicines and treatments are ineffective. Some of them may work very well—but good luck figuring out which ones!

The problem is that we can't know for sure which drug or treatment works as promised until it is proven with proper scientific testing, but if an alternative medicine or treatment is shown to be effective by credible scientific testing, then it would become part of modern medical science and no longer be an alternative medicine or treatment. Therefore, alternative medicine in the context of healthcare really means nothing more than "unproven" by the standards of proper science. In order to be profitable and to survive in the marketplace, alternative medicines and treatments must rely, not on cold hard data, but rather on tradition, marketing, and anecdotal evidence (individual stories and word-of-mouth referrals).

None of this is meant to suggest that medical science and evidence-based healthcare get it right 100 percent of the time. Far from it. Some "tested and proven" medicines are not tested properly or well enough, are not prescribed safely, or were compromised by incompetence or profit/ethics issues and end up causing more problems than they solve. Every year, tens of thousands of people are injured or die of complications from *evidence-based*, *tested*, and *regulated* drugs and treatments. But this is not a sensible reason to reject medical science and turn to CAM. If anything, I would hope that one would view the very serious problems with modern medical science as only more reason to avoid the largely unregulated and fraud-riddled world of CAM. If we can't even trust *tested*, *regulated*, and *scientific* health treatments all the time, why would anyone choose to risk his or her safety with *untested*, *unregulated*, and *pseudoscientific* health treatments? It should be no surprise that medical science and mainstream healthcare fall far short of perfection. How could it not with humans involved? What is important to understand, however, is that there is at least a good chance, a reasonable hope, that a particular drug or treatment on the evidence-based side of healthcare was produced by a process based on science, was tested, and is more likely to be safe and effective. Across the border, over in the land of alternative medicine, the risks of a treatment being ineffective or dangerous skyrocket because anything goes.

I have always found the popular attraction to unproven and unregulated medicine odd. If more people understood what it means for a drug or a treatment to be outside of modern medical science, I suspect that CAM might lose a lot of fans. I wouldn't want to drive a car or eat a candy bar that wasn't tested and regulated in some way. I certainly don't want to put untested and unregulated medicine into my body when I'm sick. Yet millions of people are willing to trust mysterious pills and potions that could have just about anything in them and do just about anything to the human body. The popularity of this stuff is staggering. A national study found that 38.3 percent of adult Americans (83 million) and 11.8 percent of children (8.5 million under age eighteen) accounted for $33.9 billion out-of-pocket spending (not paid by insurance companies) for CAM treatments and consultations with practitioners in 2007.[3] Nearly three billion dollars alone was spent on homeopathic medicines that many doctors and scientists view as nothing more than very expensive water.[4]

WHY PEOPLE TRUST ALTERNATIVE MEDICINE

Natural. Alternative medicine is often touted as being "natural." The assumption being that this means they are safer and more in tune with your body than chemicals in plastic capsules that were manufactured in a grim, windowless factory somewhere. But the reality is that the claim of "natural" by itself means nothing in the context of medicine. "Natural" does not necessarily mean a treatment works or is safe for you. Rattle snake venom is natural, but I wouldn't want to drink it. Malaria is natural, still not much fun. Water is natural but it doesn't cure diabetes. One should also be aware that many of the top-selling CAM products are not exactly cooked up in grandma's kitchen using the finest natural ingredients. They come from big factories too.

Tradition. Many alternative medicines tap into our fondness for the good old days, simpler ways of doing things, and especially the lure of ancient wisdom. But, just like the "natural" claim, old or traditional does not necessarily mean safe and effective. Bloodletting was a traditional treatment, but I wouldn't recommend giving up a few precious pints if you come down with the flu.

A study of women in Pakistan who delayed seeking treatment after being referred for care after lumps were discovered in their breasts found that 34 percent of these women delayed seeking proper medical treatment *because they relied on traditional treatments*. Of these treatments, homeopathy was most common (70 percent), followed by "spiritual therapy" (15 percent). "The delay results in significant worsening of the disease process," concluded the researchers.[5] CAM kills.

A study in Africa found that a significant number of acute renal failure cases (rapid kidney failure) are *caused by reliance on traditional or folk remedies*. This is another example of death by alternative healthcare: "In conclusion, ARF [acute renal failure] occurring after use of folk remedies in South Africa is associated with significant *morbidity and mortality* [italics added]. . . . Significantly, although a proportion of patients have underlying systemic or renal conditions that may contribute to renal dysfunction, in the majority of patients, folk remedy use appears to be the most likely proximate cause."[6]

Clearly traditional medicine fails millions of people when modern, science-based medicine could have given them a better chance.

Cheapness. Not that all CAM treatments are inexpensive, but many are and this is explains some of the appeal. When a poor person is sick, an *unproven* $5 herbal "cure" is likely to sound a lot more attractive than a $75 visit to a doctor who may prescribe a *proven* $100 medication. If people were more informed and understood that wasting $5 only leaves them with the same problem they started with, they might make better choices. In many circumstances, however, people simply can't afford evidence-based medicine, or it's not available to them where they live. In these cases the particular society and the world at large have a moral issue that needs to be addressed. It's bad enough when people choose snake oil because they are not skeptical and don't think critically. It's far worse when the conditions they live in leave them no choice.

Fear factor. A huge advantage CAM has over science-based medicine is that it's not nearly as scary. If I'm sick or injured, I will choose to go with modern medical science, but there is no denying that a visit to a hospital can feel a lot like being sent to a torture chamber. In hospitals people jam needles into your skin; they suck fluids out of your body; they probe and peer into very personal places using scary metal instruments; sometimes they even cut you open and take things out of you. Yes, these procedures are great because they save lives all around the world every day. But they horrify most people who experience them, nonetheless. Now, consider what is entailed with virtually all alternative medical treatments: you pop a sugar pill, drink a shake made from some plant, or maybe just wear a magnetic bracelet. It's the path of least resistance, far more appealing and comfortable than cold surgical instruments, chemotherapy, or drugs with severe side effects. Well, more appealing right up until your untreated illness kills you, I suppose.

Safety. It may be a surprise to some, but in many cases alternative medicines really are safer and do have less side effects than science-based medicines. But the reason for this is because many alternative medicines contain no significant ingredients and don't do anything! For example, a typical homeopathic medicine truly is safe and really doesn't cause side effects. But that is only because it's water! One wonders if there is any limit to the absurdity of alternative medicine. I fully expect to see somebody selling pills filled with air one day soon. They will say the natural and safe air in the capsules is magnetized or some nonsense and claim that it cures cancer, Alzheimer's, and other diseases. And people will buy it. Wait and see.

Trust issues. A general mistrust of governments and corporations probably adds to the appeal of alternative medicine. A 2010 Gallup poll found that nearly half of Americans rate the work of pharmaceutical companies as no better than fair or poor.[7] That's fertile ground for CAM to work in. Everyone knows that modern medicine in the United States is big business and treating human beings often seems like a trivial concern next to profit. The mystery for me in this is, however, why anyone who finds it difficult to trust the for-profit evidence-based medical industry would run into the open arms of the for-profit unscientific CAM industry. They're big business and they care a whole lot about profits too. Given the frauds, dubious claims, and dishonest advertising that is rampant in the alternative healthcare industry, why in the world would anyone trust them more than evidence-based healthcare?

Frustration. There simply is too much frustration with modern healthcare. Some of it is justified. Some of it is unjustified yet understandable. If someone is suffering, she wants relief and she wants it fast. Many parents of autistic children, for example, are unwilling to wait on science to catch up to their personal and immediate crisis. In 2011, as many as 75 percent of autistic children were receiving unscientific alternative treatments, most of which were obviously bogus.[8] Understanding the desperate circumstances some people face, however, doesn't excuse anyone from forsaking reason.

Too many patients today feel rushed, confused about their illness and treatment, and even dehumanized by their journeys through healthcare systems. This is a serious issue with significant implications for the CAM problem. While promoting critical thinking and skepticism are important, of course, I suspect that a large part of the solution to the problem of unscientific healthcare, fraud, and medical quackery has to come from improving the overall experience of today's science-based healthcare. We are human beings, and as such we bring all kinds of fears and emotional needs with us into the doctor's office. We are not robots in need of an oil change, and the health workers who make up modern healthcare have to keep this in mind. Modern medicine can learn a thing or two from the CAM culture. Clearly alternative medicine must be providing something for all these people who keep throwing away billions of dollars and risking their health on it year after year. Maybe the modern-day shamans and snake oil salesmen simply do a better job of inspiring hope and connecting with sick people as "people" rather than something called "patients."

Effectiveness . . . sort of. Undoubtedly CAM maintains much of its popularity and generates most of those word-of-mouth endorsements via the placebo effect and fortunate timing. It's well known that taking some form of treatment, no matter how impotent it is, often helps a significant percentage of patients even though the inert medicine could not have done anything to directly treat the ailment. One may argue that alternative medicines are worthwhile, even if most of them can do no better than placebos, because that's still better than nothing. But this is not necessarily true. A placebo might make one *feel* better, for example, but it's not necessarily going to help with the actual problem in every case.

No doubt fortunate timing and faulty analysis explains many CAM success stories too. Imagine a sick person who takes some alternative medicine and then, after a week or so, has a full recovery. Chances are the person will credit the alternative treatment *even though their body probably recovered on its own naturally*. CAM believers also are likely to credit alternative treatments even when they were taking science-based drugs or treatments at the same time and could not possibly know which of them was responsible for their recovery.

GIVE CREDIT WHERE CREDIT IS DUE

Imagine if we could compile a list of every person whose life was saved over the last fifty years by vaccines, antibiotics, and other science-based healthcare. Imagine a headshot next to each name. I suspect that just one glimpse of such a massive compilation might change the minds of many people who trust in unscientific treatments. It seems few people today appreciate the amount of suffering and shocking death rates of the past. Everyone should do a tour of the developing world for a different perspective. Those of us fortunate enough to live in the wealthier societies today are shielded from diseases in ways we take for granted. Those who sell, buy, and defend alternative medicine seem to think they have no need of science-based healthcare. I certainly harbor no such illusions. I'm a huge fan of modern medicine—primarily because it has probably saved my life at least twenty times by now. I've stayed in some hotels that were so infested with germs, plants, spores, and critters that I swear the walls and floor seemed to move. I've eaten more than my share of weird and risky meals on the road (some of them seemed to be moving as well). My coddled-American immune system has also been up close and personal with all kinds of people in all kinds of places. If not for vaccines and antibiotics it's possible, if not likely, that I would have been whacked by a germ long ago.

Several years ago my son had to be hospitalized because he became dehydrated while fighting a stomach virus. The doctor assured me that everything would turn out fine, and it did. Nonetheless, I was deeply moved by the stressful event and dedicated one of my newspaper columns to it. I described my gratitude for being fortunate enough to live in a time and place that enabled me to take my sick son to a hospital that practiced medical science rather than to some hut filled with nothing more than chants, smoke, and superstition, or to a gilded temple that could offer my son nothing more than prayers and promises. I saw like never before how nice it is to have a clean, well-stocked modern hospital to turn to in a moment of crisis. Had I been less aware of the advantages of science over pseudoscience and relied on alternative medicine, or was so poor that I could not afford access to evidence-based medicine, my son might have died from that illness.

We all should give credit where it's due. Medical science is the reason many of us alive today have a fair chance of making it to see an eightieth or ninetieth birthday. Vaccines, antibiotics, X-rays, scanning technology, the double-blind method of testing drugs and treatments have extended life and reduced suffering. There was a time when forty was a very old age to be, and in some places most babies never made it to year one. Medical science changed that. What have herbal drinks, homeopathy, and touch therapy done by comparison?

While some of the reasons people are drawn to alternative medicine are understandable, to a degree, none of the reasons should be allowed to overshadow the most important point of all: an alternative medicine is called "alternative" because it has not been shown to work

POINTS TO REMEMBER

- Alternative medicine means unproven medicine.
- "Natural" does not mean safe or effective.
- Anecdotal evidence (mere stories) is not good evidence. One can find a story to support just about anything.
- When struck by illness or injury, the human body is usually capable of healing itself given enough time. But people often give the credit to alternative medicines unjustly.
- Many people mix science-based treatments with unscientific alternative treatments. But when their condition improves, they may give credit only to the alternative treatment.

by the tried-and-true methods of modern science. If it had been, nobody would call it an *alternative* medicine. It would simply be called *medicine*.

GO DEEPER . . .

Books

- Barrett, Stephen, and William T. Jarvis, eds. *The Health Robbers: A Close Look at Quackery in America*. Amherst, NY: Prometheus Books, 1993.
- Bausell, R. Barker. *Snake Oil Science: The Truth about Complementary and Alternative Medicine*. Oxford: Oxford University Press, 2009.
- Singh, Simon, and Edzard Ernst. *Trick or Treatment: The Undeniable Facts about Alternative Medicine*. New York: W. W. Norton, 2008.
- Sharpiro, Rose. *Suckers: How Alternative Medicine Makes Fools of Us All*. London: Harvill Secker, 2008.
- Wanjek, Christopher. *Bad Medicine: Misconceptions and Misuses Revealed, from Distance Healing to Vitamin O*. New York: Wiley, 2002.

Other Sources

- Quackwatch, www.quackwatch.com.

Chapter 24

"HOMEOPATHY REALLY WORKS, AND NO SIDE EFFECTS!"

> **Homeopathy is utterly impossible. Homeopathic preparations are so thoroughly diluted that they contain no significant amounts of active ingredients and thus can have no effects on the patient's body. . . . Those who believe it works either do not understand the science, or are simply deluded.**
>
> —Dr. Kevin Smith, bioethicist[1]

After talking to people who sincerely believe in homeopathy, laboring through a towering stack of books on the subject, talking to pharmacists, and loitering in the isles of alternative medicine shops, I have come to the surprising conclusion that homeopathy is not all bad. There is virtually no possibility of harmful side effects, for example, much different from most science-based medicines. No one can ever overdose or become physically addicted to homeopathic medicines either. There are some real advantages to the stuff.

Unfortunately, the reason that homeopathic medicines are so safe and agreeable to the human body is the same reason they are rejected by medical science: they are mostly water and not medicine. A traditionally prepared homeopathic solution is diluted to extreme levels. So diluted, in fact, that there is usually nothing left of the original active ingredient. It's water! But this minor technical detail doesn't stop the homeopathic medicine industry from selling cures for just about everything to millions of people. According to information posted on the website of one prominent homeopath and author, homeopathy can treat trauma, sprains, cuts, bruises, burns, acute pain, headaches, colds, influenza, ear pain, inflammation, chronic pain, joint disorders, chronic fatigue, migraines, stomach disorders, insomnia, frequent urination, eczema, psoriasis, PMS, menopause, post-partum depression, hormonal mood swings, hot flashes, memory loss, hormonal headaches, hypo- and hyperthyroid complaints, adrenal fatigue, depression, anxiety, phobia, grief, Attention Deficit Disorder, eating disorders, immune system problems, "and much more."

Wow, an impressive list, to say the least. Doesn't this sound like the perfect medicine, far better than the expensive and often-dangerous stuff doctors and pharmaceutical companies keep pushing on us? Homeopathic medicine can treat virtually everything and has no downside—unless we run out of water, of course. Many celebrities, from movie stars to Prince Charles, swear by the stuff, and it's sold over the counter in major pharmacy chains. So where do these wonder drugs come from and how are they supposed to work?

LESS IS MORE

Homeopathy traces its origin to the late eighteenth century, when a German doctor named Samuel Hahnemann came up with the strange idea that using a substance that caused symptoms similar to the symptoms of a particular disease would cure the disease. This is the "like cures like" claim of homeopathy. This may hint at the way vaccines work, but it shouldn't because it's not the way vaccines work. Of course, Hahnemann knew better than to load up patients who are nauseous and feverish with something that causes vomiting and fever. His solution was to dilute the "medicine" so that it would only be enough to stimulate the body's natural defense mechanisms into action to deal with the disease. But "dilute" is an understatement. This is where it gets really weird.

Homeopathic medicine is diluted to mind-boggling extremes. In fact, solutions are usually diluted to a point where there is none of the original ingredient left in the remedies—not even a single molecule! But this doesn't make the medicine weaker or less effective, say believers. In some sort of upside-down, mirror-universe logic, homeopaths claim that their medicines get more potent the more they are diluted. So a homeopathic solution that is pure water would be immensely more powerful than one that contains 50 percent, 10 percent, or 1 percent of an active ingredient. Of course this doesn't seem to make sense. How can a medicine work if there is no medicine in it?

According to homeopaths, it works because water can "remember" the molecules it comes into contact with. I'm sure this sounds incredible to readers who are unfamiliar with homeopathic medicine and always assumed that it was simply some kind of "natural" medicine, not fundamentally different from herbal concoctions. But it's true; homeopathy defenders actually say this and seem to believe it. Practitioners and proponents claim that the water somehow retains an imprint of the active ingredient when the homeopath preparing the solution bangs the container a certain number of times. So even though no molecules of the supposed "medicinal" substance (which was itself unscientific in the first place) are likely to remain, the solution is supposed to work anyway. One skeptic estimates that a person would have to drink twenty-five metric tons of a typical homeopathic solution in order to have even a remote chance of swallowing just one molecule of the original substance.[2] Dr. Ben Goldacre, another critic of homeopathy, calculates that many of these popular potions are almost incomprehensibly weak: "At a homeopathic dilution of 200C (you can buy much higher dilutions from any homeopathic supplier) the treating substance is diluted more than the total number of atoms in the universe, and by an enormously large margin. To look at it another way, the universe contains about $3X10^{80}$ cubic meters of storage space (ideal for starting a family); if it were filled with water and one molecule of active ingredient, this would make for a rather paltry 55C dilution."[3]

This claim of water memory is extraordinary. So how do homeopaths know that water can do this? Hahnemann may have imagined or guessed it somehow, but he certainly didn't "know" it in any scientific sense back in the eighteenth century. Obvious questions are, How does water know to remember what the homeopath wants it to remember but not remember all the other stuff that has been in contact with it? (A horrifying thought, considering the places water has been! What if it remembers the bus station toilet bowl it was in last month?) And have scientists confirmed this claim since Hahnemann came up with it? Of course not. Homeopaths say that water remembers the desired active ingredient because it is shaken vigorously during the dilution process and this creates the imprint. To date there is no credible science that has proved this claim. I'm willing to concede that it's possible, only because things get so weird at the atomic level that nothing would surprise me. Read up on entanglement and you will understand what I mean. But wild possibilities are never reason to believe wild claims that are not supported by evidence.

In fairness, many mainstream medical treatments in the 1700s and 1800s were brutal, dangerous, and often ineffective. Just as many mainstream treatments today will be viewed in the future, no doubt. Many patients died not from their illnesses and injuries but from treatments administered by doctors. In many cases patients probably would have been better off doing nothing or taking a water/homeopathic treatment than facing the wrath of their time period's medical care. Fortunately for us, however, medical science has come a long way since Dr. Hahnemann's time. The scientific method, double-blind tests, and a willingness to change by accepting things that work while rejecting things that don't have given us treatments that are effective and save millions of lives every year. Evidence-based medicine is far from perfect at this time, but it performs measurably better than anything else by far.

Two things should give one reason to pause before spending any money on homeopathy or trusting one's health to it: (1) No one has been able to prove that it works even though it's been around since the late eighteenth century, and (2) it can't work, at least not according to the official laws of physics and the unofficial rules of common sense. But people do believe, nonetheless. Homeopathy is a massive industry in India, for example, where there are more than three hundred thousand "qualified homeopaths" and three hundred homeopathic hospitals! More than half of the people in Belgium buy the stuff and 36 percent of the French population are users. In the United States, homeopathic medicine almost died out early in the last century but has roared back bigger than ever in recent decades to earn big profits. Today it rakes in more than one billion dollars per year.[4] This impressive popularity is maintained despite wide condemnation from scientists and doctors in many countries. In 2010 the British Medical Association, for example, called homeopathy "witchcraft" and voted overwhelmingly to ban the practice, as well as to stop placing trainee doctors into any programs that promote homeopathy.[5]

As understanding as I am about how well-meaning people come to sincerely believe in paranormal and pseudoscientific claims, I confess to having very negative feelings about homeopathic medicine. The claim is fraudulent in the way its proponents usually portray it as scientific even though it stands in opposition to science and they have failed to demonstrate that it can do the things it is supposed to be able to do. I have more respect for witch doctors because they are far more honest about what they are selling. Some argue that homeopathy is a safe alternative to science-based treatments and if it gives some people satisfaction, what's the harm? The harm is that it puts people at risk by misleading them into trusting a vial of expensive water or a sugar pill over real doctors and real treatments. Anyone who cares about the health and safety of others has a moral obligation to oppose homeopathic medicine.

I asked a senior pharmacist at a CVS drugstore in southern California how she felt about homeopathic medicines being sold over the counter not more than twenty feet from the pharmacy counter she worked at. I suggested that the juxtaposition of science and pseudoscience was improper and that selling homeopathic treatments in this context suggested that it was legitimate medicine. It's presence next to a pharmacy counter likely gives it unjustified credibility in the minds of many customers. I pointed out the odd contrast of her having worked to earn professional qualifications as a pharmacist in order to distribute the best help medical science can offer to people in need, while a few steps away magic water was being touted for the treatment of everything from anxiety and asthma to vertigo and warts. It would be no less incongruent and improper if somewhere in the National Academy of Sciences Building there were an office dedicated to tarot card reading. She replied that she only works in the pharmacy and has no control over what is sold in the store. "But does it bother you?" I asked.

"Yes, I understand what you are saying," she said. "But what can I do?" She seemed to acknowledge that it was a waste of money and her facial expressions suggested that she agreed the stuff had no business being in a respectable drugstore. However, "it's not my department" was the best she could come up with.

HOMEOPATHY FAILS IN THE UK, TOO

Homeopathic medicine is even more popular in the United Kingdom than in the United States. A House of Commons Science and Technology Committee conducted an in-depth inquiry in 2009 and 2010 and came to the conclusion that homeopathy is unproven, unscientific, unlikely to work better than placebos, and unworthy of government funding or endorsement. Some highlights from the official report of the proceedings:

- We [House of Commons Science and Technology Committee] consider the notion that ultradilutions can maintain an imprint of substances previously dissolved in them to be scientifically implausible.
- In our view, the systematic reviews and meta-analyses conclusively demonstrate that homeopathic products perform no better than placebos.
- To maintain patient trust, choice, and safety, the Government should not endorse the use of placebo treatments, including homeopathy.
- We conclude that the principle of like-cures-like is theoretically weak. It fails to provide a credible physiological mode of action for homeopathic products. We note that this is the settled view of medical science.
- We do not doubt that homeopathy makes some patients feel better. However, patient satisfaction can occur through a placebo effect alone and therefore does not prove the efficacy of homeopathic interventions.
- There has been enough testing of homeopathy and plenty of evidence showing that it is not efficacious. Competition for research funding is fierce and we cannot see how further research on the efficacy of homeopathy is justified in the face of competing priorities.
- Professor David Colquhoun, professor of pharmacology at University College London: "If homeopathy worked the whole of chemistry and physics would have to be overturned."

(Source: House of Commons Science and Technology Committee, Evidence Check 2, www.publications.parliament.uk/pa/cm200910/cmselect/cmsctech/45/45.pdf)

If I were a pharmacist in this situation, I would attempt to unite with my colleagues and send a strongly worded message to company headquarters requesting/demanding that they stop selling bogus medicines to people, many of whom are seeking treatment for serious health problems. Yes, the aisles near the pharmacy counter are technically a different domain, but I don't think that is how it is perceived by many people. To many customers, that general area of the store is "the pharmacy" and pharmacists often blur the line further by coming out from behind their counter to assist and answer questions about products found on the shelves. Maybe if indifferent medical professionals paused and actually thought about the real harm trusting in homeopathic claims over evidence-based medicine can cause, they might take it more seriously.

DEATH BY HOMEOPATHY?

Consider the case of Gloria Mary Sam, a baby who died in 2002 in Australia of septicemia (blood infection) after eczema ravaged her tiny body for months. Gloria's parents—both college educated—were convicted of manslaughter because they refused to seek medical treatment for the suffering child. They received repeated advice to seek proper care for her from science-based medical professionals but did not. Why not? The reason was that the father, Thomas Sam, a "practicing homeopath," was confident that his child did not need any treatment other than homeopathic concoctions. The *Sydney Morning Herald* reported on the trial:

> They [parents] watched her skin bleed, her hair go white and her small frame shrink as their baby girl fought to battle her eczema. They watched her constantly scream out in pain, her sores weeping through tears in her skin, and the corneas in her eyes melt. And in the end, by not seeking proper medical treatment until it was too late, a homeopath and his wife watched nine-month-old Gloria Mary die.[6]

Don't think for a second that this was some bizarre once-in-a-million-years case. There are too many examples of homeopathy distracting from, hindering, or blocking medical science to the detriment of someone's health. In Ireland, Mineke Kamper, another homeopath, was allegedly linked to the deaths of two of his patients. He reportedly told one to stop taking her asthma medication and she promptly died of an asthma attack in 2001. Kamper also used homeopathic medicines to treat another patient who had a treatable (with medical science) cancerous tumor. That patient died in 2003.[7]

Russel Jenkins, fifty-two, stepped on an electrical plug at home in 2006, cutting his foot. When the wound failed to heal, he sought treatment from a homeopath and later died from complications. A vascular surgeon said Jenkins would have had a 30 percent chance of survival if he had sought proper medical treatment *even just two hours* before he died. His mother was devastated: "To lose my son is devastating, absolutely. But the way he died. I just can't come to terms with it when I know all it needed was a phone call for a doctor or ambulance, for antibiotics and my son would be here today."[8]

In 2009, a baby died in Japan from a subdural hematoma due to vitamin K deficiency. The mother sued the midwife who had been caring for the baby because the midwife allegedly had taken it upon herself to give the baby a homeopathic supplement rather than the vitamin K supplement she was supposed to administer.[9]

In the United Kingdom, homeopaths have drawn fire from the medical community for recommending their water over conventional antimalarial treatments. "We've certainly had patients admitted to our unit with falciparum, the malignant form of malaria, who have been taking homeopathic remedies—and without a doubt the fact that they were taking them and not effective drugs was the reason they had malaria. . . . [T]hey could die," said Dr Ron Behrens, director of a travel clinic at the London Hospital for Tropical Diseases.[10]

Medical science has come far in recent centuries. The human life span has been extended significantly as a direct result of science-based healthcare. Magic water may seem appealing at first glance. Look more closely, however, and it becomes clear that it's a waste of time and money. Yes, modern medicine is imperfect, but for those who value proven results it's the only game in town.

GO DEEPER . . .

- Ernst, Edzard. *Healing, Hype, or Harm? A Critical Analysis of Complementary or Alternative Medicine*. London: Imprint Academic, 2008.
- Goldacre, Ben. *Bad Science: Quacks, Hacks, and Big Pharma Flacks*. New York: Faber and Faber, 2010.

Chapter 25

"FAITH HEALING CURES THE SICK AND SAVES LIVES."

One of the saddest lessons of history is this: if we've been bamboozled long enough, we tend to reject any evidence of the bamboozle. We're no longer interested in finding out the truth. The bamboozle has captured us. It's simply too painful to acknowledge, even to ourselves, that we've been taken. Once you give a charlatan power over you, you almost never get it back.

—Carl Sagan

The best thing about being a writer is that you have a good excuse to place yourself in unusual places with interesting people. Tonight I'm a mere twenty feet away from Benny Hinn, perhaps the most popular faith healer in the world today. I'm surrounded by thousands of his devoted fans in an open-air football stadium. They have come expecting this little man with a big stage presence to call down miracles from heaven, and he's about to deliver.

Hinn stalks from one side of the stage to the other. The white suit and white shoes glow against the black night sky behind him. Hinn's overengineered hair never falters in the strong breeze. He closes his eyes tightly and begins to sing:

"Hallelujah . . . Hallelujah . . . Hallelujah . . . Hallelujah . . ."

The slow, one-word song is remarkably dramatic. Each word seems to penetrate the believers who begin praying, mumbling, chanting, and screaming. Many reach up to the sky, some flail at it almost as if trying to claw their way up to heaven. Hinn continues the slow-paced one-word song:

"Hallelujah . . . Hallelujah . . . Hallelujah . . . Hallelujah . . ."

The effect is stunning. It feels hypnotic—maybe it is. Some of the people in the audience begin to twitch. One woman falls to her knees and begins to convulse as if she is receiving jolts of electricity. Although I'm a skeptic filled with doubt about Hinn's ability to provide or facilitate supernatural cures, I can't deny the impressive emotional power of his presentation. It's a potent psychological cocktail of music, religious belief, and hope. It's not difficult to understand how people are swept up in the excitement of Hinn's performances.

Suddenly Hinn freezes on stage and whispers into the microphone: "The Holy Spirit is with us now. It's time to claim your healing." Random shrieks rise and fall in the audience. Several believers "speak in tongues," the incomprehensible "language" some say God inspires them to speak. Suddenly Hinn screams: "God heals today!" His eyes close tight and he extends an open hand toward the audience. "Some will feel electricity. Some will feel heat all over their body. Some will feel vibrations. Take your healing now."

This seems to me like a clear case of the "power of suggestion" in action. Hinn wants the evening to go a certain way, so he tells his primed and willing audience precisely what they are supposed to experience. And they do as they are told, so far as I can tell. An interesting aspect of a Benny Hinn faith-healing event is that the "healings" take place at this point, not later when believers are sent up on stage to be touched by Hinn. According to Hinn, God actually heals people while they are still in their seats. They only come up on stage in order to get some sort of final dose of the Holy Spirit. Most outside observers, I would think, miss this important distinction. On his television show, for example, it is presented as if the healing happens on stage when Hinn touches people. Of course this works well for Hinn because it allows him to describe himself as little more than a humble master of ceremony, while at the same time everyone sees people collapsing and writhing on stage after his "anointed touch" which suggests that he has great healing powers.

For those who have never seen Hinn's "touch" in action on his television show or clips on YouTube, it's a sight to behold. It shouldn't surprise anyone that his ministry rakes in many millions of dollars per year because he really does put on an unforgettable show. After the "healing" by God takes place, Hinn calls for those who were healed to come forward to possibly appear on stage. This is key to the process because those who come forward are self-selected believers who are already at a high level of excitement and satisfaction because they think they have just been healed. This virtually guarantees that Hinn won't get any duds on stage. As people arrive at the side of the stage, Hinn's workers conduct quick interviews to determine who gets on stage with Hinn and who doesn't.

The crowd's excitement peaks when people are sent out to be touched. Hinn asks most of them what God has healed them of. On this night a wide variety of ailments were wiped away, including AIDS, cancer, arthritis, diabetes, chronic eczema, and one heart problem. The workers feed believers to Hinn rapid-fire from both sides of the stage alternatively. They do this at high speed, as if there is some time limit or deadline to meet. The frantic pace adds to the excitement. A woman tells Hinn she has been addicted to drugs for many years. Hinn touches her face. Immediately she shakes violently and collapses into the arms of one of the "body catchers" who stand behind the believers. "She has been freed of the addiction demon!" declares Hinn. The crowd explodes with cheers. It's odd that the woman would have such a dramatic reaction to Hinn's touch because the supposed actual cure from addiction already took place fifteen or twenty minutes ago. It's also interesting to note that anyone who displays an above-average physical reaction to Hinn's touch is immediately lifted back up on their feet so that the preacher can knock them down again with his powers—or, technically, God's power that travels through Hinn's fingers. The audience goes wild over these double and triple doses of the "Holy Spirit."

Amid the falling bodies and frenzied crowd's screams, Hinn repeatedly states that he is not doing the healing, that it's all God. Of course, this does little to diminish the obvious image being sold here, that, no matter what he says, Hinn is very special. I don't know about you, but when I touch people they don't go into convulsions, faint, and become twitchers on the ground. But I'm sure this professed deflection of credit and humility plays well with the audience who might otherwise suspect that he is a little too godlike for their tastes. "I have nothing to do with this," says Hinn. "I am even more amazed than you are." Then he promptly returns to dropping people on stage with his fingers. Sometimes he knocks believers down with a mere wave of his hand from several feet away.

I carefully observed Benny Hinn in action and saw nothing to suggest that he might really be a conduit to a god or otherwise possess magical healing powers. Everything I witnessed can be explained as nothing more than the expected behavior of people who were predisposed to play along with a carefully choreographed performance that has all the tried-and-true elements of faith healing that have been in practice since the first prehistoric shaman hung out a shingle. Hinn is very good at what he does and he is able to tap a deep well of religious passion and conviction that is thousands of years old. The audience was a self-selected group who wanted to see Hinn. Then a small subset of that crowd was selected for the onstage antics. They came with expectations and Hinn was smart enough to cater to those expectations precisely. Did anyone get healed of a serious medical condition? Of course not, and I suspect Hinn knows this. If he (working as God's humble middleman) really did have a hand in curing people of AIDS, cancer, arthritis, and diabetes that night he would have demanded, and deserved, front-page coverage in the *New York Times* the next morning. He also would have clinched the Nobel Prize in Medicine.

With such an ability to heal, Hinn no doubt would feel morally obligated to visit hospitals day and night. I hear they have lots of sick people in need of healing in those places. Faith healers often say or suggest—cruelly, in my opinion—that if a healing doesn't occur, it is the fault of the sick person because she or he lacks sufficient faith. I heard Hinn say more than once that one's faith is the key to successful healings. But if Hinn is worried about wasting his gift on the undeserving, there's an easy way around the problem. He can simply share his anointed touch with the more than twenty-five thousand babies who die in extreme poverty every day in the developing world. If his supernatural delivery system of divine healing can overcome AIDS and cancer, then malnutrition and dysentery should be easy. Surely neither God nor any preacher would begrudge lifesaving healing from a one- or two-year-old baby for lack of faith, right? But no, Benny Hinn seems to prefer to hop from stadium to stadium where he is greeted by thousands of preconditioned believers who know the routine—and have money to give.

So what is the reality behind this popular claim? Can prayers or the touch of some special holy person actually heal? Yes and no. There may be some positive benefit experienced by some people as a result of faith-healing efforts. But there is no reason to jump to the conclusion that a god or supernatural forces are the reasons. It can more sensibly be explained as the placebo effect or the excitement that makes people feel better temporarily. Sometimes it is likely a coincidence that people get better soon after they participated in a faith-healing event.

Remember, we all recover from the majority of injuries and illnesses we get in our lives—whether or not a preacher touched us. The placebo effect still is not fully understood, but some people do improve simply because they believe that they are receiving treatment for their problem even though they are not. It's worth noting, however, that the placebo effect does not restore missing limbs or instantly close wounds. It is remarkably similar to faith healing in this regard. By the way, why don't gods and professional faith healers heal amputees, anyway? It's a fair question, especially given the many thousands of children who have lost limbs in car accidents or by stepping on landmines. There is even a website dedicated to this mystery.[1] No, faith healing only seems to work on internal problems we can't easily see or types of illnesses that often heal independently with or without faith healing.

Faith healing almost certainly does not work the way millions of people believe it does. If prayers and faith healers had really been curing serious health problems for thousands of years, we would be certain of it beyond all doubt by now. It should be simple to measure: Are those people who are touched by faith healers in these ritualized events healthier and do they live longer lives than those who are not? To date, no one has ever been able to prove it.

Keep in mind when thinking about faith healing that it's nothing new. Faith healing has been around at least since the Stone Age. The only significant differences between Benny Hinn and a Paleolithic shaman are clothing allowance and a television show. Faith healing is a claim we can confidently reject as almost certainly not valid because it is so audacious in its claims and so utterly lacking in proof. If one healer, one specific type of prayer, or one religion really and truly healed people as dramatically and consistently as preachers like Hinn claim, then humankind certainly would have noticed it by now and all seven billions of us would have rushed to it. We don't see millions of desperately ill Muslims, Hindus, Buddhists, and animists flocking to Christianity because Christian faith healers claim they can cure cancer, AIDS, and other serious illnesses. And we don't see Christians and Jews with cancer running to animism or Islam because of the long traditions of supernatural healing associated with those belief systems. As with many other unproven beliefs, proponents of faith healing have unjustified confidence in their claims because they ignore the equally passionate and sincere claims of people in contradictory belief systems.

In addition to the big Benny Hinn production, I also attended two small faith-healing services. One starred a woman who said she met Jesus, who gave her special powers and sent her on a mission to heal people. She "laid hands" on many people that night, including a very sickly old woman whom she declared to be healed. I photographed this and wrote an article about it for the newspaper I worked for at the time.[2] I learned that the old woman died a week later, but I was instructed by my boss at the newspaper not to do a follow-up story.

Another healing event I attended was by far the worst. After a lengthy sermon that was nothing but a series of the most shallow clichés imaginable punctuated by random Bible quotes, the preacher called for anyone to come forward who was suffering back pain. A few people raised their hands, and the preacher chose one. The man walked to the front of the church without limping and he did not show any signs of being in pain. The preacher promptly sat him down and declared to the rest of us that the man's back hurt because he had one leg that was shorter than the other. It just so happens that I had read about this common scam. I was shocked because I was under the impression that it was outdated and nobody used it anymore. Apparently they do, however, as the preacher lifted the sitting man's feet, shook them and then either aligned them in a way that made his legs appear uneven by about an inch or two or pulled one shoe off just enough to make the feet misalign. This is not difficult to do, no matter how long or short the legs are. The audience was invited to come up close and inspect the asymmetrical legs. The preacher then shook them and realigned them again so they were even. I couldn't believe what I was seeing. It was so transparent and embarrassingly bush league. This guy was no Benny Hinn. But, to his credit, he pulled it off. The audience *oohed* and *awed* as if they had just seen Lazarus raised from the dead. That experience made me wonder just how bad a faith healer has to be before people refuse to believe. But it turns out that the most enlightening faith-healing session I would ever witness would be the one in which I was the one getting healed.

THEIR PRAYERS LIFTED ME OFF THE PAVEMENT

Cruising fast through the streets of the Cayman Islands on my bicycle, I had a Bob Marley tune playing in my ears and no worries on my mind. Without warning, however, a careless driver pulled out of a parking lot and we collided. I don't remember the launch or the landing, but I do recall the flight. It may be my edited and embellished memory speaking, but I really do remember feeling like a superhero as I soared over the hood of the car. I managed to plant a hand on the hood, which stabilized my flight just enough to allow me to execute a fast one-handed cartwheel. It was a smooth move, to say the least. An Olympic gymnast would have been impressed. I only had to flip my body around, stick the landing on the other side of the car, and take a bow. Unfortunately, my moment of supreme physical grace degenerated into the dorkiest belly flop in recorded history. My feet had rotated too far under me during the cartwheel, and I crash-landed hard. To this day I cringe thinking about it.

After a brief nap on the warm asphalt, I looked up through a haze to find several very large Jamaican women hovering above me. They circled around like a tight formation of Zeppelins, blacking out the stars like an eclipse. And then it got weird. They began shouting:

"Lord, heal dis boy!"

"Come Jesus and help 'im now, please."

"In Jesus name we pray!"

Apparently the women had rushed over from their evening service at a tiny church across the street and were eager to practice what they had just heard preached. Still flat on my back, bruised and bleeding, I began to feel special. Everybody likes to feel loved. Even though they didn't know me, these kind women came to my rescue. They even brought a god with them to help. I was flattered.

As they continued to pray for my divine healing with increasing passion and volume, I became highly motivated to get up on my feet and reassure them that I was going to be OK. So I did. I hopped up, dusted myself off and thanked them for their concern. I didn't realize it until later, but this gave me deeper insight into the way many people likely feel when they stand before Benny Hinn or some other faith healer. People aren't stupid; they know how the script is supposed to go. Who wants to be the party pooper, the creep who disappoints everybody by not getting healed? This becomes especially powerful if a preacher ties one's level of conviction to the success of the healing. It puts the blame on the person rather than the god. I am sure that this "pressure to perform" plays a big role in faith healing. I'm not suggesting that everyone who takes part in a faith-healing session is faking, but the urge to go along and not let people down is powerful. I know because I felt it that night in the Cayman Islands. There is also the inevitable adrenaline surge most people will get in such a situation. That alone can be enough to make a bad back feel better momentarily, for example. I recall interviewing a World War II P51 Mustang ace who told me that adrenaline in combat could always be counted on to make any pain or discomfort vanish—at least for a while. In short, until someone can prove that faith healing works, it makes no sense to believe it does because there are many simple explanations for what is really going on. Unlike some other irrational beliefs that might only waste time and money, this one is a clear danger to people. Sure, it may offer comfort, maybe even temporary relief, to some who believe in it. But is it worth the risk? Terence Hines, a professor of neuroscience, offers a stern wakeup call for anyone who may think faith healing is harmless:

> One point about faith healers cannot be overemphasized: They kill people. Convinced that they are cured when they are not, they may be dissuaded from seeking legitimate medical help that could save their lives. For example, many kinds of cancer are now treatable, if treatment begins early enough. . . . How many of the largely elderly and poor members of [a typical faith healer's] audience go home to great pain and even die because they have thrown away the medicine that was really treating their health problems? . . . In many states, laws covering child abuse and neglect contain specific religious exemptions. These permit a parent to withhold medical treatment from a child if the parent is a member of a religious group that believes in the power of faith healing or in the power of prayer to heal. Such exemptions have resulted in the death of many children whose lives could have been saved by legitimate medical treatment.[3]

GO DEEPER . . .

- Brenneman, Richard J. *Deadly Blessings: Faith Healing on Trial.* Amherst, NY: Prometheus Books, 1990.
- Randi, James. *The Faith Healers.* Amherst, NY: Prometheus Books, 1989.
- Sloan, Richard P. *Blind Faith: The Unholy Alliance of Religion and Medicine.* New York: St. Martin's Griffin, 2008.

Chapter 26

"RACE-BASED MEDICINE IS A GREAT IDEA."

How accurate can "race" be in determining genetic links to disease and health conditions when the definition of race is one that eludes most researchers?

—From the report *Geneticizing Disease: Implications for Racial Health Disparities*

In 1998, the American Anthropological Association (AAA) issued a formal statement on the concept of race. It opens with these words:

> In the United States both scholars and the general public have been conditioned to viewing human races as natural and separate divisions within the human species based on visible physical differences. With the vast expansion of scientific knowledge in this century, however, it has become clear that human populations are not unambiguous, clearly demarcated, biologically distinct groups.[1]

The AAA's statement goes on to explain that there is more diversity *within* racial groups than there is *between* them. It also points out that humans have mated whenever and wherever they came into contact throughout prehistory and history and that this "continued sharing of genetic materials has maintained all of humankind as a single species." The AAA communicated it loud and clear: Traditional racial groups such as "black" and "white" are not natural or consistent biological categories. Races are cultural categories of our own design. We invented divisions between people and then proceeded to blame nature for it. While the AAA's important statement is direct and readily available to anyone, there is a significant problem with it: Virtually no one read it. Including, it seems, many medical doctors and drug company executives, because currently there are hundreds of race-based drugs in development at major pharmaceutical companies. Many scientists warn that this is a mistake that could have severe consequences for patients.

What's wrong with racialized healthcare? The more personal and precisely targeted a drug is, the better it should work, proponents say. So why not race-based medicine? It seems logical that a drug engineered for black people would work better for a black person, for example, than a drug designed for anyone and everyone, right?

No. Drugs designed and prescribed according to race are not a good idea. It's a terrible idea for the simple reason that race groups are nowhere near the biologically distinct and unique subsets of humanity that most people think they are. The last thing you want is to be prescribed a drug based on some doctor's spontaneous assumptions about your current genetic makeup and your ancestry stretching back several thousand years, all based on how you look or identify yourself culturally. No one can glance at another human being and know such things with certainty. Consider Barack Obama; he self-identifies as "black" or "African American" and, based on his appearance, people who have been encultured to think in line with traditional American race rules would identify him as such. Imagine, however, if he was not the president of the United States and instead was some relatively unknown lawyer. What would a doctor probably assume his race to be? "Black" or "African American," most likely. While this may be culturally accurate, it would be far off the mark biologically. And if Obama took a drug specifically designed and tested for African Americans—who knows?—it could be ineffective or even harmful to him. The reason Obama's genome would likely be significantly different from most African Americans is because they derive their recent African ancestry from West Africa. Obama's recent African ancestry comes from his father, who was an East African Kenyan. Ironically, it is his recent African ancestry that puts the most genetic distance between Obama and most other African Americans. His "white" mother is likely the closest kinship tie that he shares with African Americans because that population has high levels of recent white European ancestry too. The reason for this is because East Africans and West Africans are not nearly as closely related as many people assume.

Africa is our oldest home, "the cradle of humanity," and our species has lived there longer than anywhere else. This is why Africans today are the most genetically diverse population. It's also why thinking of all dark-skinned Africans as simply "black people" that fit within a single biological race is nonsense scientifically—and risky when it comes to medicine. Obama may have skin color that is similar to many other African Americans, but this does not justify lumping him in with them when it comes to the serious business of prescribing drugs that were designed based on genetic and ancestral factors. The same applies to all of us. No doctor can know the complexities of your genome and ancestry stretching back across thousands of years by simply looking at you or noting which box you checked on a form.

In my experience discussing the concept of race in a formal setting or casual conversation, the question of "racial diseases" comes up virtually every time. It seems that most people are under the impression that many diseases afflict one race and not others and that this proves both the existence of races as well as the need for racialized medicine. The reality, however, is that commonly cited examples such as sickle-cell disease as a "black person's disease" and Tay-Sachs as a "Jewish disease" are not proof of anything. Sickle-cell disease is not a "black disease." Many black people are in no danger of it and many non-black people are in danger of it. Sickle-cell trait is the result of people in some regions undergoing evolutionary adaptations in the presence of the malarial protozoa over thousands of years. It has nothing to do with our illogical concept of race. "Sickle-cell is a disease of populations originating from areas with a high incidence of malaria," states race researcher Kenan Malik. "Some of these populations are black, some are not. The sickle-cell gene is found in equatorial Africa, parts of southern Europe, southern Turkey, parts of the Middle East, and much of central India. Most people, however, only know that African Americans suffer disproportionately from the trait. And, given popular ideas about race, they automatically assume that what applies to black Americans also applies to all blacks and only to blacks. It is the social imagination, not the biological reality, of race that turns sickle-cell into a black disease."[2]

Sickle-cell disease is a problem for many Hispanic people, people in northwest India and people throughout the Mediterranean region, many of whom have light-colored skin. "The label 'black disease,' however, rendered the distribution of sickle cell anemia invisible in other populations, leading to erroneous understanding of the geographical distribution of the underlying genetic variants," writes Charles N. Rotimi in *Nature Genetics*. "This is one reason why many people, including physicians, are unaware that the town of Orchomenos in central Greece has a rate of sickle-cell anemia that is twice that of African Americans and that black South Africans do not carry the sickle-cell trait."[3]

Similarly, Tay-Sachs should not be thought of as a "Jewish disease" because it is also a problem for non-Jewish French Canadians. False perception often trumps reality when it comes to race, and this is a mistake no one should be willing to tolerate in healthcare.

"We know race is a terrible proxy for genotype," explains biological anthropologist Jonathan Marks. "The most relevant examples would be something like a drug working in 75 percent of Africans but 25 percent of Asians. First, within each category, there will have to be huge variation, patterned the way we know human diversity is patterned: the African Ethiopians are more likely to cluster with the Asian Pakistanis than with the African Ghanaians. So the continental average value is hardly of use. More importantly, the drug won't work in 25 percent of Africans—and might sicken them—and will work in 25 percent of Asians [who won't receive its benefits because it won't be given to them]. What I'm getting at is that the test has to be made at the individual genotype level; prescribing a drug based on the census category of the patient is a medically very bad idea."[4]

Dr. Marks is correct, of course. Races as biological categories of humans are far too illogical, inconsistent, and untrustworthy to base things as important as medical diagnoses and treatments on. I certainly don't want a doctor looking at me, seeing a "white man," and making her decisions accordingly. My personal history could be a lot more complicated than that. Like it is for most people, my family history blurs after only a few generations and becomes a total mystery after six or seven. If I have no idea who the majority of my ancestors are, I know my doctor doesn't. What we need is not racialized medicine but *individualized* medicine. Fortunately, it's on the way. In 2003, the US government produced a rough draft of a human genome for about $2.7 billion (1991 dollars).[5] By 2010, a genome could be sequenced for a few thousand dollars (though most companies charge a lot more to do it). Soon, experts predict, the cost will be less than $1,000.[6] That's great; in the meantime, however, be weary if offered medical treatment that is based on race. Ask to be viewed by your doctor as a unique individual with an ambiguous ancestry spanning many thousands of years—because that is precisely who you are.

It is important to understand that rejecting the existence of biological race categories and criticizing racialized medicine does not mean that race has no place in discussions about health and healthcare. Cultural race groups are real and meaningful because we made them so. They can have deadly biological consequences and therefore can't be ignored. "Black" or "Native American" may be nonsense as biological categories, for example, but they still have to be considered because membership to a group can sometimes have very serious health consequences. This is because people often are treated differently, make unique choices, or experience different environmental conditions (stress, toxins, nutrition, violence, and so on) as a direct result of belonging to a particular cultural race group. This can cause confusion on the issue of race for the general public and many doctors because racial health disparities do exist within some societies. This leads some people to conclude that biological race is the cause, when in reality *cultural* race is to blame. The Pima Indians are a good example of this.

Nearly half of the adult Pima Indians in Arizona have type 2 diabetes, one of the highest rates in the world. No doubt it would be automatic for many people to see this as a biological race problem in need of a biological race solution, perhaps a race-based drug. But hold on, there is a community of Pima Indians just across the border in Mexico, and they have a type 2 diabetes rate of approximately 7 percent.[7] This is essentially one population of people, with the same ancestry, split into two by an imaginary line. If race is all about kinship and ancestry, then it makes no sense that this could be a racial problem in need of a racial solution. Clearly there is something about living conditions on one side of the border (diet perhaps?) that created this health disparity. The Pima Indians in Arizona don't need a new pill designed and marketed for their "biological race." What they need to do is address whatever it is within their specific culture or environment that is causing the high diabetes rate.

Another example of a health problem that can be misinterpreted easily is the relatively low birthweight of black babies born in the United States compared with white babies born in the United States. Low birthweight is a serious matter, as it corresponds with many health risks for infants. But where many might reflexively imagine a biological race problem in need of a race-based cure, a long-term study of the problem points directly to social causes. Researchers found that black African women who had immigrated to America had babies that were the same weight as white American babies. But the daughters of these African women who grew up in America later had babies that weighed, on average, half a pound less than white babies. Furthermore, the researchers discovered that infants born to black Caribbean women who had immigrated to America weighed more than the babies of black women who had been born in the United States. If this problem were tied to biological race, then it would not spare black women from Africa and the Caribbean. It seems clear that there is something about living your entire life as a black woman in America that puts your infant at a relatively high risk of low birthweight. This is a cultural challenge.[8]

The fact that biological race categories do not exist in a consistent or logical sense means that we cannot safely rely on them to diagnose and treat people with health problems. *Cultural* race groups may be relevant and important to determining the best course of action to help people in some cases. But it's vital that we keep it clear in our minds that there is a difference between biological race and cultural race. The lay concept of race can easily trick us into making false assumptions about our genes, ancestry, and medical needs. For this reason, belief in biological races clearly has no place in a doctor's office, in hospitals, or in the development of drugs.

GO DEEPER . . .

Books

- Graves, Joseph L. *The Emperor's New Clothes: Biological Theories of Race at the Millennium.* Piscataway, NJ: Rutgers University Press, 2003.
- Harrison, Guy P. *Race and Reality: What Everyone Should Know about Our Biological Diversity.* Amherst, NY: Prometheus Books, 2010.
- Montagu, Ashley. *Man's Most Dangerous Myth: The Fallacy of Race.* Lanham, MD: AltaMira Press, 1997.

Other Sources

- *Unnatural Causes . . . Is Inequality Making Us Sick?* www.unnaturalcauses.org/media_and_documents.php.

Chapter 27

"NO VACCINES FOR MY BABY!"

> **Vaccines can be credited with saving approximately nine million lives a year worldwide. A further sixteen million deaths a year could be prevented if effective vaccines were deployed against all potentially vaccine-preventable diseases. In all, vaccines have brought seven major human diseases under some degree of control: smallpox, diphtheria, tetanus, yellow fever, whooping cough, polio, and measles.**
>
> —UNICEF

I'm feeling uneasy as the nurse prepares the syringe. Not that my precious little baby daughter or anyone else would ever know it. I'm upbeat, all smiles and jokes on the outside. But there is no denying the resentment smoldering within. I'm bothered because I'm trapped in a time period when the best medical science can do is stab my baby with a sharp piece of metal in order to get disease-preventing vaccines into her little body. Of course, I suppose I should be grateful that I'm not stuck in the time before vaccines. That was when half the children never made it to the age of five. I rub my daughter's head and force a smile as the nurse jabs a needle into her. "There, sweetheart, all done."

I can't remember if it was the DPT shot (for diphtheria, pertussis, tetanus) or MMR (for measles, mumps, rubella) that she received that day. I do, however, recall feeling troubled by more than just her temporary pain. It was 2002, and as a journalist who reads science news incessantly and often writes about science topics, I was well aware of the growing concerns and protests over vaccines. I understood that the claims of an autism-vaccine link were unproven, but I watch talk shows on TV too. I had heard the claims from mothers, some tearful and some enraged, that vaccines were responsible for their children's autism. I know the difference between anecdotes and credible scientific studies. I also know the difference between a Hollywood celebrity and a scientist. Still, I found it impossible to forget those passionate warnings when it was my child about to get *the shot*. "What if they're right? What if there really is something wrong with vaccines and the scientists just haven't figured it out yet? What if I am about to make the biggest mistake of my life? What if I let this happen and it condemns my little girl to a diminished life?"

Fortunately, I reacted to the scary questions in my head by *thinking*. Emotions are wonderful; I love them. But when it comes to important decisions, I prefer analysis and reason to hunches and fears. I may have *felt* the possible threat of an autism-vaccine link, but I *knew* the reality of how vaccines protect children. There is no doubt that vaccines save the lives of millions of children each year and prevent many millions more from having to suffer through painful illnesses. I want my children to be on that side of the fence. How could I possibly send my daughter out into a dangerous world without protecting her from known killers such as measles and diphtheria? In the end, I decided it is far safer to go with the known and the proven over the unknown and the unproven—especially when my daughter's safety is the issue. A few celebrities and one or two renegade doctors don't match up favorably against the world's top epidemiologists. There are no guarantees, of course, but it would have been reckless and irresponsible of me to withhold a very important protection from my child based on unproven claims.

The decision to vaccinate my children has been vindicated by very solid science in the years since that uncomfortable moment with a nurse and my baby back in 2002. Since then, vaccines have continued to do their amazing work, quietly and without much fanfare. Every day, vaccines prevent the deaths of countless children worldwide. Vaccines may have saved my children's lives as well. Unfortunately, irrational fears about vaccines have not gone away.

Anyone who researches the supposed link between vaccines and autism is likely to be surprised by how one-sided the controversy is. I certainly was. Mainstream medical science has evidence and credible studies to show that vaccines are not linked to autism. Meanwhile, the antivaccine activists fight on, continuing to rely on nothing more than fear and misplaced rage.

It is difficult to criticize parents of autistic children who condemn vaccines when already they are so burdened by challenges. One might feel the urge to give them a break and let them do or say whatever they want. But they need to hear the truth too. And their protests against vaccines, no matter how well intentioned, can lead to the deaths of children. It's also tragic that some parents compound their problems by blaming medical science and then running into the clutches of quacks and con artists who are all too willing to sell them the latest flavor of snake oil. By blaming vaccines—without proof—parents are most likely pursuing a dead-end path and discouraging other parents from vaccinating their children. I'm willing to give them the benefit of the doubt and believe that their hearts are in the right place, but their minds clearly are not. Attacking vaccines is not something to take lightly. Vaccines have probably saved more lives than any other form of medicine in history. One would think it would take a mountain of damning evidence to turn people against something with a track record like that. But apparently it takes only a little misinformation and a lot of fear because increasing numbers of parents are choosing not to vaccinate their children. Today an estimated four out of ten parents in America choose not to give their children one or more recommended vaccines.[1]

DEADLY CONSEQUENCES

Heated debates that pit science against pseudoscience—evolution versus creationism, for example—rage on and on. But few of them rack up casualties and have the potential for mayhem like the anti-vaccine controversy. This particular clash between reason and irrational belief is literally killing children right now. Vaccination rates have plunged in parts of America and the United Kingdom because of misinformation and unjustified fears. According to the United Kingdom's Health Protection Agency, a drop in vaccination coverage levels has again made measles endemic in the UK after it had already been wiped out by vaccines decades ago.[2]

Much of the fears were stirred up in 1998 when British doctor Andrew Wakefield published research claiming that the measles vaccine causes autism. He said the vaccine inflamed intestines, causing harmful proteins to leak out that then made their way to the brain, where they caused autism.[3] This generated considerable coverage in the mainstream media which, of course, sent waves of fear straight into the hearts of millions of parents. Many of them made the decision not to vaccinate their children as a result. Predictably, this was followed by outbreaks of preventable diseases that killed children. Soon after Wakefield's announcement, MMR vaccine rates dropped from nearly 90 percent to as low as 50 percent in some areas of London.[4] Now comes the kicker: It turned out that Wakefield's research is garbage. Other scientists could not confirm his findings. Something was wrong, very wrong. But not only has his work been deemed scientifically flawed, it has ethical problems as well. Investigative journalist Brian Deer reported that Wakefield's study was funded by a lawyer who also was representing five of eight children used in the study for a suit against pharmaceutical companies. In 2010, the *Lancet* medical journal formally retracted Wakefield's study that they had published, and the General Medical Council removed Wakefield's name from the medical register. He can no longer practice medicine in England.[5]

In the late 1990s, antivaccination activists set their sights on a preservative used in some vaccines called thimerosal. No studies suggested that thimerosal might cause autism, but pharmaceutical companies removed it as a precaution anyway. Now, years later, autism rates have continued to rise. "After all the research," writes Michael Specter in his book, *Denialism: How Irrational Thinking Hinders Scientific Progress, Harms the Planet, and Threatens Our Lives*, "thimerosal may be the only substance we might say with some certainty *doesn't* cause autism; many public health officials have argued that it would make better sense to spend the energy and money searching for a more likely cause."[6]

Multiple studies have failed to find evidence of an autism-vaccine link. In Japan, the feared MMR "vaccine cocktail" was withdrawn and replaced by single vaccines. A study of thirty thousand children there found that autism rates continued to rise even in MMR's absence.[7] Other countries, including the United Kingdom, Canada, and Sweden removed thimerosal from vaccines only to see autism rates continue to rise. Meanwhile, researchers in Finland looked for an autism-vaccine link by analyzing the medical records of more than two million children. They found nothing.[8]

It seems to me that vaccines are victims of their own success. People who are fortunate enough to live in countries with strong vaccination programs have been lulled into a false sense of security. Diseases once feared are not so scary anymore. Measles, for example, does not strike fear in the heart of the typical American. But it's not a disease we should take lightly. It causes brain swelling and high fever and is often fatal. In the past, measles killed millions in Europe and America. It still kills more than *one million children per year* in the developing world today.[9] Nevertheless, many parents are being scared away from the measles vaccine by warnings with no credible science behind them. The percentage of unvaccinated children in the United States has doubled since 1991.[10] This is as infuriating as it is absurd. We are moving backward.

Dr. Paul Offit, chief of the Division of Infectious Diseases and the director of the Vaccine Education Center at the Children's Hospital of Philadelphia, is one of the world's leading experts on vaccines. He is also currently waging a professional war against the antivaccine movement. But it is also clearly personal for him. His frustration and concern for children are often readily apparent when he describes the irresponsible decision to deny vaccines. "The problem with waning immunization rates in the United States isn't theoretical anymore," he told me. "Recent outbreaks of measles, whooping cough, mumps, and bacterial meningitis show a clear breakdown in population immunity. Children are now suffering the diseases of their grandparents. It's unconscionable."[11]

MISPLACED FEARS

"Caught in the middle are children," Offit writes in his book, *Deadly Choices: How the Anti-Vaccine Movement Threatens Us All*. "Recent outbreaks of measles, mumps, whooping cough, and bacterial meningitis have caused hundreds to suffer and some to die—die because their parents feared vaccines more than the disease they prevent."[12]

Maybe this would not have happened if more of us were aware of the constant assault we are under from dangerous viruses and bacteria. Microscopic monsters have killed far more people than all of history's wars combined. Some of the most important events in history involved disease. European contact with the New World, for example, is a story of germs and disease as much as or more than anything else. One thing is certain: too many people in the United States are unaware of their own recent history. Offit writes:

> In the early 1900s, children routinely suffered and died from diseases now easily prevented by vaccines. . . . Americans could expect that every year diphtheria would kill twelve thousand people, mostly children; rubella (German measles) would cause as many as twenty thousand babies to be born blind, deaf, or mentally disabled; polio would permanently paralyze fifteen thousand children and kill a thousand; and mumps would be a common cause of deafness. Because of vaccines, all these diseases have been completely or virtually eliminated. But now, because more and more parents are choosing not to vaccinate their children, some of these diseases are coming back.[13]

One likely reason for belief in an autism-vaccine connection is unfortunate timing. The first signs of autism often appear just around the time when children are getting routine vaccines. Parents, understandably, search for a cause for their child's disturbing symptoms, and vaccines seem a likely culprit. As often happens, however, correlation is easily confused with causation. Offit relates a story about a man who took his baby to get the DTP vaccine but after waiting in line for a time became tired and went home without ever getting the child vaccinated. Several hours later, the father discovered the baby had died in its crib, apparently of Sudden Infant Death Syndrome. "One can only imagine what the father would have felt if [the baby] had received DTP several hours earlier. Presumably, no study would have convinced him of anything other than that the vaccine had killed his son.[14]

As a parent who once worried about vaccinating my own children, I sympathize with mothers and fathers who are not 100 percent sure what to do. I would advise parents who are worried about vaccines causing autism to play it safe and go with common sense and reason.

Of course playing it safe means getting your children vaccinated. And it's only common sense and reasonable to protect your child from as many of the numerous diseases lurking out there as you can.

"THEY'RE NOT STUPID, JUST IGNORANT"

Nurse Shawn R. Browning is in the trenches on the frontlines of this issue. She has nearly two decades of experience in the medical field, most of it working with the US Navy. She regularly administers vaccines to military personnel and their families. She also has been involved with immunization education for many years. Irrational fears about vaccines are nothing new to her.

"I have had plenty of parents and patients that are misinformed about vaccines," she said. "When they tell me they don't want to get a particular vaccine, the first thing I ask them is, 'why'? I have heard everything from the thimerosal content is bad for you, vaccines cause autism—particularly the MMR vaccine—and everything in between. By law I give them the VIS [vaccine information statements], but in addition I also educate them on the pros of receiving the vaccine versus not. What I have learned is that more times than not, people are willing to get the vaccine once it is explained to them in words they can understand and relate to. They're not stupid, just ignorant. They have listened to their neighbors, the media, and everyone else and have formed an unjustified opinion. Drives me crazy! Many parents and patients have expressed their gratitude that someone has taken the time to explain things instead of just sticking a needle in them without any explanation. I think our particular patient population is more vaccine hesitant than antivaccine."

Like most healthcare professionals, Browning is concerned that this reluctance to vaccinate might lead to major outbreaks of preventable diseases:

> The biggest fear is that preventable diseases will rise to epidemic proportions again. Infants and children are going to die or be disabled because adults are ignorant and won't vaccinate themselves or their children. The outbreak of pertussis [whooping cough] is the latest. People think that since they are adults, they don't need a vaccine. Yet how many die from complications from the flu every year? [Influenza virus, the flu, kills as many as five hundred thousand

> people each year worldwide, according to the World Health Organization.[15]] It's very scary. We also have an obligation to get vaccinated to protect those [who] can't be vaccinated due to various reasons [such as immune system problems].
>
> There was this mom [who] came into our clinic a little more than a year ago to get her one-year-old daughter her immunizations. The corpsman that brought them back to the room started to explain the vaccines the child would be getting and their potential side effects to the mom. The mom politely interrupted the corpsman and proceeded to explain that this child was not her first baby. She had once been "one of those moms" who didn't believe in vaccines, and her first little girl had died when she got the measles. Just how do you respond to that? Your heart breaks.[16]

Offit adds, "The science is largely complete. Ten epidemiological studies have shown MMR vaccine doesn't cause autism; six have shown thimerosal [preservative once used in vaccines] doesn't cause autism; three have shown thimerosal doesn't cause subtle neurological problems; a growing body of evidence now points to the genes that link to autism; and despite the removal of thimerosal from vaccines in 2001, the number of children with autism continues to rise."[17]

In 1997, 4,138 children entered California kindergartens without being vaccinated because they had exemptions. By 2008, that number had more than doubled. Parents citing religious or philosophical objections to having their children vaccinated are putting not only their own children at risk but the lives of many others as well.[18] Babies who are too young to be vaccinated can be infected and die. Children who have immune system problems and cannot be vaccinated have to rely on others around them to be vaccinated in order to keep the diseases at bay. When vaccination rates drop, danger to these vulnerable groups increases. According to a Centers for Disease Control and Prevention (CDC) expert, a parent's decision to refuse vaccination means his or her child is thirty-five times more likely to get measles and twenty-two times more likely to come down with pertussis (whooping cough). Please don't think for a second that this is exaggeration or fearmongering. Children are paying a price for this madness in small pockets across America now, and the potential for much greater suffering is real. In April 2011, for example, a private school in Virginia had to close because half its students were infected with pertussis. None of the children had been vaccinated. Many of the parents had obtained religious exemptions that officially sanctioned their negligence.[19] News of several recent infant deaths in California due to pertussis either had not reached those parents or failed to impress them.

Why subject children to this unnecessary danger? To protect them from autism? Very large, thorough, and expensive scientific studies did not find any reason to conclude that vaccines cause autism. Therefore it simply makes no sense to withhold such important protection from a child.

GO DEEPER . . .

Books

- Mnookin, Seth. *The Panic Virus: A True Story of Medicine, Science, and Fear*. New York: Simon and Schuster, 2011.
- Offit, Paul A. *Autism's False Prophets: Bad Science, Risky Medicine, and the Search for a Cure*. New York: Columbia University Press, 2010.
- Offit, Paul A. *Deadly Choices: How the Anti-Vaccine Movement Threatens Us All*. New York: Basic Books, 2010.
- Offit, Paul A., and Charlotte A. Moser. *Vaccines and Your Child: Separating Fact from Fiction*. New York: Columbia University Press, 2011.

Other Sources

- The Centers for Disease Control and Prevention maintains an informative collection of articles and fact sheets about vaccines at www.cdc.gov/vaccines.

LURE OF THE GODS

Chapter 28

"MY GOD IS THE REAL ONE."

If God has spoken, why is the world not convinced?
—Percy Bysshe Shelley

Some books that promote science and skepticism dodge claims about gods being real. Not this one. While I understand that challenging belief in gods can be seen as rude or out of bounds, I feel that it makes no sense to take on UFOs and psychics when belief in gods makes a far greater impact on the world than those things. Unlike religion, for example, hatred, wars, and terrorism are not often inspired by belief in Bigfoot and the Bermuda Triangle. Yes, belief in gods is deeply important to billions of people. Yes, one ought to be aware of and respect, to a point, the emotional attachment many people have to their belief in the existence of a god or gods. But it only makes sense to try and ensure that something taken so seriously by so many people is actually valid in the first place. This is not, or should not be, a question for skeptics alone. Don't believers also want to know if their gods actually exist or not?

I hope readers who believe in a god or gods will not be offended by ideas raised in this chapter. Nothing here is meant to insult, only to provoke thought. Believers who respect truth and place a high value on reality should find nothing upsetting here. A common misconception is that nonbelievers claim to have disproved the existence of all gods. This is not accurate. I can't even imagine how one could do such a thing. For example, short of exploring the entire universe and all possible dimensions, how can anyone ever really know if the Aztec god Quetzalcoatl exists or not? The best a skeptic can do is point to the absence of proof and go from there.

It might help if believers recognized the common ground they share with nonbelievers. When it comes to gods, *everyone* is skeptical and *everyone* is a nonbeliever. It's just a matter of degree. I have met Muslims in the Middle East who were hardcore skeptics—about the claims of Hinduism and Buddhism. They were quick to point out many sound reasons to doubt the more extraordinary claims of those belief systems. I know a Jewish person who dismisses Mormonism as a collection of totally unsubstantiated claims put forth by a questionable source. South American Christians tell me that Islam fails the test of analysis and reason. Catholics have explained to me that Protestants are way off course and have failed to prove their claims. Some Protestants say the same about Catholics. Yes, when it comes to religion, every believer is a skeptic—almost. The only difference is that they stop analyzing, doubting, and asking questions when their thoughts arrive at the doorstep of their own religion. That's where critical thinking ends and faith begins.

THE GODS MUST BE LAZY

If only the gods had made the effort to clearly establish their existence and communicated their desires to everyone on Earth. How hard could that be for a god? But either the gods are not real or they have chosen to be elusive and utterly confusing, leaving us to spend thousands of years doubting, bickering, and slaughtering one another over religious differences. This conflict is impossible to escape given the state of religion: Millions of gods have been said to be real and none of them proven to exist. And there can be no compromise. At the very least, everyone must reject most gods. There is no person on Earth—alive today or at any time in the past—who would or could believe in every god. It is impossible to believe in all gods because nobody even knows all the gods. There are simply too many of them to keep up. People have declared so many gods to exist over the last several thousand years that nobody has been able to keep an accurate count, much less list names and attributes. These gods, and the hundreds of thousands of religions that proclaimed them to be real, are far too contradictory to be reconciled under one roof. Jesus either is the only way to heaven, or he is not. Allah is the one true god with no son, or he is not. There are millions of Hindu gods, or there are not. Ramses II either was a god, or he was not. Zeus was top god and really did hurl lightning bolts down from Mount Olympus, or he did not. (This could go on for thousands of pages, but I'm sure you get the point.) One or some of these gods may be real. Maybe none are real. What we can be sure of is that they cannot all be real. No wonder belief in gods has been a source of constant conflict throughout history.

What we can conclude from the multitude of claims for very different gods is that, at the very least, most believers must be wrong. This is just the way it sorts out and there is no getting around it. Do the math, either polytheists are wrong or monotheists are wrong, for example. The fact is, the majority of people today are missing the mark when it comes to gods and most people in the past got it wrong too. If Christianity is correct and Jesus/God the Father/The Holy Spirit are real, then that would mean the majority of people alive today and the vast majority of people who have ever lived were wrong. If Islam is true and Mohammed was correct about the Koran and Allah, then it would mean that the majority of people alive today and a majority of the people who have ever lived were wrong on the god issue. The same is true for every religion and every god claim. If somebody's hell turns out to be real, it's going to be awfully full. And it will be filled mostly with religious people who got in line behind the wrong god.

Many people mistakenly believe that the popularity of god belief in general somehow validates their belief. Not so. In fact, the conflicting claims of so many beliefs casts suspicion on all of them. If my neighbor can be wrong, perhaps I can be wrong. Christianity, for example, is currently the world's most popular religion. That sounds impressive, but some 70 percent of the world's people are non-Christians. Muslims are a minority, too, at 21 percent. Hinduism is a major world religion, but only 14 percent of all people are Hindus. These conflicting minority belief systems do not validate or support one another.

In my book *50 Reasons People Give for Believing in a God*, I analyzed the most common justifications for belief that I heard from people in various religions around the world. I found it interesting that people defending very different belief systems would almost always rely on the same handful of justifications. For example, I have been told that answered prayers, miracles, divine healings, and feelings of joy "prove" the existence of a god. I have been told these things by Christians in Europe and the Americas; Muslims in Syria, Jordan, and Egypt; and Hindus in India and Nepal. Once again, *somebody* has to be wrong here. If a Hindu who successfully prays for a favor from Ganesha proves that Hindu gods are real, then why aren't the answered prayers of a Muslim proof that there is only one god? If a Christian and a Sikh pray for better jobs and then both get better jobs, whose prayer should we consider to be proof that Jesus is or is not the only way to salvation? Meanwhile, we have to consider why the past prayers of ancient peoples such as the Greeks and Romans are not compelling evidence for the existence of their long list of gods. They said prayer worked too.

It is the same with faith healing. I have spoken with a variety of believers in a variety of unique religions who assured me that their god or gods must be real because of some supernatural recovery from illness or injury that they experienced or witnessed. But I have attended faith healing services and was not impressed. When you consider the global/historical context and recognize that claims for divine healings have been taking place for thousands of years within numerous contradictory religions, it becomes clear that this is not good evidence for the existence of a god or gods.

No one claim for a god or gods holds a decisive advantage over all the others. Sure, many people within each belief system will say that theirs is true and all others false, but they are in a poor position to judge. It must be difficult to observe the religious landscape as it really is while standing inside the high walls of just one of them. From the nonbeliever's vantage point outside the walls, however, it easily comes into sharp focus. The great number of gods that we humans have confidently believed in since the dawn of history and probably deep into prehistory, suggest only one thing: we are a god-inventing species. We see divine beings everywhere and then imagine that we know their desires. The fact that there has never been agreement on who the real gods are and what they want of us hints to the likely source of our tales. The gods have not spoken to us. Most likely it is we who are simply speaking to one another, in their name.

GO DEEPER . . .

Books

- Chaline, Eric. *The Book of Gods and Goddesses: A Visual Dictionary of Ancient and Modern Deities*. New York: It Books, 2004.
- Harrison, Guy P. *50 Reasons People Give for Believing in a God*. Amherst, NY: Prometheus Books, 2008.
- Hemenway, Priya. *Hindu Gods*. San Francisco, CA: Chronicle Books, 2002.
- Jordan, Michael. *Dictionary of Gods and Goddesses*. New York: Facts on File, 2004.
- Kurtz, Paul. *The Transcendental Temptation: A Critique of Religion and the Paranormal*. Amherst, NY: Prometheus Books, 1991.
- Mills, David. *Atheist Universe*. Berkeley: Ulysses Press, 2006.
- Sagan, Carl. *The Varieties of Scientific Experience: A Personal View of the Search for God*. New York: Penguin, 2007.
- Thompson, J. Anderson, and Clare Aukofer. *Why We Believe in God(s): A Concise Guide to the Science of Faith*. Charlottesville, VA: Pitchstone Publishing, 2011.

Other Sources

- *The Atheist Experience*, www.atheist-experience.com/archive/.
- *Letting Go of God* (DVD), Julia Sweeney and Indefatigable, 2008.

Chapter 29

"MY RELIGION IS THE ONE THAT'S TRUE."

> **Modern science is beginning to understand the neurological mechanisms that give rise to the religious experience of the believer. Given these results, the skeptic may present the believer with a simple question: How do you know that your religious experience is not a simple trick of your brain—the unfolding of a perfectly natural temporal lobe transient? How can you trust such an experience when, through science, we can convincingly mimic the face of God?**
>
> —David C. Noelle

One of the world's most common and enduring beliefs is that one particular religion is true while all the many thousands of others are wrong. Of course, those who hold this view are always sure that it's *their* religion that happens to be the one that is correct. This is another belief that can be difficult to address due to a force field of traditional respect, legal protection, and the outright threat of violence that often surrounds it. Reaction varies by religion, context, time period, and society, but one might be considered rude for challenging the concept of religious favoritism, arrested or even killed for it. It is usually considered good manners, and safer, to simply duck this one while repeating the live-and-let-live cliché. Unfortunately, however, total confidence in one religion over all others encourages many bad things, from the Crusades to discrimination to suicide bombers. This makes religions fair game for all those who care about such things as peace and human rights.

So what is wrong with the claim that one religion is true and all others false? Three points reveal the problem with religious isolationism. First, we need to look at how people come to follow one religion over thousands of others. Second, we must explore the religious landscape as it really is, not as people tend to imagine it is. Finally, we need to address the problem of religious illiteracy. How can people judge their religion to be the most sensible and accurate of all when they know virtually nothing about any others?

RELIGIOUS INHERITANCE

How do people choose a religion? They don't! The dirty little secret about religious belief is that it's imposed, not chosen, in almost every case. Very few believers voluntarily and consciously select their particular religion. The religion usually is introduced to a child by family members—without debate, questions, or consent—and then reinforced by the immediate social setting. This is clearly the case because we can look at the geography and family patterns of religious belief and see that the best predictors of a person's religious belief are what their parents believe and where they live. So if a person was born to Muslim parents and raised in Egypt or Syria, for example, the odds are very high that she or he will be a Muslim in adulthood. If a child has Buddhist parents and grows up in Thailand, it's likely that he or she will end up a Buddhist. It's nearly certain that a person raised by Christian parents in Mississippi will be a Christian. If one is born in a small village in Papua New Guinea, most likely she will not be a Scientologist or Raelian. If one is born and raised in Pakistan, chances are not good for becoming a Baptist. What this shows is that very few of the world's people are doing much thinking, if any, when they first become tied to a religion. There is virtually no comparison shopping going on when it comes to the adoption of religions throughout the global population. There is no weighing of evidence and assessing of arguments. There is no time given for fair hearings of alternative beliefs or counter explanations for religious claims. In almost every case religion is a family and social inheritance that the individual has little say in. For typical believers, religious loyalties develop early in life and within the context of trusting authority figures. These beliefs are able to grow deep roots in relative isolation, safe from challenge. Then the confirmation bias protects the imposed belief, as observations and arguments that seem to support their religion are embraced while everything that supports rival religions or casts doubt upon all religion is ignored or trivialized. There may be movement within religions by individuals, from Catholic to Protestant or from fundamentalist Muslim to casual Muslim, for example. But allegiance to the original primary religion does not change for most people.

This religious enculturation, some call it indoctrination, should concern those who grew up within the psychological cocoon of one particular belief system and now find themselves confident that this religion happens to be the one that is true above all others. Let's not forget that Muslims, Christians, Jews, Hindus, Sikhs, Buddhists, and all other believers are humans first. And as humans, they are vulnerable to the same errors in reasoning. For example, it is not difficult for me to imagine what likely would have happened if I grew up in India, was raised by loving Hindu parents who taught me all about the gods, lore, and rituals of that particular belief system. By the time I reached adulthood, it is very likely that I would have a strong bias favoring belief in Hinduism. I probably would view Hinduism as the best religion, even if I knew little about all the others that exist today and existed in the past. Maybe I would one day embrace skepticism and begin to have doubts. If that happened, however, my struggle likely would be between Hinduism and atheism with little or no thought given to Christianity, Islam, or any other religion.

I believe that it would be constructive for all believers with total confidence in their religions to imagine what their mind-sets might be if they had been born into different societies, with different parents, and raised to be loyal to a different religion. Hopefully people can recognize that simply being taught one religion and no others early in life is not a reasonable justification for believing that your religion is best and makes more sense than all the others. The logic is not so different from a sports fan loving a hometown football team he or she grew up with and supporting it through the years, no matter what happens on the field each season. Such feelings and actions are more about geography and childhood loyalties than any rational or logical decision making.

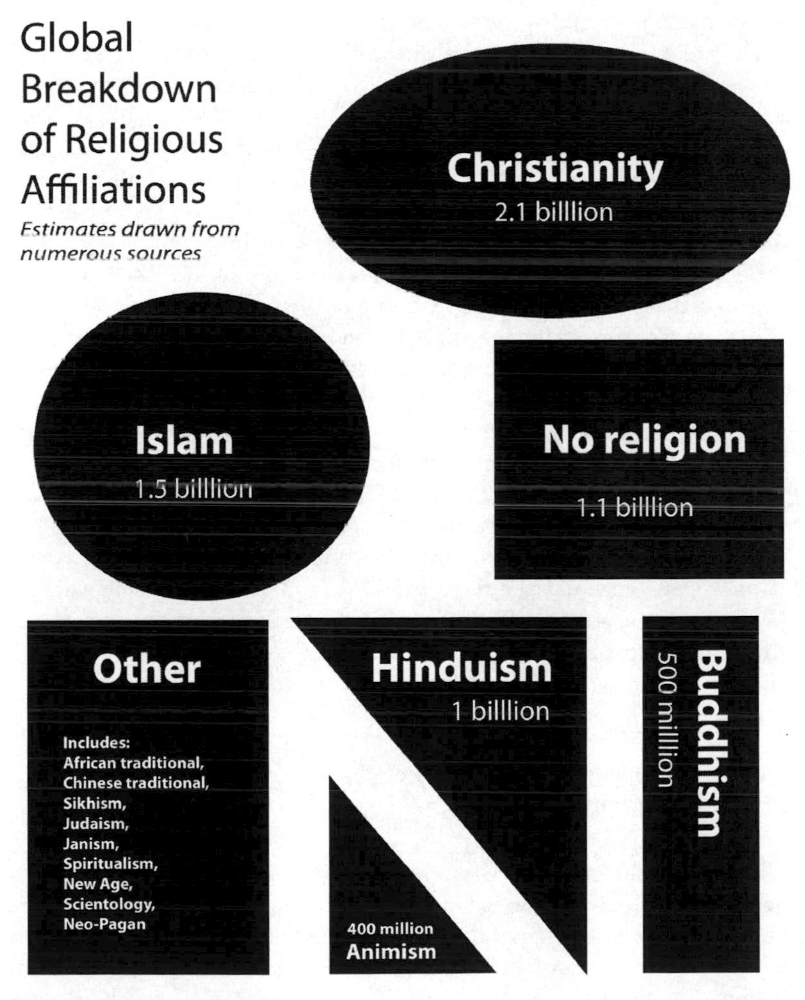

Figure 5. Illustration by the author.

THE REAL RELIGIOUS LANDSCAPE

A second key reason that many people believe their religion is the right one is the failure to see religions as they are and not how they are imagined to be. I am convinced that if people were aware of the basic numbers, structure, and history of contemporary religions, they would find it much more difficult to dismiss or look down on rival belief systems. I have interviewed and had conversations with a variety of believers around the world who are under the false impression that the general popularity of belief in gods somehow gives credibility to their particular religion over all others. "Most people are religious, therefore my religion must be true," is the popular idea. But this makes no sense. First of all, reality and truth are not determined by popularity contests. Even if every human who ever lived believed in fairies, it would not mean they are necessarily real. To know if something is real, we can't rely on a show of hands. We have to assess the evidence. An abundance of believers in a god or gods does not mean any gods necessarily exist. Yes, religion has been near universal throughout history. Humans have created hundreds of thousands of religions and claimed the existence of hundreds of millions of unique gods. But no proof for gods is to be found in these numbers. If anything, it is a compelling argument for the likelihood that all gods were invented. More to the point, key contradictions between those hundreds of thousands of religions do not help make the case for any single belief system being correct. One billion Hindus believing in millions of gods does not reinforce the claims of 1.5 billion monotheistic Muslims, for example. The disharmony among believers today and throughout the past suggests just one thing: we look very much like a species that loves to make up gods and invent religions.

WHY DO RELIGIOUS PEOPLE KNOW SO LITTLE ABOUT RELIGION?

Ignorance is the greatest reason so many people are able to confidently declare that their religion is true while all others fall short. What could be easier than to feel superior about your belief system when you know little or nothing about all other belief systems? It would be easy to believe hamburgers were the best food in the world if you had never tasted any other food. To be clear, this is not about anyone's intelligence. This is about curiosity, awareness, and educational opportunities. We may live in a world teeming with religion, but most people do not know basic facts about the world's current most popular religions, and even their own religion in many cases. Strangely, religion is hailed by billions of people as something very important and necessary in our lives. Believers say it has profound implications for life, death, and even eternity—yet almost no one understands it.

Boston University religion professor Stephen Prothero addresses the problem in his book *Religious Literacy: What Every American Needs to Know—And Doesn't*:

> Today religious illiteracy is at least as pervasive as cultural illiteracy, and certainly more dangerous. Religious illiteracy is more dangerous because religion is the most volatile constituent of culture, because religion has been, in addition to being one of the greatest forces for good in world history, one of the greatest forces for evil. Whereas ignorance of the term *Achilles' heel* may cause us to be confused about the outcome of the Super Bowl or a statewide election, ignorance about Christian Crusades and Muslim martyrdom can be literally lethal.[1]

Former president George W. Bush demonstrated the dangers that can come from religious ignorance in a religious world. Bush said shortly after the 9/11 attacks that he would respond to the terrorists by launching a "crusade." Apparently he was unaware of the religion-based ill-will and suspicions that come when a Christian leader uses the term *crusade* when talking about Muslims. Bush then ordered the invasion of Iraq without knowing anything about the historic tensions between Sunni and Shia Muslims. In fact, he did not seem to have ever heard of Sunni and Shia Muslims. Even the most basic superficial knowledge about Islam and its history might have meant better planning and fewer lives lost in postinvasion Iraq.

So just how bad is the problem of religious ignorance? In 2010, the Pew Forum on Religion and Public Life conducted a survey of Americans' knowledge of religion. Here is some of what was revealed:[2]

- About half of Protestants (53 percent) cannot correctly identify Martin Luther as the person whose writings and actions inspired the Protestant Reformation, which made their religion a separate branch of Christianity.

- More than four in ten Catholics in the United States (45 percent) do not know that their church teaches that the bread and wine used in Communion do not merely symbolize but actually become the body and blood of Christ.
- Roughly four in ten Jews (43 percent) do not recognize that Maimonides, one of the most venerated rabbis in history, was Jewish.
- Fewer than half of Americans (47 percent) know that the Dalai Lama is Buddhist.
- Fewer than four in ten (38 percent) correctly associate Vishnu and Shiva with Hinduism.
- Atheists and agnostics scored the highest on a general religion knowledge quiz, outperforming believers.
- Only about a quarter of all Americans (27 percent) correctly answer that most people in Indonesia—the country with the world's largest Muslim population—are Muslims.

A disturbing picture to be sure, and not exclusively an American problem. During my travels I encountered stunning religious ignorance in virtually every society I visited. The problem is global, and the cause of it seems obvious to me. There is an arrogant confidence that seems inevitable when a person is immersed in one religion from childhood and constantly assured by authority figures that it is the true one. This process likely discourages investigation and curiosity toward other belief systems and nonbelief. There also is far too little competent religious education for young students in most schools. Rare is the school that presents even the basic facts and history of a variety of religions. In most societies such classes are usually limited to more sophisticated high schools or offered as university electives. The world's children need to be taught early on about religions in an unbiased and academically competent manner. They should get some exposure to the scholarly history of today's more popular religions as well as some religions of the past. This is vital to having a chance of a clear and sensible worldview. One can never really understand world history or current events very well without a minimal understanding of religions given their impact on the world.

A final point for readers to consider is that if one particular religion were really right and all others wrong, then it seems reasonable to think that this particular religion would stand apart from its rivals in some obvious way. Wouldn't there be a global rush toward this one belief that everyone could see was delivering the supernatural goods? But nothing like this has ever happened and it's not happening now. Even Christianity, the current most popular religion, is fractured into tens of thousands of contradictory versions and the majority of the world's people are non-Christians. The fact is, not one religion has ever produced sufficient evidence or compelling arguments to attract most people. Imagine if the claims of one belief system were confirmed in the real world in a way that all of us could see and know to be credible. What if one religion really did boast meaningful and unambiguous predictions that anyone could clearly see came true? What if it could point to numerous scientifically confirmed faith healings, or miracles like lost limbs regenerating in seconds, or corpses coming back to life after being dead for years? If a religion could point to answered prayers that occurred at a rate that could not be explained by chance or faulty interpretation, that religion would win out overnight. Such a religion likely would leave all its rivals in the dust and become the dominant, if not the only, remaining belief system in the world within a generation or two. Every sane person on the planet would run to it. The fact that this has not occurred is a very good indication that no one religion is obviously true or superior to all the others.

GO DEEPER . . .

Books

- Daniels, Kenneth. *Why I Believed: Reflections of a Former Missionary.* Duncanville, TX: Daniels, 2009.
- Dawkins, Richard, and Dave McKean. *The Magic of Reality: How We Know What's Really True.* New York: Free Press, 2011.
- Epstein, Greg. *Good without God: What a Billion Nonreligious People Do Believe.* New York: Harper Paperbacks, 2010.
- Farr-Wharton, Jake. *Letters to Christian Leaders.* Queensland, Australia: Dangerous Little Books, 2011.
- Harris, Sam. *Letter to a Christian Nation.* New York: Vintage, 2008.
- Harrison, Guy P. *50 Reasons People Give for Believing in a God.* Amherst, NY: Prometheus Books, 2008.

- Head, Tom, ed. *Conversations with Carl Sagan*. Jackson: University Press of Mississippi, 2006.
- Loftus, John. *The Christian Delusion: Why Faith Fails*. Amherst, NY: Prometheus Books, 2010.
- Prothero, Stephen. *God Is Not One: The Eight Rival Religions That Run the World—And Why Their Differences Matter*. New York: HarperOne, 2007.
- Prothero, Stephen. *Religious Literacy: What Every American Needs to Know—And Doesn't*. New York: HarperOne, 2008.
- Wilkinson, Phillip. *Myths and Legends*. London: Dorling Kindersley, 2009.
- Zuckerman, Phil. *Society without God: What the Least Religious Nations Can Tell Us about Contentment*. New York: NYU Press, 2010.

Other Sources

- *Free Inquiry* (magazine).
- *Religulous* (DVD), Lions Gate, 2009.
- The Non Prophets (podcast), www.nonprophetsradio.com.

Chapter 30

"CREATIONISM IS TRUE AND EVOLUTION IS NOT."

> **There are now tens of thousands of hominid fossils in museums around the world supporting our current knowledge of human evolution. The pattern that emerges from this vast body of hard evidence is consistent across thousands of investigations. All models, all myths involving the singular, instantaneous creation of modern humans fail in the face of this evidence.**
>
> —Dr. Tim White

> **Reality is that which, when you stop believing in it, doesn't go away.**
>
> —Philip K. Dick, *How to Build a Universe That Doesn't Fall Apart Two Days Later*

One of the most difficult challenges many religions face is figuring out how their supernatural claims can be aligned with the natural world in a way followers can accept as sensible. The relentless progress of discovery and science has made this increasingly difficult. There was a time, for example, when one could get away with pointing at the ball of fire in the sky and declare it to be a god. That's not so easy to do these days because science has revealed, based on evidence, that the Sun is a giant ball of fire fueled by the fusion of hydrogen and helium atoms—nothing supernatural about it. Over the last few decades or so this struggle between religious claims and the natural universe that science continues to reveal has been most visible and most contentious in the evolution-versus-creationism conflict.

Creationism is most often defined as the religious belief that the Judeo-Christian god created the universe, Earth, and all life as described in the Genesis story. It has to be made clear, however, that while this version of creationism may get all the attention in the United States, and Europe, it is in fact only one of many. Through prehistory, history and up until today there have been many thousands of unique creation stories—none of which can be fairly judged to be superior to the others based on evidence. It is also important to note that creation stories are not scientific and were not discovered by scientific means—although many people claim otherwise. They come to us, as most believers say, by divine revelation in a book or in some other form of spiritual enlightenment.

Creationists traditionally have claimed that Earth and all life on it are less than ten thousand years old, usually around six thousand years old. Today some creationists have retreated a bit and admit that this is obviously wrong and accept the evidence-based or scientific age of Earth, which is around 4.5 billion years. They are called "young Earth creationists" and "old Earth creationists" respectively. Many old and young Earth creationists, however, believe that their religious origin story should be taught in schools, masquerading as science. Of course this presents a problem in the United States because the Constitution forbids the government from promoting religion, and teaching the Genesis story to children in government schools is obviously the promotion of religion.

The specific conflict with evolution comes into play since creationists claim that God created all life in just one week and in present form. According to them, there were no dinosaurs evolving into birds, no long road to a planet teeming with biodiversity, and absolutely no australopithecines evolving into modern humans. That's not how the

Torah, Bible, or Koran says it happened so that's not how it happened, they declare, and evidence to the contrary be damned.

I asked paleontologist Jack Horner, the famed dinosaur hunter who has made some of the most spectacular paleontological discoveries ever, why the acceptance of evolution continues to be a sticking point for so many people. Rather than blame parents, teachers, and preachers, Horner looks in the mirror.

"I think it is a problem for the scientists more than those opposed to evolution," he said. "Scientists have done a bad job in teaching people what evolution is. I think it's about time for us to figure out how to teach it to people. [Evolution] has to do with similarities. We look at the human being and we look at the ape and we can see that they share more common features with each other than they do with anything else. So, just like with a brother and a sister, we can assume that they have a common ancestor. That's really all there is to it.

"Now the proof of evolution is the mere fact that you are different from your parents. That's all evolution is about," Horner continued. "Charles Darwin's theory is descent with modification—and selection. If you are different from your parents, then you cannot argue with evolution. And if you can see that we can select characters [or traits] and the environment can select characters, and you can start with a wolf and end up with a Chihuahua, then you believe in selection and you can't argue with Charles Darwin's theory."[1]

While some who appreciate the value of modern science find it difficult not to give up on creationists, I try to be more understanding. In my opinion, creationism is not dependent upon a lack of intelligence. Nothing more than reason gone a bit astray combined with a little confirmation bias allow it to thrive even in very bright minds. Given the right circumstances, it can happen to just about anyone. Creationists have allowed religious belief to cloud their judgment about reality and, given the emotional attachment people often have to their religion, this should not be surprising to anyone. I never assume that creationists are chronically dim because I have met too many of them who are obviously highly intelligent. The key problem is that creationists fail to recognize or choose to ignore the difference between science and pseudoscientific claims pushed by people who are not experts and have no evidence behind them. Further, most creationists have been bamboozled by appeals to their religious loyalty into seeing only an unnecessarily restrictive version of their religion and nothing else.

Many creationists have pinned themselves under a false choice that says they must choose between their god and modern science. Of course this is not necessarily the case, as proven every day by the many millions of Christians, Muslims, Hindus, Sikhs, Buddhists, Jews, and so on who are able to accept evolution without abandoning their religion. Many religious leaders have openly stated that adhering to antiscience creationist claims is not necessary for one to be sincerely religious. You really can have your religion and keep your mind when it comes to the evolution issue. Even the late pope John Paul II declared in 1996 that evolution was obviously true. He certainly was not suggesting that anyone had to stop believing in God in order to better align him- or herself with modern science. He was saying, correctly, that one can believe in God *and* accept evolution. There was, after all, a time when it was a life-threatening heresy to accept that Earth was not the center of the universe. In time believers got over it, of course, and today the location of Earth in space is not seen as a litmus test for belief in a god. So it likely will be with Darwin's theory of evolution one day. At some point in the future the obvious fact that life evolves will have no bearing on religious belief and everyone will accept it.

A large part of the problem is that creationists misunderstand what the theory of evolution is and is not. Evolution is the description of how life changes over time, primarily because of genetic mutation and natural selection. The theory of evolution is not and has never been a declaration that no gods can possibly exist. The theory of evolution has nothing to say about gods, only that natural processes, given enough time, can bring about profound changes in life-forms and produce a richness of biodiversity such as we see here on our planet. This point has frustrated me greatly over the years, as I have interacted with creationists of various religions around the world. It seems to me that creationists do not really have a problem with evolution, even if they do understand it. They don't reject it because the theory, the fossils, and the genetic evidence fail to convince them. They refuse to accept evolution—or even give it a fair hearing—because they mistakenly believe doing so would mean they have to give up their religious belief. This is profoundly untrue. Admittedly, those who believe Earth is six thousand years old and all species were created at the same time would have to make some adjustments. In most cases, however, one can accept the findings of modern science and still keep one's religious beliefs. Religions adapt to change all the time, anyway. No matter how rigid and entrenched they may pretend to be, the reality is that religions are infinitely flexible. That's why we see tens of thousands of versions of Christianity today, for example.

Donald Johanson, discoverer of the famous Lucy fossils, says believing in a particular creation story depends on where, when, and by whom one is educated into the world. Evolution, however, is evidence-based and therefore, universal.

"There are two very different ways to try [to] explain our human existence," Johanson told me. "One of them is the faith-based endeavor and that depends exclusively on how one is raised. If you are raised as Hopi Indian, then you will learn the myths of creation of the Hopi Indians. You will believe that creation story is the true story. If you are raised in a tribe in South America, you will believe that story is the truth. And if you are raised as a Catholic, then you will believe that is the true answer. All of that is based on experience, how you grow up, and how you are taught. It is based on faith. We don't subject religious ideas to the same rigorous investigation as scientific issues. Regarding evolution, we are looking at the scientific evidence for how we came to be who we are today. It is a fact of the natural world that all animals, plants, and insects have gone through a process of evolution by means of natural selection. We don't look at gravity, which is a fact, and ask if it is moral or immoral. It is simply a fact."[3]

Creationism makes the extraordinary claim that a god magically created all bacteria, algae, fungi, plant, animal, and virus species instantly and in present forms. This would mean that no species have ever evolved or are evolving currently. Observation and evidence do not back this up, of course. In light of the fossil record, genetic discoveries, and real-time observations of evolving microbes, plants, and insects, creationism is not so much wrong as utterly bizarre. Once again, to be clear, this does not necessarily mean that anyone's god does not exist. But the specific creationist description of life is about as far off base as claiming the world's geologists are all wrong and Earth is actually flat. The disagreement between creationism and modern science is not even close enough for compromise, truce, or reconciliation. The sides are just too far apart for an amicable settlement. I suspect that the majority of creationists just do not realize how far out of line with the scientific evidence one has to be to hold such a position. Consider that a creationist must effectively reject all modern biology because evolution is the central theory of that entire discipline. The study and understanding of everything from microbes to whales to redwoods is fatally compromised if evolution is omitted. It would be like trying to explain libraries without ever mentioning books. It's impossible. But it's not just biology that must be sacrificed. If evolution goes, you can also forget about zoology, anthropology, comparative anatomy, marine biology, entomology, herpetology, microbiology, embryology, paleontology, and so on. And don't forget that evolution is the foundation of modern agriculture as well as vital areas of modern medical science. In order to have a good chance of working, for example, vaccines and antibiotics have to be formulated with a close eye on *evolving* viruses and bacteria. Furthermore, in order to believe that Earth is less than ten thousand years old, one must reject fundamental core discoveries and conclusions of geology, astronomy, cosmology, physics, and archaeology.

Regarding the time factor, young Earth creationists would do well to pause and consider just how six thousand years compares to 4.5 billion years. Claiming that Earth six thousand years old is like saying that the distance from Earth to the Moon is a few inches. It's not just wrong, it's *outrageously* wrong. Contrary to creationist propaganda, there is no debate about whether or not life evolves within the scientific community. This entire creationism-versus-evolution thing is a culture/religious war, which explains why its battles are fought in courtrooms, political campaigns, and school board meetings rather than at science conferences, in laboratories, and in academic journals.

THE GREATEST STORY NEVER TOLD

Overall I was a very good student through twelve years of public school in Florida. I graduated from high school with academic honors, a varsity letter in track, and no felonies on my record. I also managed perfect attendance six of those twelve years. The point is, I showed up and I paid attention most of the time. But never once in all those classes with all those teachers did I hear or read about evolution. Not once. No teacher so much as whispered to me or hinted anything about *Australopithecus*, Louis Leakey, or the HMS *Beagle*. No Neanderthals or trilobites ever crept into my curriculum. Fortunately, I was personally curious about science so I read a lot independently. I also went to a good university where I ended up learning all about the foundational theory of life that I am a part of. But what about my peers back in high school, the ones who maybe weren't as curious or never took college biology and anthropology courses? What happened to them?

Today an alarming 40 percent of Americans believe that Earth is less than ten thousand years old and all life was created instantly in its present form.[3] Statistics like that disturb paleoanthropologist Tim White. He worked with Don Johanson on the Lucy fossils and has made his own key discoveries in the field of human evolution, including Idaltu Man, the oldest anatomically modern human found to date. I sensed that he is both puzzled and frustrated by creationism's popularity.

"It frightens me more than disappoints," White said. "Understanding evolution, understanding that the biological world that includes us evolved, is essential. This is true in virtually every field of inquiry, fields as disparate as medicine, agriculture, and sociology. Education is the key to improving awareness."[4]

As important as general education is to this problem, I think that basic awareness of how evolution is integral to modern biology is the first step. I have encountered too many people who wrongly think that one can reject evolution while still being proscience. But this just isn't possible. Trying to embrace modern biology while rejecting evolution is like driving a car while rejecting rubber tires and the internal combustion engine.

IS EVOLUTION EVIL?

Another huge, though totally unnecessary, problem is that creationists have been misled into believing that evolution is an evil philosophy that leads to individual and social ruin. Merely teaching evolution in a high school science class, they claim, leads inevitably to drug abuse, teen pregnancy, violence, and a general collapse of morality. Somehow, this biological theory is a green light for people to run amok and knock down the pillars of civilization. Apparently it is fine to be related to Caligula and Stalin but not Neanderthals and *Homo erectus*. No matter how many times this is explained to me by creationists, I still struggle to understand how they can blame teaching the theory of evolution for crime and immorality. They say that it places us with the animals and thereby makes us no better than wild animals. But this makes no sense because evolution is a scientific explanation of the world around us. It's based on sound reasoning, numerous lines of evidence, and verified predictions. It just happens to be the way life works. How does learning about it give us permission to do evil and tear down civilization? This creationist charge is not only illogical but it is absolutely unproven, too. I'm pretty sure that the crime rate of the world's professional evolutionary biologists is pretty low. I'd be willing to bet that it's lower than preachers and priests. Nobody knows evolution better than the people who research it, write about it, and teach it full-time, yet the vast majority of them seem to live remarkably decent and quiet lives. I'm also guessing that the ratio of prisoners who are behind bars due to some fateful day that they picked up a Richard Dawkins book and read about evolution is probably small to nonexistent.

Yes, the Nazis had unscientific ideas about racial superiority linked to flawed notions of "survival of the fittest." However, nothing they did was morally justified or excused by the theory of evolution. Creationists should recognize that people were quite capable of killing and destroying long before Charles Darwin was even born. We decide for ourselves whether or not to do bad things and we have to take responsibility for our actions. Not only is a connection between understanding evolution and immoral behavior unproven, it is also irrelevant to the natural world. Reality is unaffected by high school curricula. Life will continue to evolve even if every criminal in the world chants Darwin's name while committing their evil acts. Evolution will not stop even if every nice person in the world rejects it. Our intellectual and emotional preferences are not as important to the universe as we might imagine. We may not think it's nice, for example, when warplanes drop bombs on civilians. But gravity will remain with us, nonetheless, even if we were to reject it and demand that an alternative theory be taught to our children in schools.

Some creationists think people who accept modern biology have adopted evolution or nature as their religion, with Charles Darwin as some sort of messiah figure. This is laughable, of course. For the record, I don't worship at the altar of evolution. I think nature as we see it work here on Earth is a horrible way to run a planet. For example, there is far too much predation—too much pain felt by too many creatures. Yes, I am awed and endlessly fascinated by the story of life and the staggering biodiversity that has evolved here. And I'm always vulnerable to nature's beauty. Whether I am scuba diving, canoeing on the Amazon River or sitting on a park bench watching a squirrel watch me, I easily become entranced by the natural world. But the overall way in which this unintelligent and indifferent nature operates does not warm my heart in the slightest. Look behind the pretty flowers and leaping dolphins and one quickly discovers a vast dungeon of horrors. Those who see the handiwork of a wise and loving god in nature need to look again. There certainly is great beauty to be found as well as countless examples of cooperation and peaceful coexistence, but the natural world is also a constant bloodbath of incomprehensible suffering. Every second of every day, pain-feeling creatures are eaten alive by parasites and predators. The fear of being stalked and eaten is always present for trillions of creatures.

One of the most powerful memories from my travels in Africa is the sight of blood-soaked lions making a ghoulish feast from the gut of a still-twitching zebra. Predators have to eat too, so no one can fault the lions, of course. But if zebras could talk, I doubt they would tell us that they think lions and all of nature were created by a just god.

If zebras don't matter to you, try watching a good zombie movie (or a bad one). Wait for the inevitable scene where a minor character is caught and eaten alive by a gang of the hungry undead. Imagine that it is you the zombies are taking apart bite by bite. Close your eyes for a couple of seconds and think about the terror and pain of losing your life, one mouthful at a time. Horrifying, isn't it? But that's business as usual, another day at the office for most life on this planet.

If I were a god with a minimal amount of compassion and empathy, I certainly would not have crafted a cruel and savage system like the one we have here on Earth. For starters, I think I might have powered all life by photosynthesis or at least made all animals herbivores. I could have tinkered with reproduction rates and natural life spans to avoid overpopulation. There are so many things I would do differently. I certainly would have opted not to create the malaria protozoa, seeing how it tortures and kills thousands of children every day. I might have left out cancer too. No, I don't have limitless love and admiration for nature as we find it here on Earth. I merely accept it as the way life works and find what beauty within it that I can.

I suppose I should add that I am aware of the Christian explanation for all the horrors of nature that we see today. Some claim that God let the world become as it is because Adam and Eve broke one of his rules in the Garden of Eden. Blaming so much pain, suffering, and death on one guy for biting one apple seems pretty unreasonable to me, however.

"JUST A THEORY"

Creationists tell me over and over that evolution is "just a theory" and that no one has ever observed it. This is another example of poor science education in schools in America and many other places around the world. Scientists do not casually toss the word *theory* around the way nonscientists do. For most people *theory* means nothing more than a hunch or a guess. For scientists, however, a theory is a formal explanation of something important that is backed up by a lot of very good evidence. It also has to be testable and make reliable predictions about the real world. Evolution passes with flying colors on all counts.

It's important for creationists to know that evolution has been observed because some of them claim that because evolution happened so long ago it can't be seen and therefore can't be proved. Not that everything in science has to be seen by human eyes to be proved (archaeologists don't see what happened at their sites hundreds or thousands of years before they started digging, but they still manage to do very good work), but I understand that observation can make it more believable to the general public. Fortunately, scientists have witnessed evolution. They have been able to do this because life is always evolving and some life-forms—bacteria, viruses, and insects, and some types of weeds, for example—have relatively fast reproduction rates. Bacteria can clock a thousand generations in a little over one month. By comparison, a thousand generations takes us more than twenty thousand years.[5] With that sort of turnover, evolutionary changes occur fast enough for scientists to observe them.

The claim that life is too complex to have "just happened" is another core argument of creationists. But complexity is not proof of their god's existence or his involvement with life on Earth, just as evolution's explanation of so much doesn't disprove the existence of gods. Our inability to fully understand life may say something about our current limitations, but it does not necessarily say anything about the processes and origin of life. If I brought a prehistoric human back to the present in my time machine and showed her a television, it probably would be a challenge to convince her that the TV wasn't magic and that I was not some sort of a god. But her failure to grasp how a television and a remote control work should not be seen by her or anyone else as evidence that televisions are supernatural devices. Given enough time and effort, "impossible knowledge" often comes into focus. Repeatedly throughout history things that were seen as too complex and forever beyond human comprehension ended up solved and incorporated as standard fodder for high school textbooks.

Perhaps the favorite argument of creationists is the old watch analogy: If the complexity of a watch reveals that it was designed by an intelligent being, then the greater complexity of a human body must mean that it was made by an even more intelligent being—a god. After all, a watch doesn't just "come together" on its own, so how can anyone believe that humans, more complex than a watch, were created naturally by luck and chance? This is a bad argument for a few reasons. First, we know who makes watches: human watchmakers. Second, we also know what shaped the human body into its current form. It was replication, variation, and natural selection over millions of years. I agree with creationists when they say they see design in the human body. And they are correct that there is a designer. According to the best current evidence, however, the designer appears to be neither intelligent nor divine. We owe our physical structure, both its wonders and its flaws, to the indifferent and blind process of evolution. It's a process that is not random or completely by chance, but there is no underlying intelligent strategy to it. It is merely life's interaction with whatever environment it finds itself in. There is no goal, no ascending ladder, in evolution. Success is defined as anything better than extinction. What appears to be the champion life-form today might vanish in an instant tomorrow when the environment changes. Maybe a god did have something to do with it, but without evidence for such an idea, how can we justify pretending to know?

People on both sides of this debate may disagree, but I believe the evolution-versus-creationism conflict has been misaligned from the start. The proper debate, if there has to be one, should be "a natural origin of life" versus creationism. Evolution describes how life changes over time. Yes, it has key implications for the natural origin of life, of course. However, it does not directly address how life started. The key moment when something that was not alive became something alive is not currently understood by science and may never be.

THE MOTHER OF ALL DEBATES

I predict that the mother of all debates is yet to come. Currently scientists have interesting ideas about how life may have started on Earth, but it's no secret that no one really knows. That may change, however, as scientists are working in labs right now to create self-replicating synthetic organisms—life. Success in this seems likely and will probably come sooner rather than later. There will always be questions, but if scientists are able to create life, then a very compelling theory of origins may follow close behind. If so, creationists will likely drop evolution as their great bogeyman overnight. They will realize that all this time they have been arguing for a divine origin of life against people who were arguing that life changes—two positions that are not in perfect opposition as most have always assumed. Natural origin versus supernatural origin is the real war. Once that fight heats up, evolution is sure to become standard fare in every science class, even in private religious schools. The outrage and fear over evolution will evaporate just like that other colossal crisis centuries ago over whether or not the Sun revolved around the Earth. Of course, when the irrational battle against evolution falls silent, creationists probably won't pause to apologize for having been so wrong, wasted so much time, and degraded the education of so many children, because they will be too busy fighting the new scientific theory of life's origin. Here we go again.

LOOK IN THE MIRROR

Evolution is a fascinating and important topic that has been woefully neglected in many schools and elsewhere because of an unfortunately and mostly unnecessary conflict with religion. As a result, few know what evolution is. Millions of people fear, hate, and reject something they don't even understand. Please don't be misled by mistaken or dishonest claims that the theory of evolution is evil, not well established, or necessarily incompatible with belief in a god. If you want to be in tune with modern science and desire a basic understanding of life on your planet, you must accept that life evolves. The evidence for this is not only found in museums and libraries around the world. It's in you. Every cell in your body is a testament to more than three billion years of life on Earth. Regardless of what you may have been told, evolution is not the enemy. Evolution is you.

CREATIONISM EVOLVES

It was long overdue, but in 1987 the US Supreme Court ruled that creationism is obviously a religious belief and as such cannot be taught in public school science classes because it violates the first amendment to the Constitution. But this did not drive it to extinction. As we shall see in the next chapter, creationism evolved to live and fight another day under a new name.

GO DEEPER . . .

Books

- Dawkins, Richard. *The Ancestor's Tale*. Boston: Houghton Mifflin, 2004.
- Dawkins, Richard. *Climbing Mount Improbable*. New York: W. W. Norton, 1997.
- Hazen, Robert M. *Genesis: The Scientific Quest for Life's Origins*. Washington, DC: Joseph Henry, 2007.
- Hosler, Jay. *Evolution: The Story of Life on Earth*. New York: Hill and Wang, 2011.
- Leeming, David. *A Dictionary of Creation Myths*. New York: Oxford University Press, 1994.
- Loxton, Daniel. *Evolution: How We and All Living Things Came to Be*. Tonawanda, NY: Kids Can Press, 2010.
- Mayr, Ernst. *What Evolution Is*. London: Weidenfeld and Nicolson, 2002.
- Miller, Kenneth R. *Only a Theory: Evolution and the Battle for America's Soul*. New York: Penguin, 2009.
- National Academy of Sciences. *Science, Evolution, and Creationism*. Washington, DC: National Academies Press, 2008.

- Palmer, Douglas. *Origins: Human Evolution Revealed.* New York: Mitchell Beazley, 2010.
- Pigliucci, Massimo. *Denying Evolution: Creationism, Scientism, and the Nature of Science.* Sunderland, MA: Sinauer Associates, 2002.
- Potts, Richard, and Christopher Sloan. *What Does It Mean to Be Human?* Washington, DC: National Geographic, 2010.
- Prothero, Donald R. *Evolution: What the Fossils Say and Why It Matters.* New York: Columbia University Press, 2007.
- Sawyer, G. J., and Viktor Deak. *The Last Human: A Guide to Twenty-Two Species of Extinct Humans.* New Haven, CT: Yale University Press, 2007.
- Scott, Eugenie C. *Evolution vs. Creationism: An Introduction.* Berkeley: University of California Press, 2009.
- Shermer, Michael. *Why Darwin Matters: The Case against Intelligent Design.* New York: Times Books, 2006.
- Smith, Cameron M. *The Fact of Evolution.* Amherst, NY: Prometheus Books, 2011.
- Smith, Cameron M., and Charles Sullivan. *The Top 10 Myths about Evolution.* Amherst, NY: Prometheus Books, 2006.
- Stringer, Chris. *Lone Survivors: How We Came to Be the Only Humans on Earth.* New York: Times Books, 2011.
- Stringer, Chris, and Peter Andrews. *The Complete World of Human Evolution.* New York: Thames and Hudson, 2005.
- Stringer, Lauren, and Peters Westberg. *Our Family Tree: An Evolution Story.* New York: Harcourt Children's Books, 2003.
- Tattersall, Ian. *Extinct Humans.* New York: Basic Books, 2001.
- Ward, Peter. *Future Evolution: An Illuminated History of Life to Come.* New York: W. H. Freeman, 2001.
- Zimmer, Carl. *Evolution: The Triumph of an Idea.* New York: Harper Perennial, 2006.

Other Sources

- *Becoming Human: Unearthing our Earliest Human Ancestors* (DVD), PBS, 2010.
- Talk Origins, www.talkorigins.org.

Chapter 31

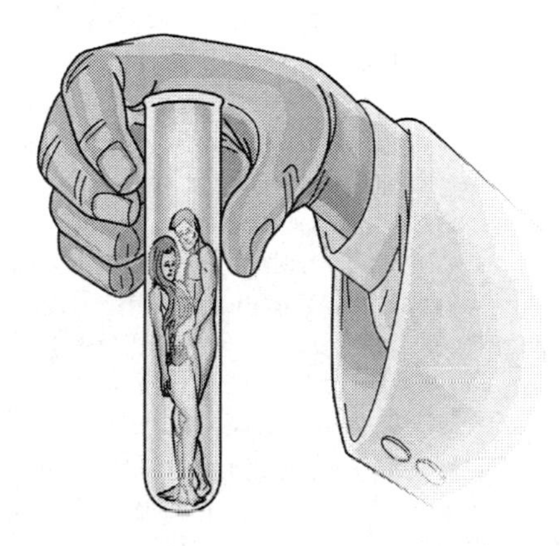

"INTELLIGENT DESIGN IS REAL SCIENCE."

Science is a philosophy of discovery. Intelligent design is a philosophy of ignorance. You cannot build a program of discovery on the assumption that nobody is smart enough to figure out the answer to the problem.

—Neil deGrasse Tyson

Even noble souls can become corrupted with wrong education.

—Plato, *The Republic*

Creationism is religion. It's about believing in a god in a particular way that forces everything else to fall in line with that belief no matter what facts, logic, evidence, and reality say. Creationism is a problem in a country like the United States because the government and its public schools are supposed to be neutral on matters of religion. The 1987 Supreme Court case *Edwards v. Aguillard* made it crystal clear once and for all that it is unconstitutional to teach overt creationism in public school science classes because creationism is religion. The ruling did, however, make the point that there is nothing to prohibit the teaching of multiple theories about the origin of humankind, so long as they are scientific and nonreligious. In hindsight, the reaction, to this ruling by determined antievolutionists was predictable. They would give creationism a makeover.

Some clever creationists figured that by repackaging it as something scientific and not blatantly religious, they could sneak it back into classrooms. They named their creation "intelligent design" (ID). It was a crafty idea for sure, but there were two big problems: it wasn't scientific and it was religion. ID is just creationism all over again. ID's claim is that life is too complex to have happened "by chance" or by natural processes and, therefore, an intelligent designer must be responsible. Of course, savvy ID proponents didn't come right out and say that the designer is their god, but that's the implication, of course.

WRONG EVEN IF IT'S RIGHT

At the core of intelligent design is the arrogant idea that life is so complex that it could not exist without an intelligent creator. It is arrogant because it assumes what we don't know now, we can never know. For this reason, embracing intelligent design is wrong even if it should one day turn out to right. Yes, I really mean that. As it exists at the moment, ID "theory" is so bad, so lame, such a pointless, defeatist, and poorly conceived concept that no thoughtful person should go anywhere near it. Intelligent design is toxic to the mind not because it seems so very wrong but because of the message it sends to people—young students in particular. ID suggests that an unanswered question is an *unanswerable* question. Everything about intelligent design screams: *Give up! Stop trying to figure out how life originated and changed over time. Just say "God did it" and shut up.* Regardless of whether or not a god or gods did create life, retreating from the challenge of biological mysteries so early in the game is neither wise nor brave. A more honest name for ID "theory" would be: "Biology is really hard so let's stop trying." Intelligent design is not science; it's the opposite of science. It's the mass marketing of intellectual surrender. ID should be rejected on principle alone. Look at it this way: if fairies were discovered tomorrow in somebody's backyard, it wouldn't suddenly mean that all the fairy believers of previous centuries were brilliant thinkers who were ahead of the curve and wiser than fairy skeptics. Maybe aliens really did build the pyramids in Egypt. But without one speck of credible evidence, it's an irrational belief one cannot intellectually justify holding today, even if it turns out to be true tomorrow. Should those who believe in intelligent design today ever be proven correct, it would not be because they were more sensible and aligned with science. They would be right for all the wrong reasons.

The ID claim that life is "too complex" to ever be explained by natural means is outrageously presumptuous. "Too complex"? Who could possibly know such a thing based on our current knowledge? The orbit of Earth around the Sun was once "too complex" for science. Should we have stopped trying to figure it out? Everything was "too complex" at some point in our past. Who can say that one day in the future we won't understand how life originates and functions in complete detail? Intelligent design proponents must know the future, for how else can they declare a cell to be forever beyond the reach of natural explanations as revealed by science? How do they know?

Anyone who looks into intelligent design will run into something called "irreducible complexity." Again, this is nothing more than jumping to extraordinary and unjustified conclusions: *we don't know how all the parts of this cell could have evolved to work in unison like this, so it must have been put together by an intelligent designer.* This is transparent nonsense. What about nonliving things that are complex? "[I]ntricacy and organization can be readily inferred even from things that were clearly not consciously designed," argues scientist Jonathan Marks. "Snowflakes are intricate, but does anyone really think that God unleashes a horde of microscopic chiselers in constructing a blizzard? I may not know much about the physics of crystal formation, but it's got to be a better explanation than that one."[1]

This is really the same old creationism. The difference is that intelligent design is packaged and marketed to appear scientific, though it is anything but. Slick promotion has fooled much of the public into thinking that the ID/evolution controversy is about fairness rather than antiscience and religion masquerading as science. "Teach both sides," is often heard. No less an authority figure than a former president, George W. Bush, said both evolution and intelligent design should be taught in science classes. One could advocate for astrology to be given equal time in astronomy classes. It would make just as much sense.

The fatal flaw of intelligent design has nothing to do with the identity of its proponents, their politics or religious beliefs, who funds it, or anything else along those lines. All that matters is that there is *no good evidence*, *rational argument*, or *valid theory* to support the idea that life was created by a god, extraterrestrial, or other intelligent being. None. Maybe it did happen that way. But until someone comes up with some real scientific evidence for it, we can't know for sure and should not pretend that we do.

The best argument one can make against intelligent design has nothing to do with DNA or fossils. Simply point to where its leading advocates choose to fight their battles. This reveals everything we need to know because they do not wage war with their arguments and evidence in academic conferences and on the pages of scientific journals. No, they do their fighting in courtrooms, school board meetings, political campaigns, on websites, and in books and pamphlets. University of Chicago evolutionary biologist Jerry Coyne describes the stark absence of intelligent design from the scientific process: "Since 1973, more than one hundred thousand peer-reviewed papers on neo-Darwinian evolution have been published. ID is represented by just a single peer-reviewed paper, and this is a generous estimate because that paper has been refuted."[2]

This is not how proper science is done. It's not how we go about determining what is real and what is not. Real science grows from the clash of ideas and claims where the victor is determined by who has the superior evidence. It's not a popularity contest. It doesn't matter if a scientific claim is presented by good people, bad people, ugly people, nice-looking people, religious people, atheists, poor people, or rich people. All that matters ultimately is who can put the best evidence on the table for all to see. Sometimes winning takes a long time. And winning is always conditional; it's never final. Some see it as a weakness, but it's the best thing about science. It's a good thing to change when change is called for. Nothing is written in stone when it comes to science. Everything is always up for grabs, open to revision, and vulnerable to attack. If I happen to have a eureka moment while I'm taking a shower tonight and figure out something that tears down the theory of evolution, my first impulse will be to write it down and send it to the journal *Nature*. My first move definitely would not be to attempt to pack some school board with people who agree with me so that high school kids can be taught my new theory—no matter what the world's scientists think of it. No, I would go the route of publication, fame, and fortune, as anyone would who had a winning theory. I suppose I could put up a slick website and print pamphlets to hand out in churches, but I would rather pick up a Nobel Prize for my Darwin-destroying theory on my way to the bank to deposit all the fat checks I would receive from book advances and speaking engagements.

Those who think intelligent design is scientific should ask themselves why the leaders of the ID movement don't act like scientists. More to the point, why don't real scientists take intelligent design seriously and embrace it? If there were anything to it, there is absolutely no doubt that many would. It is ludicrous to suggest that they choose not to rock the boat in the evolution-loving subculture of science or *can't* due to censorship. Even if mainstream secular universities were silencing scientists who possess powerful new evidence capable of sinking evolution, it doesn't explain why we do not see game-changing ideas and data emerging from religious universities. Surely Oral Roberts and Liberty Universities would not censor their professors or students if any of them could make a compelling case for intelligent design. There is also the opportunity for intelligent design proponents to write their own books or start their own science journals. If they knew something significant, they could and would make it public. And, if there were anything to it, the world's scientists would not ignore it. We can only conclude that ID proponents have no case to make beyond the political campaigning, lobbying, and public sales pitches that we have seen.

"What is really going on in the ID movement is that highly educated religious men are justifying their faith with sophisticated scientific arguments," explains Michael Shermer, publisher of *Skeptic* magazine and the author of *Why Darwin Matters*. "This is old-time religion dressed up in newfangled language. The words change but the arguments remain the same. As Karl Marx once noted: 'Hegel remarks somewhere that all great world-historical facts and personages occur, as it were, twice. He has forgotten to add: the first time as tragedy, the second time as farce.' The creationism of William Jennings Bryan and the Scopes trial was a tragedy. The creationism of the intelligent design theorists is a farce."[3]

GO DEEPER . . .

Books

- Brockman, John, ed. *Intelligent Thought: Science versus the Intelligent Design Movement*. New York: Vintage Books, 2006.
- Dawkins, Richard. *The Blind Watchmaker: Why the Evidence of Evolution Reveals a Universe without Design*. New York: W. W. Norton, 1996.
- Dawkins, Richard. *The Greatest Show on Earth: The Evidence for Evolution*. New York: Free Press, 2009.
- Forrest, Barbara, and Paul R. Gross. *Creationism's Trojan Horse: The Wedge of Intelligent Design*. Oxford: Oxford University Press, 2004.
- Humes, Edward. *Monkey Girl: Education, Education, Religion, and the Battle for America's Soul*. New York: HarperCollins, 2007.
- Margulis, Lynn, and Dorian Sagan. *Microcosmos: Four Billion Years of Microbial Evolution*. Berkeley: University of California Press, 1997.
- Smith, Cameron, and Charles Sullivan. *The Top 10 Myths about Evolution*. Amherst, NY: Prometheus Books, 2006.
- Soutwood, Richard. *The Story of Life*. New York: Oxford University Press, 2004.
- Tattersall, Ian, and Rob DeSalle. *Bones, Brains, and DNA: The Human Genome and Human Evolution*. Piermont, NH: Bunker Hill Publishing, 2007.
- Weinberg, Steven. *Facing Up: Science and Its Cultural Adversaries*. Cambridge, MA: Harvard University Press, 2003.
- Young, Matt, and Taner Edis. *Why Intelligent Design Fails: A Scientific Critique of the New Creationism*. Piscataway, NJ: Rutgers University Press, 2006.
- Zimmer, Carl. *The Tangled Bank: An Introduction to Evolution*. New York: Roberts, 2009.

Other Sources

- *Judgment Day: Intelligent Design on Trial* (DVD), *Nova*.

Chapter 32

"THE UNIVERSE AND EARTH ARE FINE-TUNED FOR LIFE."

If things were different, things would not be the way they are.

—Robert L. Park,
Superstition: Belief in the Age of Science

A common justification for believing in a god or gods that I have heard from people around the world is that the universe and Earth are so perfectly tuned for life that their god (always *their* god over millions of others) must have designed and created it. All of this around us could not possibly have just happened by chance. This must be a very attractive idea to religious people because it's repeated in the same way by many followers of many contradictory belief systems. Like any idea, however, popularity doesn't necessarily mean that it makes sense.

If the extent of one's awareness about animal life stops at a dog, a goldfish, or an occasional visit to a zoo, then I suppose it may be understandable if one doesn't appreciate how incredibly difficult, violent, disgusting, and unfair existence is for most life on our planet. Right this moment, as you are reading this sentence, millions of creatures are being pierced, clawed, snapped in half, chewed and swallowed—while still alive. A constant and incomprehensible flow of pain and suffering is standard operating procedure, just the way life goes on this planet. If our world is finely tuned for life, then why is the overall extinction rate more than 95 percent? If this planet is intelligently designed specifically for life to flourish, then why are struggle, disease, agony, misery, and failure the norm?

If Earth had significantly more mass, or less mass, if there were no water, or if the atmosphere were different, then we couldn't exist. So what? If the conditions were so different that we could not exist—then we would not exist. Nothing should be so difficult to understand about that. There is life here and no life on the Moon, for example. That's just how it is. If the Moon had an ocean and an atmosphere like Earth's, then maybe there would be life on it. But since there is no life there, should we conclude that the Moon was intelligently designed to exclude life? By this sort of logic, any kind of environment can be attributed to an intelligent designer no matter if it can support life or not.

Those who believe the Earth is tuned for life get it precisely backward. *Life has been tuned for the Earth*. And the tuning was/is done by the natural process of evolution. This makes more sense than anything else, especially in light of all the dead ends and unintelligent designs that are found in many life-forms, including us. A basic understanding of evolution makes it easier to grasp why there are so many oxygen-breathing animals on a planet that has lots of oxygen. It explains why there are so many life-forms with fins and gills on a planet that has so much water. And it makes sense for all these sun-dependent plants to exist on a sunny planet. If life fits on Earth, it's because it evolved here.

Look around the Earth. Some creatures evolved in extremely hot environments; others evolved in extremely cold environments. If the entire Earth had been freezing cold for the last one hundred million years, then hot-weather creatures would either not be here today or they would be very different. Should we conclude, then, that the Arctic was "designed" for polar bears because that's where polar bears live? Isn't it more likely that polar bears have evolved to survive in cold environments and that's why we find them in a cold place like the

Arctic? If they hadn't evolved for cold weather, they either wouldn't exist or they would have evolved to live someplace else. What we can say for polar bears on one part of this planet, we can say for all life on this planet. If conditions were different, life would be different—or simply not exist.

Many people say the entire universe, like the Earth, is fine-tuned for life. But this claim falls far short too. Of course there are key qualities to this universe that we rely upon to live, but most likely that is only because we exist in *this* universe and we evolved under *these* conditions. If our universe was fundamentally different—say atoms didn't exist and matter was constructed of something else—then we would either be made up of something other than atoms or we would not exist. It's as simple as that.

Another problem with this claim of a universe designed to cater to our needs is the fact that the universe doesn't seem very hospitable to us! Most Earth life, with the exception of some microbes, would not last long in space or on any other known celestial bodies. We certainly can't survive without making a significant effort to fend off the hellish environment that seems to be standard throughout virtually all the universe. Space is too cold, too hot, and has too much radiation for most life-forms we are familiar with. It's also mostly empty, hardly ideal for building a home and raising a family. Overall, the universe does not appear to be the most hospitable place one could imagine, or design and create for humans if one had such powers and such a desire. Earth is also far from ideal and certainly does not seem to be intelligently designed and built specifically with human needs, comfort, and safety in mind. If it were, why are there hurricanes, earthquakes, tsunamis, supervolcanoes, the Ebola virus, and so on? Any designer with an ounce of compassion would have left those out of the recipe, right?

I suspect that this common belief of an Earth and an entire universe created and catered for us is nothing more than human arrogance. Don't forget, we had to be dragged, kicking and screaming, away from believing that the Earth was the center of the universe. Now many of us are clinging to the belief that we are still centrally *important*, if not centrally located. "The universe was made just for me," sounds a lot like the imagination and insecurities of a self-centered child. But in fairness, who knows? Maybe the universe really is custom-made for us by divine design. But until we learn a lot more about this universe we find ourselves in, it is preposterous to make such a bold assumption without evidence.

GO DEEPER . . .

- Ferris, Timothy. *Life beyond Earth*. New York: Simon and Schuster, 2001.
- Ferris, Timothy. *The Whole Shebang: A State-of-the-Universe(s) Report*. New York: Simon and Schuster, 1998.
- Greene, Brian. *The Elegant Universe: Superstrings, Hidden Dimensions, and the Quest for the Ultimate Theory*. New York: Vintage Books, 2000.
- Greene, Brian. *The Hidden Reality: Parallel Universes and the Deep Laws of the Cosmos*. New York: Knopf, 2011.
- Hawking, Stephen. *The Grand Design*. New York: Bantam, 2011.
- Potter, Christopher. *You Are Here: A Portable History of the Universe*. London: Hutchinson Radius, 2009.
- Sagan, Carl. *Cosmos*. New York: Ballantine Books, 1985.
- Stenger, Victor J. *The Fallacy of Fine-Tuning: Why the Universe Is Not Designed for Us*. Amherst, NY: Prometheus Books, 2011.

Chapter 33

"MANY PROPHECIES HAVE COME TO PASS."

I have not sent these prophets, yet they have run with their message: I have not spoken to them, yet they prophesied.

—God, quoted in Jeremiah 23:21

For as long as I can remember I have heard talk of prophecies. The gods, it seems, are always sharing with us key information, critical facts, important predictions, dire warnings, and spectacular insights about what is happening and about to happen to us. I have had Nostradamus experts and pulpit-pounding preachers declare to me with absolute confidence that earth-shattering events were close. Over time, however, I began to notice that the end was always *near* but never quite *here*. While such talk may have been mildly concerning the first few times I heard it, I'm a bit more skeptical these days. I have since learned that supernatural prophecies are not the rock-solid predictions of future events they are so often made out to be. Nonetheless, claims of "fulfilled prophecies" serve as powerful anchors of faith for many millions of believers who follow many different religions.

All around the world, prophecies are cited as justification for believing in the accuracy and validity of entire belief systems. In the Middle East, for example, some Muslims have told me with deep conviction that the Koran contains a prediction that foretold the Apollo Moon landings: "The Moon has split and the hour has drawn closer."[1] They explained to me that "the Moon has split" refers to the return to Earth of Apollo astronauts with lunar rocks and "the hour has drawn closer" refers to the fast-approaching end of the world. They insist that a seventh-century book accurately predicting an important twentieth-century event is ironclad proof that Islam is the one and only true religion. I'm not sure how you feel, but I'm not convinced that this was an accurate prediction of the Moon landings. I am sure, for example, that if the Moon cracked in half tomorrow, many Muslims would then say that the "Moon has split" line referred to that and not astronauts taking away rock samples. This is typical of the problem with most religious predictions. They are too vague and can be interpreted by believers in ways that may make them appear to fit actual historical events when they really don't. Christianity has its prophecy of a messiah that is predicted and fulfilled all within the same limited collection of sources. That doesn't seem fair, or very convincing. Imagine if I showed you a book that described in chapter 1 a magical city made of gold rising up in Australia ten thousand years ago. Then, in chapter 20 of the same book, the city was said to have vanished forever without a trace five thousand years ago. Would you be impressed if I then declared that the city of gold is a fulfilled prophecy and proof that my book is true? Didn't think so.

Rather than list a long string of prophecies that are flawed in one way or another, let's look closely at one major prophecy that most people are familiar with. I feel like I know this one very well because I have seen it with my own eyes.

VISIT TO A PROPHECY

I'm not sure why, but I reach out and touch the yellow stone of Jerusalem's famed Western Wall. I do not believe there is any point to this other than to make a tangible connection with a place millions of other people view as extremely important and intensely sacred. I suppose maybe I am trying to channel some of that human passion, to see if I can feel some of what they feel, if only for a moment. Or maybe I'm just nervous and trying to blend in with the believers around me. The surface of the gigantic wall is old and rugged. Folded papers of various sizes fill virtually every crack in the wall. These are written prayers, stuffed into a portal to the divine by believers who hope that a god will read them. On both sides of me, Jewish men with gray beards and black hats rock back and forth. Prayers stream out of them in the form of endless rhythmic chants. I can't understand any of it, but there is an obvious commitment to each word.

Although there is no agreement about whom he favors, what he desires of his followers, or even if he exists or not, the God of Abraham certainly has made an impression upon humankind. Approximately half of the world's population currently believes he is real. Numerous versions of Judaism, Christianity, and Islam declare him to be not only the creator of Earth and the universe, but also relevant and involved in the affairs of society today. If Yahweh/God the Father/ Allah had an address, I'm guessing it would be right here in Jerusalem somewhere.

Earlier I spent hours exploring the Church of the Holy Sepulchre, just a short walk from the Western Wall. Run by rival Christian sects, the site contains an odd forced collaboration of shrines. The building stands over the precise patch of land where Jesus is believed to have been executed some two thousand years ago. Obviously it is profoundly sacred to many Christians. The body of Adam, believed to be the first human, is also said to lie beneath the floor here. Christian or not, one cannot help but be captivated by the bustling mix of sweaty tourists, dazed pilgrims, shuffling monks, pickpockets, fanatics, schizophrenics, and genuinely nice people who share the dim hallways. But Christians are not the main attraction in Jerusalem. Jews and Muslims dominate my attention here.

I see a Jewish man carrying a baby. He has an assault rifle slung over his shoulder, a reminder of all the hate and tension that partly define the Holy Land. Of those people who can be identified by their clothing, I see only groups of Jews and groups of Muslims. Never do I see a mixed group strolling along in casual conversation. It is odd to see a society ripped apart believe in the same god. I wonder what the future holds for Jerusalem. Will these people go on killing one another at a measured pace, or will the streets be turned into rivers of blood again as happened during the Crusades centuries ago? Will a day come when the fiery blast of a nuclear warhead finally quiets the hate once and for all? Many believers in the God of Abraham say they are confident that the world will end in a chaotic bloodbath of destruction very soon. Of course, people have been saying this for many centuries so it has lost much of its shock value. But ideas about "God's plan for the world" are still profoundly interesting in Jerusalem. Armageddon just feels closer here—maybe because it is. Hundreds of millions of people believe that their god is coming soon and this is the place where he will first appear. This will be the first battlefield in the final war. But that prophecy concerns the future, what about something that has already happened?

The entire nation of Israel and its capital, Jerusalem, are a fulfillment of divine prophecy, according to millions of Christians and Jews. The very existence of this nation is seen by many as conclusive proof that Judaism and Christianity are accurate religions. I think about this as I walk the streets of this young country that feels so old. Are my feet treading on magical dust? Was this nation really created by a god in order to keep his promise to the Jews? Maybe, but I don't think that would be the best available explanation we have. I know another, far less extraordinary, story of how Israel was created. And this one does not require a god, magic, or even faith.

For many centuries Jews throughout the world held a strong belief that this land had been promised to them by their god and that it would one day be a nation for Jews. For many this was not seen as a mere wish or hopeful thinking. They viewed it as a dream they had to make come true by any means necessary. The creation of a Jewish nation in Palestine *would* happen someday, somehow. The constant persecution inflicted by Christians upon Jews in Europe was no doubt additional motivation for them to dream of their own country. But not all Jews were content to hope, pray, and wait. Many of them worked to make it happen. Zionism became an international movement to create a Jewish homeland. Men like Theodor Herzl (1860–1904) dedicated much of their lives to founding a new nation for Jews. Many wealthy and influential Jews around the world pushed hard for it. Over time, many Christians supported the goal. They too saw it as something God wanted. The horrors of the Holocaust helped make the newly formed United Nations and the governments of the United Kingdom and United States agreeable to the idea of a Jewish state, and by 1948 it came to pass.

This event was historic and important to world affairs, no doubt, but the important question here is why a god would be given credit for it when humans clearly did all the work every step of the way. Perhaps, as some claim, people were the instruments of God in achieving the goal. But there is no evidence for this. Having been to Israel, I can attest that all the buildings, homes, and streets appear to be human-made. Nothing I saw suggested that any government buildings or flagpoles miraculously sprung from the soil. Then US president Harry Truman was a Christian who was well aware of the idea that God had promised the Jews a homeland in the Middle East. Are we to believe that he was under the spell of God when he decided to officially recognize the new state of Israel? Isn't it far more likely that he simply was influenced by the geopolitics of the day and perhaps his own personal religious beliefs?

MY PROPHECIES ARE BETTER THAN YOUR PROPHECIES

The biggest clue that something is wrong with prophecy-as-proof arguments is that they only impress people who are already committed to the particular religion making the claims. We don't see Jews flocking to Islam the moment they hear about the "split Moon" prophecy. Hindus are not converting to Judaism and Christianity by the millions upon hearing that the creation of Israel was a divine promise fulfilled by the Abrahamic god. Most Sikhs are not won over by claims that the life of Jesus was accurately predicted within the Bible. If there was one religion that really did have an important prediction, made many centuries or thousands of years ago, and it specifically described an important event, with great accuracy and detail, then this religion would have an overwhelming advantage over all of its rivals. Imagine if seventh-century Islamic scrolls were discovered this year and included the following prediction:

> In a year to be called 1969, a chariot of metal will fly to the Moon in three days carrying three men named Neil, Buzz, and Michael. Two of the men will walk upon the surface clad in gleaming white suits. They all will return home and safely splash upon the sea.

Now, if the scrolls' age could be verified and it was determined that they had never been known to the modern world (so they could not have influenced events), we would have a very impressive prophecy. We might all be Muslims by the end of the week. But we have never seen anything remotely like this. No religion today, or ever, has been able to dominate the world's believers. There has always been disagreement and splintering of the mass of believers because no religion has an advantage on evidence, logical arguments, *or prophecies*.

A key reason that "fulfilled" religious prophecies are still embraced by many people around the world is that most are still un-

PROPHECY CHECKLIST

Keep the following points in mind when thinking about supernatural predictions:

- Was the prediction made before the event? Sometimes they aren't. Psychics and Nostradamus enthusiasts have been known to pull this trick. One always has to check.
- Was the prediction specific enough to make it clear that it could only mean one thing, or is it so vague that it could apply to many things and be interpreted in many ways?
- Is it a prediction of something that people could have accomplished by natural means without a god's help?
- Are the odds favorable for the prophecy to come true no matter what? A prediction of victory in a war between two sides, for example, is not impressive because there is a fifty-fifty chance of being right.
- Could it just be a coincidence or something that was bound to happen anyway? Predictions of economic hard times, natural disasters, wars, and so on mean nothing because those things have been occurring over and over for millennia.
- Keep in mind that when many predictions are made, some are bound to come true by chance alone. Remember the old saying: even a broken clock shows the correct time twice per day.

familiar with the simple ways in which one can cut through fog and spin to recognize empty claims for what they are. These "fulfilled prophecies" are nothing more than vague predictions that can be easily analyzed and shown not to necessarily rely on supernatural events. When confronted with one of these divine-prediction-come-true claims, always be sure to poke it and prod it with an inquiring mind to see if it's worthy of your trust.

GO DEEPER . . .

Books

- Allan, Tony. *Myths of the World: The Illustrated Treasury of the World's Greatest Stories*. New York: Duncan Baird, 2009.
- Dennett, Daniel. *Breaking the Spell: Religion as a Natural Phenomenon*. New York: Penguin, 2007.

Other Sources

- *Humanist* (magazine).

Chapter 34

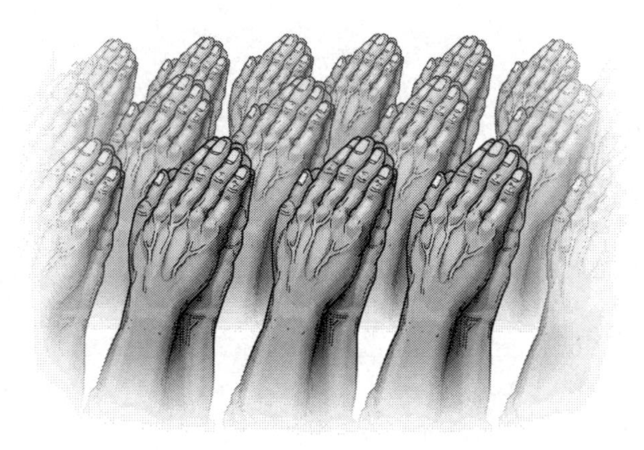

"PRAYER WORKS!"

> **If you believe, you will receive whatever you ask for in prayer.**
>
> —Matthew 21:21

> **Why won't God heal amputees?**
>
> —WhyWontGodHealAmputees.com

During my travels on six continents I always made sure to ask people about their religious beliefs. Religion fascinates me and I understand its importance to history and contemporary culture. The many conversations I had with believers taught me a lot about how they think about their particular religions and what they feel is important. A statement I heard over and over was: "I pray." The world's believers pray to their gods often and with great enthusiasm, it seems. They may worship different supernatural beings, revere contradictory holy books, and follow different rules, but they all pray. So what is this thing called prayer, why is it so popular, and, most important, does it work?

Like everything else in religion, prayer means different things to different people but most commonly it is nothing more than an attempt to establish some kind of a communication link with a god. It can be in the form of a request for something like healing from illness, world peace, or a new bicycle for Christmas. Prayer can also be a "thank you" or an unselfish form of worship, a way of showing respect and reverence to a god or gods. But when people say that prayer "works," they usually mean that they asked for something and their god delivered it. Based on my interviews and research, I believe that this aspect of prayer—asking and getting—is one of the most popular justifications for religious belief worldwide. When I ask Catholics, Protestants, Muslims, Hindus, Sikhs, Mormons, Jews, and so on to explain how they know their gods are real, an answer I hear almost every time is, "my prayers were answered."

The amazing thing about this common confidence in prayer is that there is no proof to back it up. There have been studies designed to test the ability of prayer to help sick people, but the results are not convincing. Nonetheless, it seems that billions of people are confident that their gods not only hear their requests but often respond to them positively. There clearly is a problem here. If billions of prayer requests are being acted on every day, then it should be easy to document and establish this as a real phenomenon. But nobody has ever been able to do it. How can this be? Surely all of these people who say prayer works are not lying. They are not insane. So what are they talking about?

First of all, I should note that challenging the claim that gods answer prayers is not necessarily a total condemnation of the act of praying. It seems clear that people can gain some psychological benefits from it. It might be helpful for some by making them feel unified with others who also pray, or it could provide a sense of peaceful solitude. Many Buddhists do not pray to a god but they do meditate, a ritual that can be very similar to prayer, and they seem to gain practical benefits from it.[1] If praying lowers one's blood pressure and heart rate, then it may be beneficial, but that is not proof of anything supernatural.

Those who are convinced from personal experience that prayers are acted upon in favorable ways by a god are most likely falling victim to confirmation bias, that troublesome habit we all have of emphasizing and remembering things that confirm what we believe while ignoring and forgetting things that contradict our beliefs. Unfortunately, very few schools and families teach children critical thinking skills or encourage skepticism. So it's no surprise that many children grow up to become adults who think that their prayers are answered when a bit of simple record keeping could prove otherwise. Claims of "answered prayers" seem to rest on the same foundation that professional psychics rely on to make a living: believers remember the hits and forget the misses. So when someone prays for a new job and then gets it a month later, he is likely to credit prayer without any thought given to his efforts to land that job as well as all the other prayer requests for things that never materialized.

The idea that prayers are answered by "my god" comes with that insurmountable contradiction problem raised in the chapters about belief in gods and religious preference. How can the claim of answered prayers be accepted when so many people in very different belief systems talking to very different gods make the exact same claim? For example, how can it be that a Hindu, Wiccan, Muslim, Christian, and ancient Roman all prayed and received positive supernatural results? One may have asked Ganesha to heal a sick aunt and another asked Allah to relieve his migraine headaches. The Wiccan wanted a tree to grow tall and strong. Meanwhile, the Christian asked Jesus to cure a dying father and the Roman asked Jupiter for a new chariot. How are we to explain it if they all claimed to have had their prayers answered? The universe is just not big enough to contain Jesus, Allah, Ganesha, and Jupiter. Based on their core claims, Christianity, Islam, Wicca, Hinduism, and ancient Roman religion cannot all be true. Maybe one of them is right or maybe none of them are right. Yet I have heard repeatedly from Christians, Muslims, and Hindus that their prayers are answered favorably. They claim to pray to their specific god in their specific way and get positive results. And, of course, this is seen by them as proof that their religion is the accurate one and their god or gods are the real ones. Some of these people, or all of them, must be mistaken. To push the point further, even if one religion is valid and all its claims are true, its followers should still be skeptical of the power of prayer and how they measure its success. They only have to look around the world and acknowledge the obvious: that just about anyone can be convinced that prayers are answered no matter what is really going on. Today's intellectually honest Christian or Muslim should be troubled by the idea that her "answered prayers" are a real phenomenon but the "answered prayers" of animists, pagans, and so many others over the last several thousand years are not.

THE MOST IMPORTANT PRAYER OF ALL

As a habit, I also observe and learn as much as I can about the poorest people in any city or country I visit. Global poverty, particularly as it relates to children, is an important issue for me. I founded a small charity that raised more than $100,000 for UNICEF, and I have written numerous published commentaries and news articles about the poorest of the poor. I have seen, smelled, and touched this jagged edge of human existence, and it is not pretty. The slums of Africa and India changed me. Actually, it's more like they wounded me. Forever. I saw the tiny bodies of child beggars twisted and tormented by malnutrition, polio, and elephantiasis. I saw the vacant stare of listless babies who were riding out a death sentence for the crime of choosing the wrong society to be born into. My small efforts to raise money and awareness for the poor have nothing to do with saving the world (it won't) or posing to appear heroic (there are easier and less disturbing ways to do that). The truth is, I do it purely for selfish reasons. It helps me push back against the sounds and images of the poor people who haunt my mind. It's a comfort thing, my simple way of feeling a little less uncomfortable for being born over on the lucky side of fate.

One thing I noticed about the poorest places on Earth is that they are also the most religious places on Earth. Atheists are scarce in those parts. The many millions of people who try their best to make do on a few dollars a day tend to be devout believers. And whether they are Muslims in Syria, Hindus in India, or Christians in Haiti, they pray frequently and they pray hard. I've seen many of them speaking to their gods and was struck by their passion. And this is where I found the most glaring problem of all with the claim that prayer works.

In the developing world, approximately ten million babies under the age of five die each year in poverty. They die because they couldn't get a two-dollar vaccine, a cup of clean water, or fifty cents worth of food. This is a stunning statistic and few people in wealthy nations know about it. Sure, everyone knows that there are suffering children around the world. But I don't think very many know just how many of them die each year, and I don't think very many people pause to consider how much suffering and agony this toll translates to every day. It's nearly a double Holocaust—every year. Don't underestimate the horror of death by dysentery or some other disease that could have been prevented. It's usually a slow, cruel death, and parents trapped in extreme poverty, with feelings of love and responsibility like any other parent—have to watch their children slip away from them.

The societies with the world's highest child mortality rates correlate strongly with the world's most religious societies. Therefore, it is reasonable to assume that virtually all of the parents of these ten million babies are believers and that they pray intensely for a god or gods to save their babies during the most desperate moments. Understand, we are not talking about praying for a promotion at work or to make an A on a math test. I would think that prayers to a god in the form of a mother's plea to save her suffering and dying child must be among the most powerful and valuable words ever uttered. In my view, this is the best test of prayer we can ever hope for. And prayer fails.

Nations with the highest rates of nonbelief—Sweden, Denmark, Finland, Japan, Canada, and the UK, for example—have the world's lowest child mortality rates. It's reasonable to assume that this means significantly fewer prayers for a baby's good health are spoken or thought by all these nonbelieving mothers. But in places where virtually all the mothers are praying like mad for a god to give their babies good health, babies suffer and die by the millions.

In 2011, the charity organization Save the Children released its rankings of the best and worst places to be a mother on planet Earth. Again, the pattern of prayer not seeming to do much for children is evident:

2011 *Mothers' Index* Rankings[2]

Top 10 Best Places to Be a Mother*		Bottom 10 Worst Places to Be a Mother**	
1	Norway	155	Central African Republic
2	Australia	156	Sudan
2	Iceland	157	Mali
4	Sweden	158	Eritrea
5	Denmark	159	DR Congo
6	New Zealand	160	Chad
7	Finland	161	Yemen
8	Belgium	162	Guinea-Bissau
9	Netherlands	163	Niger
10	France	164	Afghanistan

*(all these nations have relatively low rates of religious belief)
**(all these nations have relatively high rates of religious belief)

A prayer advocate's response to the challenge of dying babies and their praying mothers might be that child mortality rates are determined by the state of local economies and are impacted by poor or nonexistent healthcare and sanitation. But that would be the secular answer for why ten million babies die in the arms of praying mothers each year. Prayer, according to most of those who believe in it, is supposed to be able to transcend details such as the availability of clean water and antibiotics. Why would a prayer have to be reconciled with earthly economic conditions and doctor-patient ratios?

When we look back over history and consider all the natural disasters, famines, wars, and other dire circumstances people have found themselves in, it is clear that countless trillions of desperate and sincere prayers for help were sent up to the heavens but never answered as hoped. I suppose one could argue that the gods of these various religions are unimaginably cruel and simply ignore such pleas. However, another possible explanation is that the gods hear no prayers for the simple reason that they do not exist in the first place.

GO DEEPER . . .

- Barker, Dan. *Godless: How an Evangelical Preacher Became One of America's Leading Atheists*. Berkeley: Ulysses Press, 2008.
- Harris, Sam. *The Moral Landscape: How Science Can Determine Human Values*. New York: Free Press, 2010.
- Harrison, Guy P. *50 Reasons People Give for Believing in a God*. Amherst, NY: Prometheus Books, 2008.
- Stenger, Victor J. *The New Atheism: Taking a Stand for Science and Reason*. Amherst, NY: Prometheus Books, 2009.

Chapter 35

"RELIGIONS ARE SENSIBLE AND SAFE. CULTS ARE SILLY AND DANGEROUS."

It's a comforting feeling to think that it could never happen to you.

—Deborah Layton, Jonestown survivor

I avoid using the term *cult* and advise others to as well. The general public and news media use this label inconsistently and illogically. The result is that unpopular, weak, or unusual groups are denigrated unfairly. It is wrong, for example, when larger and older religions slander smaller and newer religions as "cults," especially when most well-established religions were once condemned as cults before achieving mainstream acceptance. But there is another reason not to use the term *cult* that is even more important than basic logic and fairness.

The primary reason the cult label is a bad one is because it too often gets in the way of opportunities to learn important lessons when bad things happen. Imagining a separate and distinct beast called "cults" suggests that we don't need to be on guard against any and all groups that exercise excessive control over members, encourage fanaticism and irrational belief, and are led by people who are abusive and power crazed. It is dangerous to believe that one only has to steer clear of "those evil, weird cults" and that it's OK to let your guard down around popular groups and organizations. The safer way is to *always* be cautious and skeptical.

THE LOST LESSON OF JONESTOWN

In 1978, more than nine hundred people died in "Jonestown," the jungle community built in Guyana by members of an organization called the Peoples Temple that was headquartered in San Francisco. Today most people think of the Jonestown event as a mass suicide by crazy, brainwashed cult members who willingly drank cyanide-laced Kool-Aid® because their leader, Jim Jones, told them they had to in order to go to heaven. The phrase "drink the Kool-Aid" even became entrenched as a pop-culture reference to blind loyalty to anything weird or untrue and is still heard often today.

Rarely, if ever, is Jonestown described as an event in which the Reverend Jim Jones, head of a popular Christian organization, forced more than nine hundred people to their deaths. But that would be a more accurate description of what happened. I learned this thanks to a long interview with a Jonestown survivor, a woman who escaped that nightmare jungle shortly before the end. She knew what was coming and tried unsuccessfully to warn the world.

"Nobody joined Peoples Temple thinking they were going to be taken to South America and killed," explained Deborah Layton, a seven-year member of the People's Temple.[1] She said that Jim Jones was a prominent and respected preacher in California and a member of a human rights commission who received positive coverage from the news media.

"Nobody joins a cult," she said. "You join a religious group. You join a political organization. You join a self-help group. Then things change gradually and at some point you stop and ask, 'What am I in?' It is OK to use the word *cult* in the dictionary, but not when discussing various religions that are active now."

Layton's perspective on the 1978 deaths at Jonestown is significantly different from the popular version of the tragedy:

> [Large] groups of people do not commit suicide. Children do not take their own lives. In Jonestown, two hundred babies and children died first. Mothers were holding onto their babies. They had no idea what else to do. [Imagine] if someone has a gun on you and your family, *you* try to think like they do and figure out how *you* can escape. Those people at Jonestown did not commit suicide. They were coerced. They were frightened by guards with guns pointing at them. [You think,] "OK, I'll run into the forest. But wait, what if I get shot in the back and my baby will be pulled from my arms, screaming and crying? Do I stay here and hold my baby? Do I give my child these last few moments with me holding him?" So many chaotic thoughts go through your head that by the time you figure it out it's too late.

Layton's view from the inside of Jonestown left no doubt in her mind that most people are vulnerable to becoming trapped by dangerous people and dangerous groups. The process of seduction and capture is deceptive. She believes it may comfort people to believe that it could never happen to them, but the reality is that it can:

> One of the ways we do a real disservice to our kids is that when something like Jonestown happens we tell them that they were just a bunch of nuts. This sets up our children to one day be in a situation that is a little bit weird and think, "Oh, it can't happen to me." Their antennas won't be up. But if we tell our kids that this kind of thing can happen to the best of us, then they will be aware.

Layton is also annoyed by what she feels is a common misconception that the Jonestown victims were "just a bunch of uneducated black people." That's just not true, she says. According to her, many of the black members were well educated. There were also highly educated white people who died, including her own mother.

WATCH OUT FOR DANGERS, NOT LABELS

Layton warns against belonging to a group or organization that discourages questions or forbids dissent. Other signs of danger she cites include being told that if you leave the group you can never come back or that you must separate from family members who do not join. "I think the most dangerous [groups] tell you that their way is the only way," she said. "I know that a lot of mainstream organizations fit this description, but it is dangerous for so many of them to be willing to just cut out so many people. To say, for example, that if you are not 'born again,' you will burn in hell. I also think that it is very dangerous when the only enlightened person is the leader of the organization and to question him or her is forbidden. That is a huge warning sign. Dissent is imperative."

Layton's comments offer an important warning for all of us. If ever you find yourself seduced by a person or an organization that has all the answers, perhaps it is only because you have failed to ask enough questions. We should always maintain awareness and caution, not only toward fringe groups that may be considered weird, but also toward powerful and popular organizations as well. Some of the most respected organizations in the United States and the world have by any reasonable measure controlled their members excessively, exploited people, stolen from them, abused them, and even killed them in some cases. When it comes to risky allegiances, it is not the cult label that matters; it's the danger. If I had to, I would choose an odd but safe cult over a respected but dangerous religious organization. Wouldn't you?

GO DEEPER . . .

- Haught, James. *Holy Horrors: An Illustrated History of Religious Murder and Madness*. Amherst, NY: Prometheus Books, 1999.
- Layton, Deborah. *Seductive Poison: A Jonestown Survivor's Story of Life and Death in the Peoples Temple*. New York: Anchor, 1999.

Chapter 36

"THEY FOUND NOAH'S ARK!"

If history and science have taught us anything, it's that passion and desire are not the same as truth.

—Edward O. Wilson,
Consilience: The Unity of Knowledge

Many years ago I remember watching a documentary about Noah's ark. The program made the claim that remains of the ark had been found on Mount Aratat in Turkey. Obviously those filmmakers were either incompetent or dishonest because no such discovery has ever been made.

In the years since that documentary aired, I have seen many more such claims come and go. Aside from shoddy documentaries, the ark is a popular topic with believers. Whenever someone tells me that the ark or parts of it have been found, I always ask them for details about the discovery. But they never have any. This is a dead giveaway. Trust me, if a team of university archaeologists, a Christian expedition, or some lost hiker ever finds a five-thousand-year-old wooden ship with fossilized dinosaur feces inside of it and Noah's name carved in the hull somewhere, you will hear about it in great detail. It will be a media sensation. The reality, however, is that the story in Genesis is all we have to go on. But that's enough for many millions of people when it comes to this popular belief.

The basics of the story are well known to almost everyone. The God of Abraham was disappointed with the humans he created, so he decided to murder them all and start from scratch. Noah was the last good man on Earth, so he and his family were to be spared. But God gave him a chore to do, a big one. Noah had to build a massive ship and take care of two animals of every "kind" onboard while God drowned the world. After forty days of rain and months of drifting, the water finally receded and the ark came to rest on top of a mountain. Noah and his family disembarked along with the animals and proceeded to repopulate Earth.

There are numerous problems with this story that should be obvious. But that doesn't stop hundreds of millions of people from believing that it's true. An ABC News poll found that 60 percent of Americans believe that it happened *exactly* as described in Genesis.[1] The Noah's ark story is one of those beliefs that easily generate endless debate over minor details that never lead anywhere. As one who has wasted far too much time discussing the lesser aspects of the story, I recommend sticking to the big issues. For example, could all the land on Earth have been completely flooded only five thousand years or so ago? Impossible, say all credible geologists, climatologists, physicists, historians, archaeologists, and marine scientists. Could Noah have saved enough animals on his ark to have repopulated the entire animal kingdom within the time frame claimed by believers? Impossible, say all credible biologists, zoologists, and geneticists. Could an eight-member family give rise to numerous cultures, civilizations, and a global population of seven billion people alive today in less than six thousand years? Impossible, say all credible archaeologists, anthropologists, historians, and geneticists.

WE'RE GONNA NEED A BIGGER BOAT

The vessel's dimensions are not known precisely, but most ark believers seem to agree that it was around four hundred to five hundred feet long. That's a fairly big boat, for sure, but not nearly large enough to house two representatives of every land species in the entire animal kingdom. Excluding plants and aquatic life, there is still a staggering amount of life Noah would have had to board and care for on his ship. In fact, no one today can even guess how many because we are still discovering new species all the time. Millions await discovery, say biologists. There may be more than twenty million unnamed species alone! Clearly we can't know how big Noah's boat would have to have been because we don't know how many pairs of species representatives would have needed a place on it. I'm guessing the ship would have to be about as big as the continents of Africa or Australia to do the job.

Speaking of continents, how did animals from different continents make it to the ship for boarding and how did animals disperse to various continents after the flood? Challenges of a peaceful and timely boarding and disembarking process aside, it seems very unlikely that two representatives of millions of mammal, bird, and reptile species could live aboard a five-hundred-foot ship for nearly a year. Many believers recognize these basic problems and don't even try to explain them. They prefer to call it a miracle and leave it at that. Supernatural herding of animals and magical violations of population breeding requirements are one thing, but everyone should be aware that the laws of nature and scientific evidence do not support this story in any way. In fact, virtually everything we know undermines it. There simply is no compelling case for the entire planet being flooded four or five thousand years ago, and all of today's land life, including humans, being descended from the passengers of a single boat.

It is important to note that having doubts or outright rejecting the flood story does not mean one must necessarily abandon one's religion. While some do argue that every detail of this story and others must be believed, the fact is there are hundreds of millions of sincere Christians, Jews, and Muslims around the world who do not accept the flood story as literally true and yet still adhere to their religion. Those believers who insist that one must believe every line of every story in a holy book in order to be a credible follower of a particular belief system are expressing an opinion and nothing more.

I have learned over the years that ark believers can be very clever in defending their claims. Most young Earth creationists say dinosaurs were on the ark, a stunning claim given the size of many dinosaur species. But believers have explained to me that larger dinosaurs such as argentinosaurus, sauroposeiden, and spinosaurus may have been taken aboard as smaller juveniles or even in unhatched eggs. A neat answer, indeed, but I'm still not convinced.

Perhaps the weirdest aspect of the ark story is the way many believers present it as a positive and inspirational event that should be admired and shared with children. I have seen many children's books, for example, that depict the flood as happy and upbeat, complete with smiling animals and cheery old Noah waving from the deck of the ark. Let's not kid ourselves, if the Noah's ark story really happened, it would have to rank as one of the most horrible events in history. According to Genesis, God flooded the entire world for the specific purpose of killing everyone on it—babies included. In addition to people, however, the collateral damage would have been horrifying. Imagine all the puppies, kittens, and cute bunny rabbits bobbing in the water, kicking frantically until they exhausted themselves and sank beneath the surface to cruel deaths. Even if every adult on every continent was sufficiently evil to warrant death by drowning, surely at least a few children here and there weren't so evil to deserve death by drowning. Imagine the screams that would have filled the air as people struggled in the raging waters before dying. Remember the terrible tsunamis of 2004 and 2011? Imagine that sort of death and destruction striking everywhere on Earth. Not a cheerful story, definitely not a children's story, in my opinion.

NOAH'S ARK DISCOVERED—NOT!

Although reports pop up from time to time about someone claiming to have found Noah's ark, no one has to date. But you might be able to see and board the famous ship anyway. A group linked to the antiscience Creation Museum in Kentucky plans to build a $125 million "Ark Encounter" theme park nearby. The centerpiece will be a $24 million full-size replica of the ship. The ark will be "perfectly proportioned" and allow visitors inside to explore where they can see both live and animatronic animals. The developers expect the attraction to draw 1.6 million people in its first year and make a $4 billion economic impact in the region over ten years. Donations to fund the project are currently being sought from the public. If interested, you can even sponsor an official wooden ark peg for $100, a plank for $1,000, or a beam for $5,000. The purpose of the ark, say organizers, is to show people how Noah lived and how he built the ark.[2] No word yet if they plan to have thousands of animatronic twitching victims in the surrounding waters and bloated corpses littered about the property to further enhance the realism.

GO DEEPER . . .

- Asimov, Isaac. *Asimov's Guide to the Bible*. New York: Wings, 1981.
- Asimov, Isaac. *In the Beginning: Science Faces God in the Book of Genesis*. New York: Crown, 1988.

Chapter 37

"ARCHAEOLOGY PROVED MY RELIGION IS TRUE."

Archaeology has confirmed many of our church's claims.
—A Mormon interviewed by the author

Thanks to a dry climate and long streak of human habitation, the Middle East offers an abundance of both historic and prehistoric treasures. Certainly no archaeologist would dispute that. People have been living in that region for many thousands of years and left behind a wealth of artifacts that help illuminate their cultures for us today. Stick a trowel in the ground just about anywhere in this region and there is a good chance you will come up with something.

During casual strolls around Cairo, Damascus, Amman, or Jerusalem I had the feeling I was at or near excavation sites at all times. Archaeology never seemed very far away in these places. The ancient greeted me every time I turned a corner. The always-present theme of the deep past magnifies the intense pressures of modern life in the Middle East as well. No matter how important it may be, archaeological work often goes unnoticed and unappreciated in most parts of the world. Here, however, an archeological dig can be seen as work toward vital confirmation of the most important truth in the universe—and even a reason to kill someone. Archaeology matters in the Middle East. There is something powerful and appealing about tangible evidence that few believes can resist. Faith is wonderful and adequate on its own, say believers. But clearly they also love the idea of evidence that can be seen and touched.

Biblical archaeology, vaguely defined as archaeology related to people and stories contained in the Bible, is a robust little international industry. Hundreds, perhaps thousands, of books have been published on the subject; seminars are held, and classes taught. There are websites, magazines, newsletters, and clubs to join. Land tours and even sea cruises are available to travelers interested in learning about biblical archaeology. There is one significant fact that seems to have become lost in all this excitement and activity, however. Nothing has ever been found—*not one artifact*—that proves any supernatural claims found in the Bible. After many years of effort, biblical archaeology has failed to deliver on the point that matters most. The same is true for all religions. No archaeological artifact has ever been discovered that confirms any paranormal/supernatural claims made by any religion ever. But this does not stop some followers of various religions from overreaching when it comes to archaeology. For example, I read the line, "archaeology testifies to this event," in an article about God's destruction of Sodom and Gomorrah in a Jehovah's Witness booklet. This is either an error based on ignorance or an outright lie because no archaeological discovery has ever confirmed that any city anywhere has ever been destroyed by a god.

While believers may point to archaeology for confirmation of their gods' activities, the truth is that this important and productive scientific discipline has not provided evidence of anything other than *human* activities. In every case, archaeology has proven nothing more than the existence of ancient human believers, but never the target of their belief. Such discoveries showing supernatural claims are possible. For example, if the fossilized skeletons of Pegasus, the flying horse, or Cerberus, the three-headed hound of Hades, were discovered tomorrow near the Acropolis in Athens, it certainly would prove that at least some ancient Greek religious claims were true. The discovery of something like the Ark of the Covenant—with demonstrable magical powers intact—would be powerful evidence for the Torah's accuracy. If the Bible is correct, this sacred box that is believed to contain holy relics such as the Ten Commandments should emit supernatural powers detectable to scientific methods. After all, the Bible claims this box made rivers go dry and delivered magical military victories for the Jews. And what about all the relics held in churches across Europe? Many believers and clergy say they have the power to heal and protect with divine magic. So why haven't biblical archaeologists been able to confirm the supernatural properties of even just one? Just imagine if a splinter from the cross Jesus was crucified on or the finger bone of a saint was found to have extraordinary powers. Something like that would be very strong evidence, possibly even proof, that the Bible's supernatural claims are valid. So far, however, we have nothing but claims and anecdotes.

The fact is, despite all of the time, money, and energy devoted to biblical archaeology, there is nothing to show for it beyond natural and human artifacts. No trace of God, angels, demons, or magical artifacts. Work described by some as biblical archaeology may have contributed to our knowledge of Middle Eastern cultures, and that's great. But it goes no further than that. Unfortunately, many believers confuse the meaning of some artifacts and discoveries. Did a certain place mentioned in the Bible really exist? Sure. Did a place named Jericho once have a wall around it? OK. Did the Romans execute people by brutal means? Absolutely. The important question, however, is whether or not there is archaeological evidence to back up any of the Bible's *supernatural* content: the big stories about angels, demons, miracles, and God. Yes, finding an ancient boat in the Sea of Galilee (the "Jesus Boat") was a big deal, but it didn't prove that Eve was made from a rib, that a snake talked, or that Jesus walked on water. According to believers, the primary point of the Bible is not to shed light on human activities. It is primarily supposed to make the case for the existence of the Jewish/Christian god and make his desires known. Therefore, the claim that archaeology has verified the Bible's accuracy is wrong because, so far, archaeology is silent on all of the important claims. Excavating ancient streets, temples, and oil lamps just doesn't cut it if one is trying to verify occurrences of supernatural events and the existence of a god. For example, I have been to Athens and Rome where I saw many artifacts in museums and touched the hard stone of many ancient ruins. They do exist. Never once, however, did I imagine that the existence of these things proves the existence of Greek and Roman gods. Those things only establish the reality of ancient Greek and Roman *people*, just as biblical archaeology to date has only been able to confirm the existence of ancient Jewish and Christian people.

I have interviewed many believers about their religions over the years, and a number of them tried to play the archaeology card with me. I don't like friendly talks to degrade into arguments if it can be avoided, so I always resisted challenging them too harshly on this. The way I usually handle it is to simply ask: "What is the single most important archaeological discovery that confirms your religion?" Blank stares and silence almost always follow. A few might say the Dead Sea Scrolls. I explain that the Dead Sea Scrolls are ancient documents that include some text of the Bible/Torah. But an old partial copy of the Bible just doesn't prove the claims within the Bible are true or that anyone other than people were responsible for producing it. I add that I've been to the Israel Museum in Jerusalem where I saw the Dead Sea Scrolls. The exhibit was impressive and it was thrilling to see something so old, but nothing about it suggested to me that it was magical or anything other than something written by people a long time ago.

I am certain that virtually everyone on Earth would have heard about it by now if there were even one archaeological discovery that confirmed one of the Bible's supernatural claims. Surely any archaeologist who discovered something so important to so many people would be a well-known figure, at least as famous as the fictional character Indiana Jones. Even Howard Carter, discoverer of the tomb of Tutankhamun, is fairly well known. Can you name the most successful and famous biblical archaeologist?

Speaking of Tutankhamun, the young pharaoh-god, those who are prone to give biblical archaeology more credit than it deserves might ask themselves a simple question: If bricks and scrolls found in Israel prove the existence of the Judeo-Christian god, then, by that same standard, wouldn't the treasures found in Tut's tomb prove that he was a god as well?

GO DEEPER . . .

- Fagan, Brian. *In the Beginning: An Introduction to Archaeology*. New York: Prentice Hall, 2008.
- Fagan, Brian. *People of the Earth: An Introduction to World Pre-History*. New York: Prentice Hall, 2009.
- Perring, Stefania, and Dominic Perring. *Then and Now: The Wonders of the Ancient World Brought to Life in Vivid See-Through Reproductions*. New York: MacMillan, 1991.

Chapter 38

"HOLY RELICS POSSESS SUPERNATURAL POWERS."

An old joke tells of a pilgrim's response to seeing a second head of John the Baptist. When he asked how this could be, he was told, "The other one was from when he was a boy."

—Joe Nickell, *Relics of the Christ*

Having successfully weaved my way through a maze of tacky gift shops, heavily armed soldiers, and parading pilgrims in Jerusalem's Old City, I finally arrived at the entrance of the Church of the Holy Sepulchre. This is the specific spot where many Christians claim Jesus was crucified. A confusing jumble of a church owned and operated by rival denominations has grown up around it over the centuries. It's fascinating and filled with intrigue, to say the least. Barely a couple of feet inside, I noticed odd behavior. A woman kneeling before a slab of brown stone sways rhythmically from side to side. Her eyes widen and seem to glow intensely. Then she shuts them tight as her hands slowly move across the brown stone as if it is precious and powerful. I learn that it is just that, thanks to a grinning pilgrim who explains to me that this is the Stone of Unction. Also called the Stone of Anointing, it's supposed to be the rock upon which the body of Jesus was laid after his execution and prepared for burial. Never mind that this couldn't possibly be true because this stone was placed here in the 1800s; it's still an important relic nonetheless. The woman before me is obviously convinced that it contains divine power, and she wants some of it. She continues to methodically sweep her hands across the surface. Then she rubs her palms on her arms, neck, and face. Amazingly, the woman is "bathing" in the stone, apparently trying to apply its magic to her skin. I'm transfixed by the scene. I don't believe there is anything supernatural going on, of course, but I do recognize a bizarre beauty of some kind within this woman's joy and passion. Magic or not, she seems close to being overwhelmed with emotion. Such is the power of religious relics.

Further on in the Church of the Holy Sepulchre, I discover a man creating sacred relics right before my eyes. With a handful of rosary beads, he crawls up to the precise rock upon which Jesus died on the cross. He then rubs the rosaries against the surface, one at a time. With the supposed magical transference complete, the man crawls back out from the altar and bags his beads. All around me, people pray, weep, and smile. Many of them touch and kiss various adornments and even the walls inside the Church of the Holy Sepulchre. Clearly there is a common desire to make contact with something, anything, Jesus or other great holy figures have touched. The believers seem to seek a physical chain of connection, a step up perhaps from the invisible and formless faith that fails to fully satisfy.

Christians are not alone in their attraction to relics. During a visit to the Umayyad Mosque in Damascus, Syria, a guide told me that the head of John the Baptist was safely inside a shrine within the compound. I couldn't see it, but he was certain that it's there. Interestingly, a few other places around the world claim to have the same head. A good portion of the Prophet Mohammed's beard and a single tooth are on display at the Topkapi Palace Museum in Istanbul, Turkey. And on a bitterly cold February day in 2011, thousands of Muslims turned out in Srinagar, Kashmir, in the hopes of catching a glimpse of a single hair from Mohammed's beard that was briefly displayed outside the shrine that houses it.[1] A temple in Kandy, Sri Lanka, claims to have a tooth from the Buddha. They say it is the only remnant of his body in existence. But I visited Lingguang Temple in China, where I was told by locals that a Buddha tooth is kept as well. I chatted with some sort of a holy man while photographing the beautiful Bodnath Stupa (Buddhist temple) in Kathmandu, Nepal. The man told me that one of the Buddha's leg bones is inside the temple and it has "great power."

After encountering and hearing about so many of them around the world, I developed an interest in religious relics and the power they can hold over some believers. I'm amazed by how wide open the field of relic veneration is. A relic can be virtually anything tangible that is connected with a person or supernatural being considered important to a particular religion. The key word is "anything." Relics range from a nail used to crucify Jesus to the toenail clipping of a saint. And, much like the sports memorabilia market is today, the religious relics market has been flooded with fakes. But few seem to care about authenticity when it comes to these objects. The truth, it seems, is less important than the claim.

Religious relics often are seen as much more than mere souvenirs or trophies. Relic believers claim that they have special powers, such as the ability to heal the sick or bring good fortune. They can also be profitable, which explains how different locations end up with the head of the same person or how several femur bones of the same saint can be housed in various churches. The desire to possess things that have some direct association with beloved figures is alive and well today and extends beyond religion. Collectors have purchased used gum chewed by Britney Spears, Cher's bra, and Ty Cobb's dentures. While I would like to think I'm above such silliness, in truth I am not. Over the years I've been fortunate enough to interview many historic figures, great scientists, and famous athletes. I often came away from those encounters with autographed photos. I've accumulated a large collection that is probably worth a lot of money. But I can't imagine ever selling them. Even though I recognize that they are just photos with names scribbled on them, they have had some form of "magic" breathed into them by people I admire and are therefore special to me.

Although the practice predates Christianity, relics connected to Jesus and the Gospels became popular in Europe during the fourth century. According to Joe Nickell, author of the fascinating book *Relics of the Christ*, remains and objects linked to martyrs and saints quickly became a very important and profitable activity. The lust for these objects drove people to open up the tombs of martyrs in order to mine them for relics. Nickell explains that relics came to be viewed as the necessary link between tombs and altars. "By 767, the cult of saints had become entrenched," Nickell writes, "and the Council of Nicaea declared that all church altars must contain an altar stone that held a saint's relics. To this day, the Catholic Church's Code of Canon Law defines an altar as a 'tomb containing the relics of a saint.' The practice of placing a relic in each church altar continued until 1969."[2]

The catalog of relics that earned some degree of credibility and made their way into churches is astonishing. There seems to have been no limit to the resourcefulness, or imagination, of the people who traded in them. Here is a sampling:

Pieces of the Ten Commandments stone tablets; a flask of the Virgin Mary's breast milk; a piece of stone that her breast milk dripped on; Mary's hair; the loincloth of Jesus; Jesus' baby clothes; Moses's staff; the skulls of the "three wise men"; Jesus' crib; Jesus' foreskin (at least six of them!); hay from the manger Jesus was born in; Jesus' baby teeth; Jesus' umbilical cord; feathers from the angel Gabriel's wings; the Shroud of Turin and many more cloths with the same claim; Mary's burial shroud; a chair an apparition of the Virgin Mary once sat in; a tear Jesus shed at the tomb of Lazarus (how was it collected?); the tail of the ass he rode into Jerusalem on; the basin Jesus used to wash the feet of his disciples; thorns from the "crown of thorns" Jesus was forced to wear; the spear a Roman soldier stabbed Jesus with; chains used to imprison Saint Peter; vials of Saint Peter's tears and some of his toenail clippings; the blood of numerous saints; nails used to crucify Jesus; fragments of the "true cross" Jesus was crucified on; and one of doubting Thomas's fingers.[3]

The collecting, selling, and displaying of religious relics became so rampant in the fifth century that even Saint Augustine became fed up with it, leading him to write about his disgust for "hypocrites in the garb of monks for hawking about of the limbs of martyrs, if indeed [they are] of martyrs."[4]

"As investigation after investigation has shown, not a single, reliably authenticated relic of Jesus exists," declares relic researcher Nickell. "The profoundness of this lack is matched by the astonishing number of relics attributed to him."[5]

ONE MAN'S RELIGIOUS RELIC IS ANOTHER MAN'S MOVIE PROP

Once again, I find myself feeling sympathetic toward those who are drawn in by things that don't seem likely to be real or true. A minimal amount of skepticism and critical thinking should be able to deflate claims of relics possessing powers or even being authentic in the first place. But maybe most relic believers just don't want to know. They like believing, so they do.

It came to me while I was looking at the display case containing the leather pants of rock god Jim Morrison. I was waiting for my lunch at the Hard Rock Café on Hollywood Boulevard, of all places, when my thoughts wandered to medieval peasants, modern-day believers, and religious relics. I thought to myself, who cares? So long as one is not wasting needed money or putting health at risk by trusting in some beard hair or toe bone, then have fun. I love the *Terminator* films and would love to keep an authentic prop used in one of the movies in my house, say a T-800 skull or maybe even a T-600 endoskeleton. It's childish, of course, but it would be great to have a tangible piece of something bigger than life and "magical" close by—not much different from the lure of religious relics, really. The key difference, however, is that I wouldn't pay more than I could afford for a relic or be sloppy enough in my thinking to imagine that it could give me luck or cure me when ill.

GO DEEPER . . .

- McCrone, Walter C. *Judgment Day for the Shroud of Turin*. Amherst, NY: Prometheus Books, 1999.
- Nickell, Joe. *Relics of the Christ*. Lexington: University Press of Kentucky, 2007.

Chapter 39

"A TV PREACHER NEEDS MY MONEY."

> **God cannot bless you until you put something into His hand.**
>
> —Rev. Benny Hinn

> **Pray over your seed and expect an abundant harvest. Know that your seed is being sown into good ground and will be used to teach the Word with simplicity and understanding throughout the world. God bless you for being obedient!**
>
> —Rev. Creflo A. Dollar, Dollar Ministries

Back in the 1980s, after Jim and Tammy Faye Bakker lost their $129 million ministry because of financial fraud and a sexual scandal, I assumed the end of the televangelist phenomenon was near.[1] Then I guessed its final days were here when Rev. Oral Roberts announced that God would kill him if people didn't donate $8 million to him by the end of the month.[2] And I was certain it was doomed when staunch moral crusader Jimmy Swaggart was caught meeting prostitutes in hotel rooms.[3]

Much to my surprise, however, the era of big-money televangelists never ended. No matter what happens, they keep begging and people keep giving. If one TV preacher should go down for good, several more are waiting to fill his $5,000 shoes. Nothing seems capable of stopping this gravy train, not exposure of their hypocrisy, sexual scandals, fraudulent miracle claims, enormous incomes, not even decadent lifestyles that would make King Louis XIV blush. It really is an amazing phenomenon. Men with a little charisma and a lot of nerve have created a unique global industry that weaves together television, religion, capitalism, charity, show biz, and astonishing greed. Many of these preachers may seem like harmless buffoons but the good ones are not dumb. They skillfully exploit thousands of years of religious tradition to rake in tens of millions of dollars, much of it from people who probably need to hold onto their money more than most. I have a sad memory of one of the older cleaning ladies who worked in my university dorm telling me that I should give money to Jimmy Swaggart like she does. When I gently suggested to the sweet woman, whom I grew to care about, that a TV star who lives in a mansion and flies around in his own jet probably can do without donations from a custodial worker and a struggling college student, she scoffed and explained that Swaggart only "holds the money for God."

GOD'S CAN'T-MISS INVESTMENT PLAN

I attended a Benny Hinn "Miracle Crusade" event and paid close attention during donation time. Hinn hammered the live audience of approximately six thousand people on the need to give him money—and lots of it. He warned them that "God knows" precisely how much they can afford to give—a brilliant tactic to put the cheapskates on notice. He also promised the crowd that paying him now would pay off for them later. "The more you sow, the more you will reap," Hinn said.

This is known as "prosperity theology" or the "prosperity Gospel," and it really does pay off handsomely, at least for the televangelists if no one else. Hinn said his ministry needed the money in order to do "God's work" all over the world. "When you give, it doesn't go to me,"

Hinn said. "All of it goes to my ministry. I do this for free. Check me out. Go ahead, check me out. All the money goes to the work of God, to get the Gospel out."

Hinn keeps his tax-free personal and ministry finances private so who knows how to sort out his minister's salary, book royalties, and donation money? What is clear based on his visible lifestyle, however, is that the reverend does pretty well for himself.

There is no denying that Hinn is a master on stage. His facial expressions, charming banter, faux humility, and rhythmic speech in dramatic moments play to the crowd perfectly. Multiple television cameras capture every step and every word. He instructs the believers to hold their donations up high while waiting for the ushers to collect them. This is another brilliant move because it publicly exposes anyone who is not prepared to give money and probably pressures them to reconsider. I think I was the only one in the crowd with my arms down.

"If you have problems, if you want to get out of debt, then give tonight. God said: 'Give and it shall be given unto you.' God cannot bless you until you put something into His hand. Don't just give," he added, "*sow*! Sow, so that you can reap a mighty harvest."

A small army of donation collectors fanned out into the crowd. No mere collection "plates" for this event, however. The men carried what appeared to be large plastic garbage cans—and the believers filled them up in short time. Some collectors were equipped with credit card scanners. Many people prayed aloud as music played during the procedure. The men loaded the treasure into a caravan of vans and SUVs that promptly sped away to the nearby airport where, I was told by a police officer friend, Hinn had a private jet waiting. The collection procedure was efficient and precise, more like a well-executed military operation than the passing-of-the-plate routine one sees in small churches.

HAVE THEY NO SHAME?

Benny Hinn won't reveal what his ministry takes in from donations, but it is estimated to be as high as $100 million per year. He told ABC News that his personal salary is "over half a million."[4] Details of his salary and personal wealth have never been publicly confirmed by a credible independent source, however. For many churches it is common for "the ministry" to pay for a pastor's house, car, and other living expenses in addition to a salary, so Hinn may be able to simply bank that big income. What is known is that Hinn dresses, drives, and flies like a very wealthy man. In 1997, he was a guest on the *Larry King Show* and a caller asked him why he takes so much money when Jesus never took any. "Jesus didn't have a TV show to run," Hinn answered.

In 2011, I saw a small-time preacher on TV who may have stooped lower than any of his colleagues ever have—and that's saying a lot: "You must give to the Lord," the preacher declared. "Give as much as you can, no matter what your situation is. The worse off you are, the more important it is for you to give. He will reward you. Even if you are homeless, you have to give something."

Even if you are homeless? How do these preachers who squeeze money out of poor people sleep at night? Yes, I know: in very nice beds with silk sheets that cost more than my car. Most amazing of all is how so many of these moneymakers are caught red-handed in money and sex scandals only to resume their careers as soon as the smoke clears. One could not be busted and hung out to dry much better than Rev. Peter Popoff was, for example. In the 1980s, prominent magician and skeptic James Randi exposed him for using a wireless device to receive personal information about people in the audience staff members had previously interviewed so that he could impress them by claiming God told him about them. Popoff confessed to the scam and soon after filed bankruptcy. So, where is he today? Cowering in shame in some cave somewhere? Working the night shift at a Waffle House on some lonely highway? No, he's back on television, of course. Popoff is "healing" people and begging for money because God always needs a little more cash to get by. His ministry reportedly brings in more than $20 million per year.[5] Currently his infomercial-style programs air in early morning slots on BET (Black Entertainment Television).

Randi makes his disgust for these people obvious. "A thing like astrology is just a slow drain on the economy," he said. "It makes the astrologers quietly and rather unobtrusively rich over a long period of time. The medical quackery and the faith-healing racket that is out there now make a very large amount of money. People who do this, like Benny Hinn, are multimillionaires many, many times over. They just have so much money pouring in every minute of every day that I'm sure they can't even keep track of it."[6]

My humble advice to people who think televangelists are God's appointed treasurers is this: please don't give your hard-earned money to anyone who is wearing a pair of shoes that cost more than your entire wardrobe. These preachers speak of sacrifice and digging deep to give money, but where is their financial sacrifice? Why aren't they "giving all they can," living in tiny houses, and making do with just one car?

If you are deeply religious and have an irresistible urge or sense of obligation to "spread the Gospel to all corners of the Earth," then why not do it yourself? Cut out these millionaire middlemen and keep your money. Why should your money support their luxurious lifestyles? And ignore their whining about needing help to pay the bills for satellite television ministries. This is the computer age; television is so twentieth century. If you absolutely must, then use the Internet to push your religious beliefs on others all by yourself. Why finance TV preachers? Why rely on them? *You* can preach too. These days anyone can convert people in faraway places via Facebook, MySpace, e-mail, and chat rooms. You don't need TV preachers.

Here's an even better idea: take whatever money you were going to send to a television preacher so that he can buy his tenth Rolex watch or whatever and send it to UNICEF instead. UNICEF directly saves lives and improves the living conditions of children and mothers in the poorest countries every day. How can it be wrong to give them your few dollars over some guy who slurps caviar in a Gulfstream jet? If the god you worship has a problem with choosing UNICEF over a TV preacher, then maybe it's time to shop around for another god.

GO DEEPER . . .

- Martz, Larry. *Ministry of Greed: The Inside Story of the Televangelists and Their Holy Wars.* Frederick, MD: Grove PR, 1988.
- Randi, James. *The Faith Healers.* Amherst, NY: Prometheus Books, 1989.
- Wilson, Bill. *How to Get Rich as a Televangelist or Faith Healer.* Boulder, CO: Paladin Press, 2008.

BIZARRE BEINGS

Chapter 40

"GHOSTS ARE REAL AND THEY LIVE IN HAUNTED HOUSES."

Human history of the past millennium is a halting march away from superstition toward knowledge. We are still far from our goal, but our species is young.

—Hank Davis, *Caveman Logic*

Science is all well and good, but when you get firsthand information from a ghost, it doesn't get much better than that.

—James Van Praagh, *Ghosts among Us*

"Ghosts are real. I know it because of something that happened to me," he said. "I know what I experienced." My friend of many years was intelligent, sober, and—as far as I could tell—completely sincere. I believe that he really did experience "something" but wasn't sure if it justified jumping to the conclusion that ghosts are real. I pushed for more details, but the apparition apparently was a deceased parent or someone close and he was uncomfortable talking about it further. That's understandable, so I backed off. That conversation always comes to mind when I discuss ghosts. The subject can seem silly with the goofy ghost-hunter television shows and the complete absence of good evidence, but I try to tread lightly and be respectful when dealing with true believers in case they connect ghost belief to people they knew. I had another similar encounter with a woman who believed in ghosts, reincarnation, heaven, and seemingly every other afterlife claim. She explained that she *had* to believe death was not the end because it was the only way she would ever see her deceased father again. Again, I eased off and did more listening than lecturing. Being respectful of people who have deep feelings tied up with ghost belief does not mean, however, that we should leave this claim unchallenged.

The number of people who believe in ghosts is significant. In the United States, 42 percent of the adult population think that ghosts are real.[1] In Great Britain, 40 percent of adults believe that houses can be haunted by ghosts.[2] One study found that women (20 percent) are more likely to say they had an experience with a ghost than men (16 percent). And a person with a college degree is significantly less likely to report such an encounter than someone with a high school education or less (13 percent versus 21 percent).[3] These statistics could be much higher, of course, if the definition of *ghost* was not left for survey subjects to determine for themselves. For example, billions of Christians, Muslims, and Hindus believe that something called the soul leaves the body at death. Because definitions are so loose, *soul* could be thought of as the same thing as a *ghost*. If so, that would raise the number of ghost believers dramatically. Further confusing the true number of ghosts believers is the reluctance I have observed by some religious people to admit that they believe in ghosts when they do. I have also seen this with astrology. Some religious people feel that ghosts or astrology are "of the devil" and are dangerous forces that one should not associate with. So in confusing "believe in" with "follow" or "participate in," they claim not to believe when they do. In any case, my travels and experiences with people around the world convinced me that ghost belief is extremely common. I would estimate the ratio of believers to be higher than 90 percent in some societies that I have visited. Globally, I would guess ghost belief to be at least 80 percent. I found belief to be distributed widely within societies as well. In the

Caribbean, for example, it is common for highly educated professionals such as doctors, lawyers, and accountants to believe in ghosts, or "duppies" as they are called there. The important question is, why? Why do so many people still believe in ghosts?

Anthropologists suggest the possibility that the dreams of prehistoric and ancient peoples may have been the catalyst for ghost belief as well as more complex notions of an afterlife, gods, and religion. We could ever know this for sure, but it is easy to imagine how big an impression a vivid dream about a dead friend or family member might make on a prehistoric person who did not know that dreams are a natural brain process. What if you were part of a small hunting and gathering clan twenty thousand years ago? Your lover dies but a week later "visits" you in a dream in which you make love to each other. Imagine the powerful impact such an experience likely would have on you. You easily might conclude that she is still alive, in some form, somewhere. Ghost belief might have begun in just that way. It would then be reinforced culturally by teaching it as fact to children, generation after generation.

Perhaps the most common flaw with claims about seeing, hearing, or sensing ghosts is that an encounter with something that can't be identified means it is just that—*unidentified*. This is the same mistake that mars so many UFO claims. It is not sensible to see something weird in the sky and conclude that it must be an alien spaceship just because you are not able to identify it. It's no different with ghost encounters. If you see, hear, or sense something strange in your bedroom or in a foggy graveyard at midnight, it's not justifiable to jump to the extraordinary conclusion that it must be a supernatural being. Maybe it's something unusual but still natural. Maybe there is an elusive but simple explanation, like a raccoon passing by in the weeds or reflected light off of a bottle. Now, if you see Blackbeard the pirate or Lizzie Borden hovering before you clearly and in great detail, that's different. However, there are possible natural explanations for these encounters too.

VISIONS OF THINGS NOT REAL

I occasionally do things that seem normal to me but not to others. I get away with it because I have an understanding wife and my kids don't know any better. A few years ago, for example, I went on a vision quest. It seemed like a fun thing to do after I had read something about this traditional practice common to many tribal cultures. I decided to skip the customary starvation, dehydration, and hallucinogenic drug consumption that are central to many vision quests. Apart from those details, however, it would be authentic. I selected a small uninhabited island in the Caribbean to live on for four days by myself. It was a wonderful experience, a beautiful break from the noise and clutter of civilization. There was nothing strange about it—at least until my spirit guide visited me one night.

I awoke confused. I felt alert and jittery but didn't understand why. Suddenly something tugged at my foot, maybe bit or clawed it. I looked down and aimed my flashlight. Staring into my beady little eyes were the large bulging eyes of a rat. It was massive, very well fed. But on what? I wondered. Coconuts? Human toes? I retracted my feet immediately and checked for wounds. I was fine. But the rat lingered. I shined the light on it and time slowed. It wasn't aggressive or afraid. Neither was I. Maybe it was my imagination, but the little beast had charisma. I liked it. This was no confrontation between a master of the Earth and a lowly human on a vision quest. It was just two mammals sharing a special moment. Eventually it exited out the hole it made and walked away. I went back to sleep—with my shoes on.

In the morning it occurred to me this rat encounter was the big moment, the peak experience of my vision quest. The primary purpose of a vision quest, according to some cultures, is to facilitate a visit from your "spirit guide," usually in the form of an animal who would pass on some wisdom or give direction and suggest a purpose for your life. I have no doubt that if I had been half-starved or on some potent drugs the rat would have had even more meaning to me. It probably would have talked to me and told me to write my mother more often or something like that. At the very least, if I had been a devout believer in ghosts and spirits, it probably would have felt "obvious" to me that it was much more than a mere rat. That experience stands out to me as precisely the sort of weird event that someone could easily have misinterpreted and injected with unwarranted supernatural meaning.

When thinking about claims of ghost encounters it makes sense to consider the massive distortions of reality that a healthy human mind is capable of producing. Most of what we see and remember are not perfect current images of reality or perfect replays of what actually happened. Our minds *construct* what we see and remember. We "see" only a relatively small percentage of what our eyes look at. The rest is *made up* or *assumed* by the mind. Our memories are edited and summarized—without our conscious consent. All this is not as crazy as it may seem because it allows us to function more efficiently in the world. Imagine if we had to constantly concentrate on each and every detail to determine if the ground is solid in front of us while we are walking on a sidewalk, or carefully look to see if a pride of lions is at a bus stop as we walk by, for example. Our brains make many assumptions so we can get things done. Sometimes this can get us in trouble—if it turns out there is a big hole up ahead or lions really are waiting to attack us—but usually we're OK. "If you tried to analyze every little thing that's happening to you, you wouldn't make it across the room when you get out of bed in the morning," says psychologist and former magician Richard Wiseman. "Optical illusions reflect our sophistication, not our idiocy. Without them we wouldn't be where we are today because we wouldn't have made so many correct assumptions. You're an effective information processor for making those assumptions."[4]

We also have to remember that hallucinations are not limited to seeing things that are not there. They also include hearing nonexistent sounds and even "feeling" physical contact with something that isn't really there to touch you. We can also easily misinterpret real input, say a gust of wind and a weird shadow for a ghost. I recently walked up an escalator that was out of service and stationary. After a few steps my mind-body coordination faltered and I had to grab the hand rail for stability. My mind assumed that the steps were moving as usual and kept instructing my legs and feet to react as if they were. It vanished when I became aware of it but then returned when I let down my guard again, just seconds later.

As bizarre as various illusions and confusions over what's really happening and not happening may seem, they are common, according to scientists.[5] In fact, the *inability* to experience illusions may be a symptom of mental illness.[6] Knowing this, it is clear that ghost encounters should be *expected* for a species with minds that operate the way ours do. This is why I'm patient and understanding with people who claim to have made contact with a ghost. Most likely they only experienced a normal human reaction to unusual thoughts, a waking dream, or real environmental conditions that were misinterpreted. It could happen to almost any of us given the right circumstances.

HAUNTED HOUSES

"I won't go in that bedroom ever again," the friendly middle-aged woman says to me. "I'll never forget it. I was terrified."

She is referring to an incident she claims occurred in the Whaley House in Old Town, San Diego. The historic building is a well-maintained "Greek revival mansion" built in 1857 by a successful businessman named Thomas Whaley. Today it's a historic site and minor tourist attraction. It's promoted as "America's most haunted house," although I question how such a thing could be measured and ranked.

"All the sudden I was freezing cold," the woman continues. "I mean, I was absolutely freezing cold. I was so scared; I remember closing my eyes and then I knew he was there. I knew there was a man in the room with me."

The man, she explains, was the ghost of Thomas Whaley, dead since 1890. She knew it was him, she says, because she could "sense his presence."

"I was still freezing cold and then I just knew he was going to pass right through me. I was so scared that I couldn't move, even though I knew something was about to happen. And then something went through my hair. I felt it touch my hair. You would never believe how it felt. It was pure death.

"There is definitely a presence in that house," she added. "The family is there. I think they are attached to the house. You feel a presence. They don't seem to ever leave."

After hearing a story like that, I couldn't resist visiting the Whaley House myself. And when I did, I was surprised to discover that even a devout skeptic like me can "feel a presence."

From the sidewalk, the Whaley House looks like a well-maintained nineteenth-century house, interesting enough for historic reasons. The red brick and clean, white wood trimming are beautiful; and thick, square porch columns give it the look of classical power. From the outside, there is nothing to suggest that there is anything scary or evil inside. None of the traditional haunted house stereotypes are visible. No cobwebs in the windows, no broken-down shutters, and no creepy tombstones in the front yard. As far as old buildings go, it could not look less haunted. Inside, however, is a different story.

The rooms inside the Whaley House are very well maintained and filled with artifacts appropriate to the time period. A large downstairs room once served as San Diego's courthouse. A portion of the house was once a general store, as well. As I continue to explore the first floor, I listen in on two women talking.

"I'm sure I heard something," said one woman. "It was a like a voice. Did you hear it?"

"No," the other replied. "What did it say?"

"I don't know. Do you think there is a vortex here?"

For some reason I don't feel the need to hear any more of their conversation and move on.

Upstairs now, I peer into one of the bedrooms. It's fascinating, filled with detail. I see a hairbrush, books, a quill pen, and so on. The current operators of the Whaley House have done a very good job of capturing and presenting the look and feel of how mid-nineteenth-century life in California was for a wealthy businessman and his family. Further along, I found a child's bedroom that was even more interesting. A painting of a young child hangs on the wall. A miniature china tea set on the dresser is supposed to be authentic to the house, once played with by the Whaley children. The room stirs my imagination and emotions. I can easily imagine the children playing on the floor. An old doll is perched on a small rocking chair. Cute at first glance, she becomes increasingly creepy the longer I hang around. Her little black doll eyes seem to stare directly at me and suddenly I realize the makings of a supernatural horror film are all around me.

Alone on the second floor of the house, I too "sense a presence." It's not a ghost (or the doll) and I'm not scared, but I have allowed my mind to run free and it's pulled me back several decades to when the house was inhabited by an 1860s family. I don't literally see or hear them, of course, but in my imagination I do. It doesn't always happen, but I have experienced this flash of heightened imagination while visiting ancient and historic sites around the world. It's a great experience when the distant past comes alive in your mind. I highly recommend it. The more you research the particular event, place, and people, the better the high. I've had special "mental reenactment moments" in places as diverse as the Palace de Versailles in France and a jungle in Papua New Guinea where Japanese and Allied soldiers butchered one another during World War II.

Reflecting on my imaginary animation of long-dead people, I wonder how much more powerful some of such moments might have felt if I had believed in ghosts. Had I been predisposed to expect a visit from a ghost, it's not difficult to imagine my emotions getting the best of me and being swept away during my brief "moment" in the Whaley House. For example, what if I believed in ghosts, experienced that brief imaginative flashback while looking in the child's room—and then, at that very moment, something unusual happened? What if a gust of wind blew a window open? What if the old building's walls or a floorboard creaked somewhere near me? What if a flash of reflected sunlight from a passing car or plane outside briefly illuminated the dark hallway? None of these things should be interpreted as ghost encounters or paranormal events. But a believer who is in the moment might feel differently.

Some of the rooms in Whaley House are blocked off by clear plastic walls in the doorways, presumably to protect the rooms from being disturbed by visitors. I immediately recognize the potential for taking "ghost photographs." Anytime you attempt to shoot photos through, or even near, a glass or plastic surface there is the possibility of a light flare that may be interpreted by some as a ghostly image. Using a flash in these situations increases the chances. It's completely natural, nothing more than light being reflected in unusual ways. I can't resist trying my luck, and in less than five minutes I have a few very good "ghost photos." Some show spherical "orbs," as ghost hunters like to call them. One photo shows a long white smoky blob "hovering" high above Thomas Whaley's bed. I wonder how many visitors per year take photos of this bedroom, only to check them later and shriek with delight/horror upon seeing the "ghost" they photographed. Apparently many do, as I discover when leaving the house. Near the exit is a photo album filled with photographs taken by visitors to the house. Photo after photo show reflected light "orbs."

Interpretation of sights and sounds is, of course, the key to the haunted houses phenomenon. Most incidents probably are mysterious noises or brief sightings of unidentified objects. None of these should be considered evidence of a ghost haunting because "unidentified" means "unidentified." If I'm spending the night in a cheap hotel in the middle of nowhere and hear a weird noise outside my door that I am unable to identify, I can't sensibly conclude that a ghost is making the sound. It could just as easily be aliens coming to abduct me, or a serial killer coming to carve me up, or maybe it's *Playboy*'s Playmate of the Year sneaking in to place a mint on my pillow. The point is, if I don't know what I have seen or heard, I shouldn't pretend to know that it's a ghost.

Nonetheless, many people do translate the unknown into the known by attributing strange noises to ghosts. John, a longtime and close friend of mine who lives in England, recalls a terrifying night he spent on vacation with his family during his childhood. He says he heard "the most unusual noises, with doors opening for no reason, and the sound of footsteps outside when there was no one there."

"I was so wet through with sweat caused by the anxiety that my parents were convinced I had wet the bed," he recalled. "It was a bit like after a normal Saturday night out on the beer these days."

Belief in haunted houses is one of the most common nonreligious paranormal beliefs in both the United States and the United Kingdom. A Gallup poll found that 40 percent of Brits and 37 percent of Americans believe that houses can be haunted. In Canada, 28 percent believe it.[7] According to that same Gallup study, belief in haunted houses by Americans is second only to belief in extrasensory perception, or ESP (41 percent) among nonreligious paranormal beliefs. For haunted houses to beat out belief in mediums (people who talk to the dead and claim to hear back from them), astrology, reincarnation, and witches indicates that something about scary and unusual things inside of houses is very compelling to many people. Perhaps the reason that it resonates with so many people is that it hits so close to home—after all, it's potentially *in* the home. For many Americans, in fact, belief in haunted houses is personal, as one-fifth say they have either lived in or visited one.[8]

This belief is found in people of all income and education levels. Consider Tia, a Princeton graduate with a very impressive résumé and job to match. She once lived in a house she believes was haunted. "The house was detached, at the end of a cul-de-sac," she explained. "There was no other house on one side, and on the other side, our neighbors were many feet away. Still, we could often hear people walking up and down the steps in our home. My mom always explained it [by] saying we could hear our neighbors going up and down their stairs. As a kid, I accepted that explanation. But now, looking back, I realize that was impossible! I got so freaked out one day, I remember, hearing the sounds of walking around upstairs when I was the only one home that I ran out of the house and sat on the porch until someone else got home. It was definitely a freaky experience."

Longtime paranormal investigator Joe Nickell has poked about in far more haunted houses than most. To date he hasn't found any ghosts, however. "Once the idea that a place is haunted takes root, almost any unknown noise, mechanical glitch, or other odd occurrence can become added 'evidence' of ghostly shenanigans, at least to susceptible people," Nickell writes in his book *The Mystery Chronicles*. "They often cite unexplain*able* phenomena, but they really mean unexplain*ed*, a condition that does not in any way imply or necessitate the supernatural. To suggest that it does is to engage in a logical fallacy called arguing from ignorance—the stock in trade of credulous paranormalists and outright mystery-mongering writers."[9]

There are many down-to-earth, natural explanations available for what may cause mysterious and creepy noises in a house. I know from experience, however, that the simpler explanations are not always satisfying or reassuring in every case. During my college days I lived in a large, old two-story house for about six months or so. Many times I was home alone studying upstairs and would hear creaking sounds as if someone were walking around on the wooden floor downstairs or in the hallway. But when I went go to see who was there, I would find the house empty. It happened so often that I learned to ignore it. My higher functioning brain reasoned that it was nothing, just the normal sounds old, drafty houses make. My reptilian brain, however, wasn't convinced. Almost every time I heard another creak or moan, I glanced up from my studies to check the doorway, as if to prepare for a showdown with the Headless Horseman or whomever. I knew there was nothing there, but some reflex within made me react anyway. Sometimes I felt uneasy, if not a twinge of fear. Whenever I encounter people who have had frightening experiences in "haunted" houses, I remember that I too was once scared by a few unidentified silly sounds. It keeps me humble.

GO DEEPER . . .

- Nickell, Joe. *The Mystery Chronicles: More Real-Life X-Files*. Lexington: University Press of Kentucky, 2004.

Chapter 41

"BIGFOOT LIVES AND CRYPTOZOOLOGY IS REAL SCIENCE!"

The sleep of reason produces monsters.

—Title of an etching by Francisco Goya

The story we're being asked to believe is that thousands of giant, hairy, mysterious creatures are constantly eluding capture and discovery and have for a century or more. At some point, a Bigfoot's luck must run out: one out of the thousands must wander onto a freeway and get killed by a car, or get shot by a hunter, or die of natural causes and be discovered by a hiker. Each passing week and month and year and decade that go by without definite proof of the existence of Bigfoot make its existence less and less likely.

—Benjamin Radford,
Committee for Skeptical Inquiry

The pilot closes the hatch and suddenly the research submersible feels small, very small. After bobbing about on the surface for a while, we finally submerge and begin the journey. It's a three-person vehicle and only the two of us are aboard but, the space available seems to shrink with every minute of our descent. Good thing I'm not claustrophobic. I'm not apprehensive about making this one-thousand-foot plunge into the Caribbean Sea, but I can't help wondering what happens if the power fails or if a colossal squid tries to eat us. The view through the curved observation window in front of me is weird to say the least. I can't see anything but dark nothingness and now the surface is out of sight. After falling for several minutes, the pilot turns on powerful external lights. A surprising abundance of tiny but visible life is illuminated as we drop deeper. So much, in fact, that it seems like the water itself is alive. I'm reminded of the late Jacques Cousteau, one of my childhood heroes. He described seawater as the "broth of life." Now I'm seeing precisely what he meant. I'm an experienced diver with numerous 130-foot dives, some of them at night, but this is different. While not scary, it is thrilling—definitely no typical Caribbean afternoon for me. I'm excited to think about all the tons of water over my head and all the mysteries out there in the deep somewhere. I wish we were going thirty-five thousand feet down instead of just one thousand.

Below seven hundred feet, the environment becomes significantly different from the coral ecosystems I'm used to diving in. The bottom comes into view. It's relatively barren and gray, like a submerged moonscape. This may not be beauty in the traditional sense, but it's impressive in its own way. Apart from the "snow flurries" of plankton and diatoms, there is no life that I can see.

The pilot kills the lights and slowly moves forward to another position. When he turns them back on, three magnificent creatures are illuminated right in front of the submersible. He draws in close. They are crinoids—big ones. Imagine a giant underwater sunflower that is actually an animal, that's what they look like. They are gorgeous. They take on a golden glow in our lights. Swaying back and forth in a gentle current, they feed by filtering out organisms that swim or drift by. Crinoids flourished in the oceans hundreds of millions of years ago and were once thought to be extinct. But here they are, alive and well, right in front of my face. I'm thrilled to see these creature up close. It's like looking back in time five hundred million years.

There was a time, back in the nineteenth century, when very smart people were convinced that the deep was a lifeless zone, nothing of interest. We know better than that now, of course, but we still understand relatively little as thousands of seamounts, canyons, and vast abyssal plains remain very much unexplored. Even my own experience interviewing scientists and explorers is telling: I've managed to talk with *nine* men who have been to the Moon but only *one*, Don Walsh, who has been to the bottom of our deepest ocean. Isn't it odd that twelve people have walked on the Moon and more than five hundred have traveled to space, while only two men to date have visited the deepest point in the ocean?

Whenever Bigfoot and other "cryptids" are discussed, I can't help recalling those beautiful crinoids I saw in the Caribbean Sea and the excitement I felt while thinking about the life waiting in the deep to surprise us. Strange, hard-to-find animals do turn out to be real. They exist. It is not necessarily crazy to believe in weird creatures and monsters because we have in fact found many of them over the years. While I don't believe it's likely that plesiosaurs evaded extinction and currently are roaming the depths of a Scottish lake or that giant primates are running around in the Pacific Northwest, I am certain that many fascinating creatures are still out there somewhere awaiting discovery. This is not even controversial. No sensible scientist believes everything has been found. There is so much unknown life right now, in fact, that the challenge of finding and naming these unknown species is immense. A 2011 article in the journal *Trends in Ecology and Evolution* estimated that it would cost US$263 billion to scientifically discover and describe the entire animal kingdom. If such a feat is even possible, I suspect that it would cost much more than that. Some habitats are incredibly challenging, perhaps impossible, to adequately study with current technology. Another reason for the high price tag is because there is so much work left to be done. The report states that only some 1.4 million of an estimated 6.8 million species are currently known to science.[1] A 2011 life census estimated that 86 percent of all land species and 91 percent of marine species have yet to be discovered and catalogued by scientists.[2] But even those figures are misleading about how little we know because they do not include viruses and bacteria. Considering all we have achieved technologically and how much we have learned about the universe, our ignorance about life right here on our own planet is stunning.

I confess to having a soft spot in my heart for cryptozoology, the "science" of hidden or undiscovered animals. The problem with cryptozoology, however, is that it's not just about "unknown animals." Cryptozoology centers on claims of large animals that are wrapped up in myths and legend or are at least spectacular animals, such as dragons or dinosaurs, that science does not validate the existence of or has declared to be extinct long ago. Is it possible that some of these sorts of creatures are real? Yes! Of course it's possible that big flesh-and-blood animals are lurking about in a thick jungle somewhere or down in the ocean depths. That possibility, however, does not justify tossing out the tried-and-true methods of science and skepticism. Cryptozoology could have been a legitimate branch of zoology if only it hadn't pitched its tent over in the pseudoscience neighborhood where anecdotes are considered evidence and possibilities pass for certainties.

My advice to cryptozoologists is to redefine yourselves and change course. Embrace science and appreciate how it produces results and eliminates false claims. Let go of the fixations with folklore and size. Every fascinating species awaiting discovery doesn't have to come with years of campfire tales hyping it up. Every fascinating species awaiting discovery doesn't have to be a giant either. A few years ago a team of scientists went into the crater of just one extinct volcano in Papua New Guinea and emerged with more than forty species new to science. I have trekked in Papua New Guinea and, given the outrageously rich biodiversity there, I have no doubt that I saw at least a few unnamed species without realizing it. This is the world we still live in. It's exciting to realize that there is so much yet to learn about whom we share this planet with. Who needs empty myths when we have so much reality before us?

Oceans cover some 70 percent of our planet yet are still some 95 percent unexplored. Marine biologists can scarcely dip a net into the ocean without discovering new life-forms. Yes, it is possible that there are some very large marine species still unknown to us, but what is absolutely certain is that *millions* of smaller unknown species are out there beneath the waves. The decade-long Census of Marine Life estimates that approximately 250,000 ocean species are known today with anywhere from a few million to hundreds of millions left to be discovered. But it's not just numbers that are exciting. The bizarre and stunning creatures that scientists are now finding and photographing in the deepest realms seem more like extraterrestrials than the life we are familiar with. To see what I mean, please find a copy of Claire Nouvian's photo-book, *The Deep: The Extraordinary Creatures of the Deep*, and explore the beautiful and bizarre collection of exotic sea life it shows. With *real* animals as freakish and mysterious as these, who has time to think about the Kraken?

Bio-pioneer Craig Venter conducted an ocean research voyage in 2003 that turned up nearly two thousand previously unknown species of ocean bacteria and viruses. The truth is, nobody really has a clue how many strange and surprising life-forms await discovery in the deepest waters. Even more tantalizing is the life that is thriving *beneath* the seafloor. That's right, underneath the bottom of the oceans there is a vast ecosystem of microbes that live without oxygen in total darkness. It is believed that they may account for as much as a third of all life on Earth! And these little creatures produce incredibly huge amounts of methane, a potent greenhouse gas. Scientists believe that leakage of this gas from beneath the seafloor in the past caused rapid changes in the planet's climate. If even a small portion of it escaped at once today, we might be hit with tsunamis, mass extinctions, and greatly accelerated global warming.[3] If you find the search for new life intriguing, then this is the right planet for you. But we don't even have to venture into thick rainforests or board submersibles at sea to find exotic life. It's much closer to home.

We only need to look at our own bodies to discover mysterious monsters and amazing creatures—if one can agree that such things sometimes come in very small packages. In case you didn't know, you are a minority in your own skin. You are a walking ecosystem of immense complexity and diversity. So much so, that the space occupied by your body is less "you" and more "other creatures." By this I mean that some ten trillion cells are "yours" in the sense that they contain your DNA, but there are more than *one hundred trillion* parasites, predators, freeloaders, and helpful cohabitants that live on you and inside of you. Think about what this means: you are 90 percent *other life-forms*. A 2011 North Carolina State University study on belly button biodiversity found some 1,400 different species of bacteria living in the navels of ninety-five people; 662 of them were previously unknown to science. So, anyone who really wants to find new life-forms ought to give up cryptozoology and try navel-gazing instead. If you have a yearning to find monsters, don't worry, they already found you.

One of my favorite little creatures is the tardigrade (also known as the water bear). It doesn't live on us or in our belly buttons, thankfully, but I'm sure it could if it wanted to. These microscopic juggernauts make Bigfoot and the Loch Ness monster seem wimpy by comparison. The tough tardigrade is known to be able to survive the following: no water for 120 years; space (some tardigrades survived in the vacuum of space on a satellite for ten days); pressure more than five times greater than the deepest ocean; freezing to near absolute zero; and levels of gamma radiation that kill most other life-forms. Then there is the Nematode worm that lives pretty much wherever it needs to: oceans, mountains, deserts, or even a mile underground. This species is so tough that some nematodes survived the high-altitude disintegration of the *Columbia* space shuttle in 2003. (They had been aboard for an experiment.) There is an entire group of life-forms called thermophiles that live in boiling water. Yes, if strange creatures and monsters are what you seek, be assured that they are all around us. We have no need for imaginary ones.

Although no Bigfoot or Loch Ness monster bodies have turned up yet, some big creatures have been discovered in recent times. The Komodo Dragon, giant squid, and coelacanth were all new finds to mainstream science. The saola, a relative of the cow, is a large mammal that lives in the dense rainforests of Vietnam and Laos. It can weigh more than 150 pounds, but no scientist had verified its existence until 1992. It may surprise some readers to know that the mountain gorilla eluded scientific confirmation until the twentieth century. The megamouth shark can grow to nearly twenty feet long and more than 2,500 pounds. But it wasn't known to science until 1976. It's not unreasonable to believe that nature might have a few more big surprises left for us. However, that belief alone doesn't make one a cryptozoologist. That requires something more.

Why isn't cryptozoology a widely accepted and respected scientific discipline like primatology, entomology, herpetology, microbiology, or zoology? One can't earn a degree in cryptozoology from universities. It's not even consistently defined. For some it's the scientific pursuit of unknown animals, not fundamentally different from what a traditional scientist does when she or he goes to do fieldwork on a remote Pacific island and hopes to find new bird, insect, or frog species. However, most casual fans and committed cryptozoologists alike probably would describe their field as the study and pursuit of *legendary large animals only*. I don't want to overgeneralize because I know there are different kinds of believers for every belief. There are some who merely let intrigue and hope carry them to the edge of science and no more than a step or two beyond. Then there are those who see dragons behind every tree. Nonetheless, cryptozoology has a severe image problem. Books on the Loch Ness monster at my local bookstores, for example, are not shelved with books on whales and sea turtles. They are placed with books about vampires and haunted houses. This might be insulting to some cryptozoology enthusiasts, but a quick glance at the content of most of these books reveals that they are shelved exactly where they should be. While I can sympathize to a point with someone who says he wants the search for famous large animals to be legitimate science, the truth is that such efforts are rarely discussed or conducted in a scientific manner. Therefore—until someone presents powerful evidence or produces the body of Bigfoot or Nessie—it won't be respected. And so long as cryptozoology in general is not conducted as a science, it won't qualify as science.

MANY FOOTPRINTS, BUT NO FEET TO FILL THEM

The biggest star of cryptozoology these days is probably Bigfoot, also known as Sasquatch. For a species that probably doesn't exist, Bigfoot has shown remarkable longevity. Somehow, say believers, a massive bipedal ape has eluded discovery in North America. In 2011—decades after Bigfoot belief soared in the 1960s and 1970s and footprints turned up all over the Northwest—the Animal Planet TV channel is airing *Finding Bigfoot*. (Spoiler alert: they don't find Bigfoot.) The National Geographic Channel's *Mysterious Science* series included an episode about Bigfoot in which skeptics and believers presented their case. (Bigfoot didn't turn up here either.) Books that are skeptical of Bigfoot claims are far outnumbered by books promoting the belief. Millions of Americans are convinced that Bigfoot really is lurking in the shadows of American forests and swamps. A 2006 Baylor study found that 16 percent of Americans believe that Bigfoot "absolutely" or "probably" exists.[4]

Professor emeritus of physical anthropology Curtis Wienker doubts Bigfoot is real and somehow managed to escape confirmation all these years. "I think it is virtually impossible," Wienker said. "Recall that almost all alleged sightings of such beasts are at night and virtually all higher primates are diurnal [active in daytime]. In recent years in remote Amazonia, a much less populated region than the Pacific Northwest, a few undiscovered species of primates have been discovered, but they are all South American monkeys related to previously known species. All of them are a branch of primates not closely related to modern humans and apes, and all are very small. Furthermore, the mythical creatures of Bigfoot, Yeti, and the 'swamp ape' are all described as bipedal, and physical anthropology data all suggest that *Homo sapiens* is the only habitual biped among the perhaps two hundred species of living primates. There is not one shred of scientific evidence, not one datum, to support the existence of such beasts. Period."[5]

The absence of good evidence is problematic for Bigfoot believers, to say the least. There may be nothing wrong with having an open mind about the possibility of Bigfoot being real, but it certainly is no justification to confidently claim that the creature's existence is probable or definite, as many do. All we have to date are eyewitness accounts, poor-quality photos and videos, inconclusive hair samples, and footprints. The oversized footprints might have been impressive evidence if not for the fact that footprints have been faked so often that they have no credibility whatsoever. The first to do it seems to have been a man named Ray Wallace who worked in road construction in the Pacific Northwest. After his death, Wallace's son went on record saying that his father was a dedicated prankster who possessed large carved wooden feet that he used to make Bigfoot trails in the dirt. "Ray L. Wallace was Bigfoot. The reality is, Bigfoot just died," declared Michael Wallace after his father passed away in 2002.[6] According to Michael, Ray first made fake prints at a Northern California logging camp in 1958. "This wasn't a well-planned plot or anything," Michael told the *New York Times*. "It's weird because it was just a joke, and then it took on such a life of its own that even now, we can't stop it."[7]

Anthropologist Cameron M. Smith lives in Oregon, so he's practically neighbors with Bigfoot. "It seems very unlikely that large primates here in the Pacific Northwest could go undetected for so long," Smith explained.[8] "Every species has a minimum viable population, a genetic barrier that it can't dip below if it will maintain genetic health. If Bigfoot reproduction is anything like, say, another giant primate, gorilla or human, then the MVP [minimum viable population] can't be below approximately five hundred individuals. So, how much foraging territory do, say, five hundred Bigfoot creatures require, considering the resources of the Pacific Northwest forests, and their daily caloric, water, and nutrient requirements, which could perhaps be modeled on similarly large primates. I don't know, but interesting to think about!"

Smith believes the best evidence we could hope to find would be DNA rather than bone.

"The techniques for understanding life-forms on the DNA level are advancing every day," Smith said. "If Bigfoot is bipedal, which everyone seems to say, its femur or other skeletal material could be significantly similar to humans or other hominids. So any bones could be fakes—just really large bones of a giant human or *Gigantopithecus* [an extinct primate that lived in Asia]. Though *Gigantopithecus* would be fossilized. However, one might say it is a fossilized, ancient Bigfoot bone! The bones alone wouldn't do it for me. I'd want the DNA and we now have DNA studies well into the twenty-thouand-plus-years-ago range, so even old Bigfoot bones would be OK. Hair or other tissue would work. At Paisley Cave, Oregon, researchers recently extracted human DNA from coprolites [dried feces] that were over fourteen thousand years old. So that might be a source if hair or other tissues could not be found."

My suggestion to cryptozoology fans has always been to keep the passion for exploration and discovery but abandon the faith-without-evidence position. If one is drawn to the idea of unknown creatures living in secluded valleys, rainforests, or in the ocean depths, then become an amateur or professional scientist and do real science. New species of animals are being discovered all the time. Most of the new species being discovered may not qualify as monsters worthy of headlines but this is very important work, nonetheless. Researchers are fleshing out the details of our home. Doesn't it make sense that we should have as complete a picture as possible of the planet we live on? It's also thrilling to find new life. Imagine the excitement of laying eyes on a creature unknown to science. I don't have to imagine. I once felt that thrill—though it didn't last.

During one of our regular expeditions out in the bush, my son Jared and I came across a bizarre bug. It looked like a cross between a scorpion, a spider, and a beetle. Because we were on a small Caribbean island that had not been thoroughly raked over by the eyes of entomologists, I allowed myself to get excited—too excited. Surely I had discovered a new species! *National Geographic* camera crews were going to be knocking on my door by the end of the week! Yahoo! Maybe the genome of this new find will lead to a cure for cancer! Just as I begin to think of a name for my new contribution to science, however, my son wanders over, takes one look at it and casually says: "Oh, you found one of those. Yeah, they're pretty cool." I tried to explain to him that I had just discovered a new species, previously unknown to humankind, and now I have to decide which shirt to wear for my photo spread in *New Scientist* magazine. For some reason, however, he wasn't impressed.

"Dad, everybody knows about those," Jared said. "They're even in one of our documentaries at home." Sure enough, when I got home and did some fast-forwarding, there it was, plain as day. It was some sort of a pseudoscorpion, well known to everyone but me, apparently. Sigh, no big discovery. The good thing to come out of that little episode, however, is that I got to experience a small taste of the electrifying thrill of finding something totally new. My little moment was fool's gold, of course, but it felt great. I can't fault cryptozoology enthusiasts for chasing that high because it's a good one. What I don't get is why anyone would feel the need to contaminate the noble pursuit of discovery with pseudoscience. If cryptozoology is simply the search for new life, then that's what scientists have been doing all along. By that definition, thousands of them are cryptozoologists. SETI is an effort in cryptozoology, they just don't call it that. NASA is doing cryptozoology when it sends probes to look for life on other worlds. Much of the scientific fieldwork conducted in the Amazon could be considered cryptozoological fieldwork. Scientists want to find new species and often do. Nobody thinks of it this way, however, because cryptozoology carries an unfortunate connotation. It's known as the club for crazies who believe in mermaids and harpies. That might not be entirely fair, but it's the reality.

Whether one is merely a fan of cryptozoology or a hardcore cryptozoologist ready to invade Tibet in search of the Yeti, it is vital to keep in mind the difference between science and pseudoscience. Science plays by a set of rules and logic that are proven to work. Pseudoscience has no rules, no logic, and the results reflect it. An important point in all of this is that it's mainstream scientists doing their homework, rolling up their sleeves, getting dirty, and working with locals who discover and catalogue the new species. Where are all the cases of cryptozoologists going out on expeditions and returning with amazing new discoveries? This lopsided score should encourage anyone with an interest in finding new life-forms to side with science. Earn a degree in biology, zoology, anthropology, botany, entomology, and then charge out into the unknown. That is the path to discovery.

Several years ago I conducted a lengthy interview with Jane Goodall, a primatologist widely recognized as one of the great scientists of our time. Her revolutionary work with wild chimpanzees in Africa famously forced us to rethink what it means to be human and a member of the primate family. I was overjoyed to be able to talk with her for more than an hour. Unfortunately, during all that time, it never crossed my mind to ask her about Bigfoot. I think I can be forgiven for the oversight. But it would have been a great question because, difficult as it may be to imagine, Jane Goodall is a believer. During a 2002 NPR interview she said the following: "Well now you will be amazed when I tell you that I'm sure that they [Bigfoot creatures] exist."[9] Goodall is not some rascal trying to drum up business at a roadside tourist trap by spreading Bigfoot rumors. She is a prominent authority on nonhuman primates and believes a population of giant ones is somewhere in North America. So what are we to make of this? Not much, I say, because just like all other Bigfoot believers, Goodall can't point to any good evidence. I'm a big fan of hers and I respect her knowledge of apes, but not so much that I would blindly follower her off into Bigfoot faith. Authority and credentials do matter in science, of course, but not so much that they can overcome the absence of evidence. Aristotle was wrong. Newton was wrong. Einstein was wrong. Nobody has ever been right about everything. And until someone finds a body, a fossil, or something else conclusive, it appears that Goodall is wrong on this point.

Two very basic problems I have long held with claims of large famous beasts such as Bigfoot and the Loch Ness monster are that almost no one considers the population sizes necessary for these creatures to survive. Unless we are talking about a magical immortal animal, there would have to be a large enough population of their kind for them to reproduce and perpetuate their species. So this would not be a case of just one or two individual creatures eluding capture and confirmation; it would have to mean a population of hundreds, maybe thousands hiding in the shadows. And yet no one has ever found a body or skeletal remains of even one. No hunter has ever shot one and showed off the body. The same holds true for other legendary animals. If the Loch Ness monster really is a holdover from the Cretaceous period, as many believers claim, then there can't just be one of them. There would have to be a significant number of them. This makes it even more unlikely that we would not have been able to positively identify one by now.

The second problem is the increasing number of people and cameras. Why aren't we seeing a flood of high-quality photos and video of Bigfoot animals these days? It doesn't make sense when we consider that the population of North America continues to grow and the number of people with still cameras, video cameras, and cell phone cameras has risen sharply in recent years. Outdoor recreation has risen in popularity as well. Opportunities for recorded sightings have never been greater, but for some reason we have virtually nothing to show for it.

Cryptozoology fans should not misperceive skepticism and the demand for proof as an outright denial of any possibility that weird unknown creatures are out there. Unlikely as it may be, nothing about the claim of some giant apelike creature living around humans contradicts the laws of the universe. In fact, we know that something like Bigfoot once existed. *Gigantopithecus* was a monstrous nine-foot tall primate that lived a few hundred thousand years ago in Asia. By the way, isn't it interesting that scientists can find convincing evidence of a giant primate that lived so long ago, but no one can find any proof for a giant primate population that is supposed to be living in North America right now? Half a million years ago *Homo erectus* clans may have encountered the massive *Gigantiopithecus* and then shared their own Bigfoot tales around the campfire. One does not have to believe in magic or miracles to imagine that some kind of ape species could still be out there on the fringes of civilization avoiding detection. I'll even go so far as to declare that Bigfoot has a much better chance of turning out to be true than most of the other popular beliefs addressed in this book. But that doesn't mean it's not an empty claim unworthy of acceptance. Yes, there are the plaster casts of footprints. But we know that Bigfoot prints can and have been faked, so they are not proof. There are also numerous eyewitnesses. But we know that eyewitnesses are not reliable. People have been wrongly imprisoned by eyewitnesses. Elvis is alive and extraterrestrials invaded Earth years ago, according to eyewitnesses. Given what modern science has revealed about the human brain—particularly how vision and memory work—there is no need to question or disparage the honesty, character, intelligence, or eyesight of people who say they saw Bigfoot. The fact that they are human beings is reason enough to withhold jumping to any extraordinary conclusions based on their accounts. And then there is the famous Patterson Bigfoot film.

THE PATTERSON BIGFOOT FILM

I don't get it. Why do so many people see the famous 1967 Patterson Bigfoot film as powerful evidence if not absolute proof? I was just a young kid when I first saw it (it was shown on a television show called "In Search Of," if I recall correctly). Even then my first reaction was that it was obviously some guy in a hairy suit. Nothing about it convinced me that Bigfoot was real. I hadn't even kissed a girl yet but I had been around the block enough times to know a fake ape when I saw one.

The brief footage, shot by Roger Patterson and Bob Gimlin in northern California in 1967, shows a husky, pear-shaped, hairy figure walking through a clearing in the woods. During the walkthrough, as if on cue, the figure turns toward the camera midway to heighten the dramatic scene. Just like in childhood, I still see a guy in a gorilla suit. Admittedly, nothing jumps out to conclusively *disprove* the claim that this is really Bigfoot, but there are many problems, as we shall see, that demand extreme skepticism if not outright rejection. (You can also search "Patterson Bigfoot film" on YouTube and judge for yourself.)

It's a subjective call, but the gait of this Bigfoot seems all wrong to me. It just doesn't move or look like a wild nonhuman primate would, in my opinion. Many believers dispute this, of course, and claim that the film is rock-solid proof. They say it shows anatomically correct flexing of the back and quadriceps muscles while walking, which means it must be real. I feel like I know how bodies move and how muscles flex pretty well. I have been a lifelong fitness enthusiast and I have photographed literally thousands of athletes in a variety of sports over many years, including at the Olympic Games. I see nothing in the Patterson film that leads me to think it couldn't have been a guy in a cheezy ape suit.

Some believers claim that the figure in the Patterson film could not be a man because no ape suits of that quality could be made back in the 1960s. This is just plain wrong. First of all, it's not that great of an ape suit. Second, ape suits were produced to that standard and better in the late 1960s. Both the original *Planet of the Apes* film and *2001: A Space Odyssey* were released in the 1960s and featured very impressive ape makeup—far better than we see in the Patterson film, in my opinion. The sequel to *The Planet of the Apes* (*Beneath the Planet of the Apes*, released in 1970 and probably shot in 1969) includes a scene with ape characters Dr. Zaius and General Ursus in a steam room with no clothes on discussing an invasion of the Forbidden Zone. The face makeup and chest area are different. But other than that, Ursus looks like he could have been wearing the same suit that might have been used in the Patterson Bigfoot film. The "Magatu," a horned gorillalike beast with white fur, was featured in the *Star Trek* episode "A Private Little War" that aired in February 1968. The Magatu's fur is the wrong color, of course, but the quality of the suit is as good or better than what we see in the Patterson film. Many forgotten B movies such as *Gorilla at Large* (1954) and *Konga* (1961) show men in ape suits that match or surpass the Bigfoot film's standard. Far more compelling than any of that, however, is the claim by Phillip Morris that he made and sold the suit seen in the Patterson film. The former magician and owner of Morris Costumes told Bigfoot researcher Greg Long that he sold one of his gorilla suits to Patterson in 1967. Back then, Morris sold the suits primarily to carnival acts and magicians who used them for stage tricks like turning a woman into a gorilla. Morris told Long that he asked Patterson what he was going to do with it and Patterson's reply was, "We're just going to have some fun."[10]

I suppose it's possible that a modern-day cousin of *Gigantopithecus* really did conveniently parade by two guys who told people in advance that they were going out to film Bigfoot that day. But the alternative possibility, that it was a hoax, seems much more likely to be true. The biggest blow to the film's credibility, however, is nothing less than a credible confession from a man who says he wore the suit! Bob Heironimus, an acquaintance of Patterson and Gimlin, says the two men agreed to pay him to wear a gorilla suit for the staged filming. Greg Long's investigative book, *The Making of Bigfoot*, presents the story behind the film in great detail and includes this confession from Heironimus: "I'm here to tell you that I was the man in the Bigfoot suit."[11]

WHY DON'T SCIENTISTS CARE ABOUT BIGFOOT?

Jeff Meldrum's book *Sasquatch: Legend Meets Science* makes the best case for Bigfoot I have encountered to date, but it's still far from convincing. Meldrum, an associate professor of anatomy and anthropology at Idaho State University, has better credentials than most Bigfoot believers but, just like them, he relies on questionable footprints, unreliable eyewitness accounts, and that Patterson film. No skulls, teeth, or bones, no conclusive DNA evidence and, of course, no body. Meldrum makes some good points about keeping an open mind, and he is right that many confirmed hoaxes do not prove that all Bigfoot evidence is faked. However, I feel that his obvious irritation with mainstream scientists goes too far: "It seems the majority of scientists are content to remain aloof, trivialize the probability of new discovery, or presume to discredit the witnesses and the evidence, leaving to others the search for proof, the definitive type specimen. They passively challenge: 'Show me the body.'"[12]

There are a few problems with Meldrum's statements. First, even if we were to allow that the majority of scientists are "aloof" and "trivialize the probability of new discovery" (which is not accurate, based on my experiences with many scientists), I am certain it would not be because they are a bunch of hopeless jerks who are too arrogant, stubborn, and entrenched to look beyond their own noses. The reason most of them have no interest in taking up the quest is because *no compelling evidence exists that would make it seem worth their time to investigate*. There is nothing to get their attention and inspire them to care. Plaster casts of footprints can be intriguing at first, but when one learns about all the footprint hoaxes that have been perpetrated over the years, they become a lot less interesting. It's like crop circles in England. They were somewhat intriguing right up to the point when the hoaxers came forward to confess and explain how they did it. Then, not so much. Eyewitness accounts are always worth listening to and considering, but what makes Bigfoot stories so special? There are many witnesses who say they have seen ghosts too. Should the world's elite scientists drop what they are doing and focus on haunted houses, based only on ghost stories? What about stories of TV preachers who routinely cure AIDS and cancer? Should the medical science field redirect the bulk of its time and money to research Benny Hinn crusades?

It's all about evidence, not aloofness or a lack of interest in discovery. I am confident that hundreds if not thousands of scientists would catch the first flight to the Pacific Northwest if solid evidence of Bigfoot were produced. What scientist wouldn't want to be in on the discovery of a new primate species, especially one with such an enduring hold on the public's interest? The payoff in terms of fame, money, and career advancement would be huge. Have no doubts, scientists would vigorously investigate Bigfoot, the Loch Ness monster, or any other fringe claim *if* something convinced them that there is at least a reasonable chance of success. In the meantime, most scientists are just not willing to waste their time on what has all the look and feel of a dead-end myth and nothing more.

Doubting the overreaching claims of cryptozoology is not a blanket rejection of everything it stands for. To be clear, I would love for Bigfoot, Yeti, the Loch Ness monster, and a thousand other mythical creatures to turn out to be real. Their existence would make our fascinating world even more interesting and exciting. Such discoveries would thrill me and every other fan of science. But I refuse to surrender my skepticism and common sense for such thin possibilities. I won't confuse hope for knowledge and I won't forget that emotional desire and scientific inquiry are two very different things.

Should people actively search for Bigfoot and other such mythical creatures? That's up to them. If they have the time and the motivation, go for it. I suspect that it's a waste of time, but if cryptozoologists do manage to find a genuine specimen that shocks the world, I will be first in line to buy their books, watch their documentaries, and shake their hands. In the meantime, however, I choose to spend my life chasing after ideas and discoveries that offer better odds for success.

GO DEEPER . . .

- Buckman, Robert. *Human Wildlife: The Life That Lives on Us*. Baltimore: Johns Hopkins University Press, 2003.
- Buh, Joshua Blu. *Bigfoot: The Life and Times of a Legend*. Chicago: University of Chicago Press, 2010.
- Dunn, Rob. *The Wild Life of Our Bodies: Predators, Parasites, and Partners That Shape Who We Are Today*. New York: HarperCollins, 2011.
- Ellis, Richard. *Monsters of the Sea*. Guilford, CT: Lyon Press, 2006.

- Jack, Albert. *Loch Ness Monsters and Raining Frogs: The World's Most Puzzling Mysteries Solved*. New York: Random House Trade Paperbacks, 2009.
- Jayawardhana, Ray. *Strange New Worlds: The Search for Alien Planets and Life beyond Our Solar System*. Princeton, NJ: Princeton University Press, 2011.
- Long, Greg. *The Making of Bigfoot: The Inside Story*. Amherst, NY: Prometheus Books, 2004.
- Nickell, Joe. *Tracking the Man-Beasts: Sasquatch, Vampires, Zombies, and More*. Amherst, NY: Prometheus Books, 2011.
- Nouvian, Claire. *The Deep: The Extraordinary Creatures of the Abyss*. Chicago: University of Chicago Press, 2007.
- Regal, Brian. *Searching for Sasquatch: Crackpots, Eggheads, and Cryptozoology*. New York: Palgrave Macmillan, 2011.
- Zimmer, Carl. *Parasite Rex: Inside the Bizarre World of Nature's Most Dangerous Creatures*. New York: Free Press, 2001.
- Zimmer, Carl. *A Planet of Viruses*. Chicago: University of Chicago Press, 2011.

Chapter 42

"ANGELS WATCH OVER ME."

More than half of Americans (53 percent) believe they have personally been saved from harm by a guardian angel.

—Christopher Bader, Carson Mencken,
and Joseph Baker, *Paranormal America*

Several years ago, an elderly woman described for me her husband's dramatic final moments in a hospital room. Understandably, she recalled the event with great emotion. "As he was passing away, he said to me, 'Do you see? Do you see the angel here in the room?' It was something else, really something else."

Then she asked me: "Do you believe in angels?"

I can't remember how I answered. Most likely I squirmed out of it by saying something kind and then changing the subject. Being asked about your position on angels can be awkward when you are unconvinced that they exist. Sure, you can explain that you are open-minded but don't know of any good evidence for angels, but that comes across to many people as just plain odd. I lived in the Caribbean for twenty years and would estimate that at least 80 to 90 percent of the population there believes in angels. But while this belief may be common, it's still quite extraordinary and should be backed up with evidence.

A typical believer thinks that angels, defined as some sort of spiritual beings with magical powers, are constantly traveling back and forth between heaven and Earth in order to deliver messages from God and save people from harm or perhaps nudge them into making the right choices in life. "I have an angel that is with me all the time," a middle-aged woman once told me. "The angel watches over me and guides me. I know it." She spoke with the sort of certainty she might have expressed if talking about the existence of her car or pet cat. Angels can also take the form of humans and blend in with us, at least that's what the Bible seems to suggest where it describes people who "entertain angels unaware."[1] One of the Bible's most interesting and dramatic angel descriptions is found in Daniel 10:

> I looked up and there before me was a man dressed in linen, with a belt of fine gold from Uphaz around his waist. His body was like topaz, his face like lightning, his eyes like flaming torches, his arms and legs like the gleam of burnished bronze, and his voice like the sound of a multitude.

Angel belief is flying high in the United States these days. According to a Gallup report, 75 percent of Americans believe angels are real and 11 percent say they aren't sure. Only 14 percent of Americans do not believe in angels.[2] A Barna Group study found that 83 percent of American teenagers think angels are real.[3] Belief is high in Canada, too, where 56 percent of adults say angels exist, with 19 percent not sure. In Great Britain, 36 percent believe in angels, with 18 percent unsure.[4]

It's easy to see why belief in angels is appealing. They are much more than human, yet not quite gods. They are best known today as messengers and servants of the Judeo-Christian-Islamic god. But angels, or similar beings, are also a part of many other belief systems today and in the past. The gods, it seems, have always had a need for middle management. While the image of what an angel is supposed to look like has become fairly standard in popular culture—winged and wearing white robes—this description is not found in the Torah, the Bible, or the Koran. The common concept of an angel, as it is widely depicted in art and imagination today, can be traced back to the Middle Ages, where it apparently was just made up out of thin air by artists of the day.[5] Some people believe that angels are spirits (another term that is not clearly or consistently defined) and as such have no physical form that can be seen by humans. Others believe they can assume human form and mingle with us in order to carry out missions for God.

Probably the primary reason so many people believe in angels is because they come standard issue with the package of religious belief. If one is a Christian, a Muslim, or a Jew, then believing in angels is just assumed to come with the territory for most. I was surprised to learn that angel belief is not just common but also carries with it intimate, hands-on relevance for millions of people. For example, the Baylor Religion Survey states that 53 percent of American adults believe they have been "saved from harm by a guardian angel."[6] This is not a vague belief in some invisible being that is out there somewhere; this is belief in direct involvement—physical intervention—by angels in the affairs of people on Earth.

Baylor researchers suggest that guardian angels are a "gateway belief" between religion and paranormal claims outside of religion. Guardian angel believers, it seems, are much more likely to believe in ghosts, psychics, Atlantis, and other such paranormal phenomenon. For example, people who claim to have had a guardian angel experience are twice as likely to believe in ghosts than people who do not claim to have had such an experience.[7]

A survey of Americans who claim to have had a guardian angel experience reveals significant differences between religious groups. Black Protestants lead with 81 percent claiming to have been saved by an angel. Evangelical Protestants are second (66 percent), followed by Catholics (57 percent), mainline Protestants (55 percent), and other religions combined (49 percent). Only 10 percent of Jews claim to have had such an experience or encounter. This is ironic, since the Jewish Torah is the oldest and perhaps most important source of the angel concept believed in by Jews and Christians. Interestingly, 20 percent of those who have no religion claim to have had a guardian angel experience as well.[8] This latter figure should not surprise anyone who has ever browsed the New Age section of contemporary bookstores. Angel belief clearly has expanded beyond the confines of traditional, mainstream religion.

Angels are interesting and popular, of course, but no one should shy away from the most important question of all: How do we know these beings even exist in the first place? The answer is that we don't know because no one has ever proved it in a scientifically credible way that others could verify. The best "evidence" for angels are a few blurry photos and eyewitness accounts, no better than the evidence we have for fairies, ghosts, Bigfoot, the Loch Ness monster, and alien visitations. One certainly can understand why people would want to believe in angels—guardian angels in particular. Who wouldn't like having an invisible flying superhero hovering just above you to provide magical protection day and night? It seems reasonable to suspect that many people who claim to have a guardian angel intervene on their behalf have chosen to inject a supernatural element into natural events based on nothing more than imagination. I have had near misses in traffic, I've been in a few very scary situations alone in far away places, and I once found my way after being lost in the wilderness. But I'm not convinced that guardian angels exist, so it never occurred to me to credit one when I got out of a tight spot and things worked out well. If I did believe in them, however, I probably would go around citing some of the more dramatic events of my life as "proof" that I have a guardian angel. Claiming that an invisible angel saved you from danger is nothing more than a matter of loosely interpreting and embellishing real events to fit into the context of prior beliefs. But not all angel believers rely solely on imaginative interpretations. Some people claim to have seen angels.

Undoubtedly people really have experienced powerful visions and even feelings of contact or intervention by something unusual. I have no doubt that many of these people are sincere about what they experienced. The problem, however, is that we can't be sure what the actual experience was, and that's why claims of contact with angels require a skeptical reaction. Given all that we know about the frailties of human vision, the reality of sleep paralysis, hallucinations, the power of suggestion, as well as the constructive and fallible nature of memory, claims of encounters with angels are just not good enough to qualify as proof. We need more in order to know for sure that angels exist.

GO DEEPER . . .

- Nickell, Joe. *Entities: Angels, Spirits, Demons, and Other Alien Beings*. Amherst, NY: Prometheus Books, 1995.

Chapter 43

"MAGIC IS REAL AND WITCHES ARE DANGEROUS."

> **Witchcraft is all around us. We must be vigilant and protect ourselves against it.**
>
> —Cayman Islands preacher

> **Magical thinking is a slippery slope. Sometimes it is harmless, other times quite dangerous.**
>
> —James Randi

I would describe myself as well above average in patience and optimism. No matter how unlikely a weird claim or story is, I'm usually able to summon up the necessary good manners and mental stamina to give it a fair hearing. I'm also optimistic about the future of humankind. There are no guarantees, of course, but my hunch is that the next thousand years will be better for most and less crazy for all than the last thousand years were. I do confess, however, to occasionally wrestling with some very negative thoughts about what tomorrow holds for humanity. There are moments, for example, when I literally feel ashamed to be associated with such a horrible and pathetic species. Sometimes I can't help doubting that we will ever free ourselves from the weight of ancient superstitions and irrational fears that have us so hobbled today. I struggled with such negative feelings a few years ago while watching a documentary about "child witches" in Africa. According to the documentary, these children are ostracized, tortured, and murdered by Christians who feared their magic. Maybe, I thought, we just aren't capable of ever escaping the pitfalls that come with these prehistoric brains we all still walk around with. Maybe science won't be enough to show us the way and clear out the cobwebs of our evolutionary past that continue to cloud our minds. Maybe the pleas of scientists and a few skeptics here and there can never win the struggle for reason. Maybe we are doomed to be *Homo irrationalist* forever.

Even after all the horrible things I have seen, heard, and read about in my lifetime, I was surprised to learn that tens of thousands of children are persecuted as witches in the twenty-first century. I wondered if the documentary makers had been loose with the facts and exaggerated the problem in order to enhance their film, so I researched the problem. What I found is that it's even worse than the film suggests. These children can be persecuted for no other reason than they have a physical or mental handicap, or simply act or look slightly different in the judgment of an accuser. Even being gifted can raise suspicions. The results of being singled out can be devastating. Parents reject their own children and turn them out into the streets. "Child witches" are more likely to be physically or sexually abused. Many very poor parents take their children to Christian preachers who promise to cure them—for a hefty fee, of course. The "cure" often includes imprisonment, starvation, and beatings. And you probably thought witch hunts were a thing of the past.

Sadly, tens of thousands of "witches"—children and adults—are still being imprisoned, ostracized, tortured, and murdered on a regular basis in many societies around the world today. Witch believers in Nigeria hammered nails into the head of a little girl in an attempt to drive out the demon of witchcraft. A Nigerian man experienced painful swelling in his legs and concluded that it must be the work of a "child witch" whom he then took to a river and tossed in.[1] In 2009, a horrifying but not uncommon incident took place in India's Jharkhand state when hundreds of people either watched or participated in a public attack on five women who were accused of being witches. In addition to beating them, the mob forced the women to parade around naked and eat human excrement.[2] In India's Sonitpur district, a father and his four children were beheaded after an unofficial witch trial held by some two hundred villagers. The mother managed to escape.[3] It was reported in December 2010 that several people accused of being witches were murdered in Haiti.[4] A family pastor in Africa accused a boy of being a witch, so the father poured acid down his throat hoping to kill the evil spirit. The boy died a month later.[5] An eight-year-old girl who liked to sleep outside of her home on hot nights because it was cooler was accused of being a witch and flying off in the dark to join a coven. In addition to being beaten with sticks, this little girl had to endure a series of expensive exorcisms that bankrupted her mother.[6] Activists estimate that in just two of Nigeria's thirty-six states some fifteen thousand children have been accused of being witches in recent years and approximately one thousand have been killed. It seems that word about the Dark Ages ending failed to reach all of us.

Skeptics point out that no one has ever scientifically proven that the popular concept of magic is a real force or that the traditional laws-of-nature-defying version of a witch exists. Nevertheless, witches and magic do matter, if only because people have believed them into relevance. I have seen the fear of witchcraft firsthand. It's a real force, even if the actual magic is not. While I lived in the Caribbean, sincere, concerned people often warned me about the dangers of "obeah," the preferred name for magic in many Caribbean societies. I recall one of my first assignments as a journalist in the Cayman Islands was to report on a frog with a padlock clipped through its mouth that someone left on the courthouse steps. Obeah believers explained to me that it was done to silence a witness in court.

Cases involving magical spells and witches make up approximately *40 percent* of the cases in Central African Republic courts. In one of that nation's districts, witch cases account for more than 50 percent of the caseload![7] Magic belief is not only dangerous, it also can be costly and time consuming. For some people magic is harmless entertainment (*Harry Potter* books and films, for example), for others it's part of a life philosophy or religion. For too many people, however, magic is about fear, exploitation, and violence. The cost is too high, in my view, for the rest of us to look the other way. Can we really afford to sacrifice so much time, money, and bloodshed over unproven magic? Can we really live with knowing that "witches" are still being burned alive?

Don't make the mistake of thinking that this particular strain of madness only involves poor people in the developing world. Watch one of the fundamentalist Christian television networks in America and sooner or later you will hear that witchcraft presents a clear and present danger to us all. Sarah Palin, the 2008 Republican candidate for the vice presidency of the United States, apparently believes that witches possess dangerous powers because she participated in a church ritual designed to protect her from them.[8] A primary problem with all of this is that the majority of the world's people accept and promote a magical worldview, even if they don't condone the witch hunts that sometimes stem from it. It would be beneficial if more people recognized that a direct line can be drawn from casual and seemingly harmless belief in miracles, astrology, ghosts, and so on to the murder of people who are thought to be witches. All these irrational beliefs grow from the same mind-set. All these irrational beliefs can be beaten back by skeptical thinking—including the one that says witches are dangerous and must be killed.

WITCHES ARE PEOPLE TOO

According to a 2005 Gallup poll, 21 percent of Americans believe in witches.[9] This is strange because it seems to suggest that 79 percent of Americans are clueless about the obvious existence of witches. Of course witches exist! They have books, websites, they appear on talk shows, and one of their belief systems, Wicca, is legally recognized in the United States. No one, especially skeptics, should ever embarrass themselves by saying that witches aren't real. I can vouch for the existence of witches because I've met a few over the years. Witches may not soar across the night sky on brooms, but they do work in shopping malls, take university classes, and attend PTA meetings. People who identify themselves as witches and adhere to Wicca, a belief system based on supernatural magic, are not significantly different from people who call themselves Christians, Muslims, Jews, or Hindus and adhere to a belief system based on supernatural beings and powers. Modern-day witches have a lot of negative propaganda heaped on them, but if one looks into what they actually say and claim to stand for, it's clear that the vast majority are nothing like the Satan-worshipping, animal-sacrificing menaces to society that so many followers of mainstream religions accuse them of being. There are no logical or fair reasons that Wiccans should not be afforded the same level of respect and their claims viewed as no more or less credible than those of people who follow mainstream religions.

Michelle Mead is a US Navy veteran and longtime witch who says she casts "spells" often but rarely "hexes" people. She told me that Wicca is positive and helps her to be a better person. "The biggest misconception about witches that people have is that we are evil and deal with the devil," Michelle told me. "The biggest misconception about Wicca is that we are a bunch of free-love types that engage in a lot of group sex. A lot of people think that we fling spells and hexes about indiscriminately. My faith provides me with some very valuable spiritual tools that enable me to consistently evolve as a spiritual being. Also, it could be said that most people think we are flakey and weird—but I can't say that's a misconception."[10]

From the skeptic's perspective, a basic problem for modern-day witches is the same one religious people face: claims of supernatural events, forces, and powers are unproven to date. Of course, this detail doesn't seem to bother most other religious people, so it probably doesn't bother most witches. Personally, I don't mind so much if someone believes in magic and calls herself a witch. But I do care very much about people—especially children—who are mistreated or killed due to the irrational fear of witchcraft. The number of people worldwide who are exploited, abused, and murdered after being labeled as witches is one more powerful argument for skepticism and critical thinking skills. The more people in the world who at least have some doubt about the reality of witchcraft, the less people there may be who think it makes sense to hate and harm fellow humans for it. Of course, we should also address the problem of witch phobia from the other side of the equation by demanding that people respect the basic right of others to think and believe whatever they wish. Christians certainly didn't like it when they were the ones being tortured and murdered in ancient Rome. Why do it to others now? Of course, respecting other people and accepting their personal choices would require humankind to grow up. Sadly, we aren't quite there yet.

GO DEEPER . . .

Who needs magic? As the following sources show, science and reality are every bit as exciting as unproven claims about supernatural forces.

Books

- Calder, Nigel. *Magic Universe: A Grand Tour of Modern Science*. New York: Oxford University Press, 2003.
- Kaku, Michio. *Hyperspace: A Scientific Odyssey through Parallel Universes, Time Warps, and the 10th Dimension*. New York: Anchor, 1995.
- Kaku, Michio. *Physics of the Future: How Science Will Shape Human Destiny and Our Daily Lives by the Year 2100*. New York: Doubleday, 2011.
- Kaku, Michio. *Physics of the Impossible: A Scientific Exploration into the World of Phasers, Force Fields, Teleportation, and Time Travel*. New York: Anchor, 2009.
- Krauss, Lawrence M. *Beyond* Star Trek*: The Physics of* Star Trek, The X-Files, Star Wars, *and* Independence Day. New York: Harper Paperbacks, 1998.
- Nelson, Sue, and Richard Hollingham. *How to Clone the Perfect Blonde: Using Science to Make Your Wildest Dreams Come True*. Philadelphia: Quirk Books, 2004.
- Panek, Richard. *The 4% Universe: Dark Matter, Dark Energy, and the Race to Discover the Rest of Reality*. New York: Houghton Mifflin Harcourt, 2011.
- Turney, Jon. *The Rough Guide to the Future*. London: Rough Guides, 2010.
- Vyse, Stuart A. *Believing in Magic: The Psychology of Superstition*. New York: Oxford University Press, 2010.

Other Sources

- *Through the Wormhole with Morgan Freeman* (DVD), Revelations Entertainment, 2011.

WEIRD PLACES

Chapter 44

"ATLANTIS IS DOWN THERE SOMEWHERE."

> **Atlantis continues to captivate people's imaginations because it offers the hope that lost ideals or some untapped human potential will someday be uncovered, not the masonry blocks of a dead civilization. Scrying for crumbled roads in Bimini or poring over the outline of some terra incognita on a forged map ignore the real Atlantis, the undiscovered country of human ideals.**
>
> —Kevin Christopher

Some people may think that the lost city/continent of Atlantis shouldn't be included in a book like this. After all, it's been discovered countless times, hasn't it? But it does belong here, of course, because—despite repeated claims of its discovery—we still don't know where Atlantis is or if it ever even existed. Not surprisingly, such pesky details have not stopped millions of people from believing it's down there *somewhere*. In fairness to the public, it's easy to be misled given the way new reports of the lost continent's discovery or near discovery keep coming, year after year. Here is a small sample of eye-catching headlines that reputable news sources saw fit to place atop reports that ultimately proved unfounded: SATELLITE IMAGES "SHOW ATLANTIS,"[1] ATLANTIS "OBVIOUSLY NEAR GIBRALTAR,"[2] TSUNAMI CLUE TO ATLANTIS FOUND.[3] But there is something interesting about such recurring reports: *Atlantis has never been found.*

The weird claims about advanced technology, crystals, and paranormal powers attached to Atlantis mean its existence is unlikely. Some believe, for example, that magical or at least highly sophisticated survivors from the sunken city migrated to other continents and founded great civilizations such as Egypt and Greece. Charles Berlitz, the same writer who stirred up belief in the Bermuda Triangle, also wrote *Atlantis: The Lost Continent Revealed.* Berlitz goes so far as to promote the belief that the people of Atlantis possessed nuclear weapons many thousands of years ago and it was a large-scale nuclear war that destroyed their culture!

The citizens of Atlantis are widely believed today to have been not only highly advanced but extraordinarily wise and peaceful as well. Atlantis was not merely a great ancient culture, it was a utopia inhabited by angelic superbeings. This directly contradicts the original source of the Atlantis story, however. Greek philosopher Plato wrote that the people of Atlantis were warlike and lacked the sense to avoid their own demise.

One version of the story is that surviving slaves from Atlantis are our ancestors. They settled around the world and attempted to recreate Atlantis culture and technology after the disaster but failed because they just weren't smart enough. The troubled and divided world we see around us today is the result of their fumbled attempts.

Despite the absence of good evidence for any of these claims, Atlantis remains a popular belief today. A 2006 Baylor study found that 41.2 percent of Americans believe in the existence of "ancient advanced civilizations such as Atlantis."[4] A study on pseudoscientific beliefs in America's classrooms revealed that 16 percent of high school *science teachers* believe in Atlantis.[5]

Many skeptics scoff at the mere mention of Atlantis and can't resist berating those who dare bring it up. I'm a bit more nuanced in my rejection of the Atlantis claim. No, I don't think the remains of a superadvanced high-tech city are submerged somewhere out in the middle of the Atlantic Ocean. Nevertheless, my mind is wide open to the likelihood of many stunning discoveries yet to come from marine archaeology. This is partly because of an enlightening interview I did with Bob Ballard a few years ago. Ballard is the underwater explorer who discovered the *Titanic* and life-supporting hydrothermal vents in the deep waters of the Galapagos Rift. He was bubbling with enthusiasm about marine archaeology and left a lasting impression on me about the promise of underwater discoveries. Ballard believes many amazing finds are likely to be made in the coming years as robot and search technology continues to improve. Imagine, for one example, finding ancient shipwrecks that still contain well-preserved human bodies from thousands of years ago, thanks to anaerobic environments in deep-sea muck. It could happen, says Ballard.[6] We have been a seafaring species for a long, long time and there is no doubt that much remains to be found beneath the waves that cover more than two-thirds of our planet's surface. But what about Atlantis? If I'm so excited about underwater archaeology, why do I stop short of believing in Atlantis?

While I have no problem with the possibility of an ancient coastal city named Atlantis being destroyed by a volcano, earthquake, or tsunami, it's an entirely different thing, however, to believe that the inhabitants of this city were aliens, were magical, or were technologically advanced to the point of possessing nuclear weapons and aircraft. A city meeting with a disastrous end is not such a far-fetched idea. It could happen. *It has happened*. There is still a problem with even this most down-to-earth Atlantis claim, however: we have no good evidence that it happened to a city named Atlantis. Apart from Plato's mention of Atlantis in his writings[7] more than two thousand years ago, supposedly based on a very old story he heard about, there is nothing else to back it up. History is silent on Atlantis. This should raise a gigantic red flag in everyone's mind. The most reasonable conclusion is that Plato was relating a fictional story as a teaching tool.

If there really was a prominent culture that boasted supernatural powers, was run by aliens, had astonishing technology, or was, at the very least, a major political and military power of the day, there should be direct and obvious references to it in other writings by other cultures. But they are not there. Archaeologist Kenneth Feder, author of *Frauds, Myths, and Mysteries: Science and Pseudoscience in Archaeology*, has considered the Atlantis claim and concludes that the cross-cultural silence about it is telling. He cites the supposed war between Atlantis and Athens that Plato described as one example:

> It is inconceivable that there would be no mention of a great military victory by Athens over Atlantis—or anyplace even vaguely like it—in the works of Greek historians who followed Plato. Yet this is precisely the case. . . . Similarly, you will not read the discourses of modern historians arguing for or disputing the historicity of *The Lord of the Rings* or Harry Potter, because these are understood, of course, to be works of fiction. In much the same way, Greek historians who followed Plato did not feel the need even to discuss his story of Atlantis; they understood it as the work of fiction Plato intended it to be.[8]

The Atlantis scenario may be plausible in some basic form, that is, a coastal or island city was hit by a natural disaster such as an earthquake or a tsunami. However, the popular descriptions of a magical, advanced, and powerful ancient Atlantis are far from convincing for anyone who demands good reasons to believe unusual claims. It's difficult, for example, to get past the idea of Berlitz's nuclear war occurring ten thousand years *before* the Dark Ages. There is also a practical problem for those who insist Atlantis was a "continent." "City" and "continent" are not interchangeable, as it might seem given the inconsistent manner in which Atlantis is described. This distinction matters. While it might be relatively easy to misplace an ancient city underwater somewhere, losing a continent is another matter entirely. I am sure oceanographers and geologists would have found it by now. Believers suggest that the continent simply flooded over. Of course that raises the question, Where did all the necessary extra water come from to do that? It might have sunk deeper into the earth after an earthquake and then been covered by water, they say. Possible, I suppose, but there is no geological evidence to support such a spectacular event occurring in relatively recent times. Keep in mind that the time frame is important because it makes it even more unlikely. We know that continents can move great distances and land can change dramatically over many millions of years. But eleven thousand years or so seems like too little a time frame to hide a continent-sized landmass from the prying eyes of modern scientists.

As for the extreme claims of magical or high-tech powers that dominate popular Atlantis belief, they are almost certainly nothing more than the products of human imagination. Where else would these ideas come from? After all, no one has ever come forward with any confirmed Atlantis artifacts. For that matter, no one has ever produced a single artifact of any kind from any ancient culture that demonstrates supernatural powers or advanced technology. But that doesn't stop some people from confidently declaring that the people of Atlantis were extraterrestrials, used crystal energy to power their civilization, could control the planet's weather, and are responsible for missing ships and planes in the Bermuda Triangle. But until somebody actually shows up at a press conference with an eleven-thousand-year-old nuclear-powered chariot that levitates and has a big *A* engraved on the side, it only makes sense to resist believing in Atlantis. Of course, this doesn't mean one shouldn't eagerly anticipate thrilling new archaeological treasures and clues to our past emerging from the depths in the coming years. I know I do.

GO DEEPER . . .

- Ellis, Richard. *Imagining Atlantis*. New York: Vintage, 1999.
- Feder, Kenneth L. *Encyclopedia of Dubious Archaeology: From Atlantis to the Walam Olum*. Santa Barbara, CA: Greenwood, 2010.

Chapter 45

"I'M GOING TO HEAVEN WHEN I DIE."

> **Here is the reality. It has been estimated that in the last 50,000 years about 106 billion humans were born. Of the 100 billion people born before the 6 billion living today, every one of them has died and not one has returned to confirm for us beyond a reasonable doubt that there is life after death. This data set does not bode well for promises of immortality and claims for an afterlife.**
>
> —Michael Shermer

> **I'm an adult. I don't need heaven to motivate me to treat others well, nor do I need it to help me cope with the realization that I will die one day.**
>
> —Anonymous

"The Holy Spirit is here," declared American evangelist Loretta Blasingame. "God spoke a word of knowledge to me right before I even stood up to speak to you. He said someone here has already been healed of cancer. Wonderful things are going to happen tonight. I just know it. God is in this place. Someone just had a tumor on their right side dissolved instantly."[1]

Blasingame was addressing a small gathering of about one hundred believers in the Cayman Islands. After relaying "God's word" of a few more miraculous healings, she told a spectacular story. "I died of a heart attack. I was dead and rose out of my body. I actually saw it lying there from above the room.

"I went to heaven," she explained. "First, I saw two beautiful gates fixed with pearls and diamonds. When they opened, I saw streets paved with gold. It was more beautiful than I could ever have imagined. I saw people eating fruit. But when they ate one piece, another one would appear. Nobody goes hungry in heaven. I even saw angels teaching a group of people how to worship. Maybe they had been saved [born again] on their deathbed, or something.

"Then I saw him," Blasingame said. "Jesus walked up to me and took my hand. This was no dream. I saw him just as I see you people here tonight."

Shrieks, prayers and moans rose from the audience. A woman near me begins speaking in tongues.

"His hair was parted in the middle and had beautiful waves," Blasingame continued. "His beard was cut perfectly. And he had the most beautiful crystal blue eyes. He spoke to me, saying he would anoint me and send me back to Earth."

With great drama Blasingame paused and stared intensely at the audience. There is absolute silence.

"I come against cancer, arthritis, back trouble!" she shouted. "Somebody has fluid on the lungs! You're healed! In the name of Jesus! Someone has female problems; you're healed!"

Several people surged forward so that Blasingame could "lay hands" on them. More shrieks, prayers, and crying filled the room. After being touched by the preacher, some believers collapsed into twitching heaps of contorted flesh. One woman told me later that she felt "electrical charges all over" when Blasingame touched her and she fell down. "I felt dizzy. When I was on the floor, I felt I was at total peace. I saw a part of Jesus reach down to me."

"I knew she meant me when she said 'somebody with female problems is healed,'" another woman told me. "When she spoke I felt God's power all over me. It was wonderful."

A believer declares that it's "obviously true" that Blasingame has been to heaven. "How else could she have this power?" I can think of a few other explanations for her "power," of course, but the story intrigues me nonetheless. Her convincing details about heaven were not really all that impressive, in my view. They seemed more like a list of clichés strung together. Even her description of Jesus felt lifeless for its lack of creativity. According to her, Jesus really does look like that classic "Head of Christ" portrait by Warner Sallman that has hung in millions of American living rooms since the 1940s. But who can say? Maybe Blasingame did visit heaven briefly after she "died." The problem for skeptics, of course, is that she has no evidence. Her faith-healing claims are not good evidence because claims of supernatural healings have been a common feature of thousands of religions spanning thousands of years. If it did not confirm the gods and belief systems of all those pagan priests and prehistoric shamans, then it does not confirm hers either. In the end, all Blasingame has is a story. Maybe if she had returned from heaven with one of those magical self-replenishing fruit baskets to show us all, a strand of Jesus' hair for DNA analysis (maybe something extraordinary would turn up). Even a chunk of gold pavement forged in heaven might have been intriguing. But that's not fair to Blasingame. Who in their right mind would swipe souvenirs from heaven? What she could do, however, is allow her healing powers to be analyzed scientifically. A double-blind test by independent researchers would do the trick. Then we might know if there is something to her claim of having gone to heaven to get empowered by a god.

Never forget that anyone can say anything about anything. Science doesn't have much use for stories alone and science is by far the best method we have for figuring things out and making discoveries. Don't misunderstand me; storytelling is still great. We love a good story. We get thrills from telling a great story and we get thrills from hearing a great story. It's an important part of being human. Good educators know that one of the best ways to get students to remember something is to present it in the context of a story. When confronted with a weird, hard-to-believe tale, one should keep in mind that our species is prone to fantasy, vulnerable to vision and hearing glitches, and human memories just can't be trusted. Often we are just plain dishonest, too. All of this means that a bizarre story needs to be supported by very good evidence. Otherwise it's in the same league with Elvis sightings, alien abductions, and every ghost story ever told.

This is the problem we are faced with when people claim heaven is a real place. There is only the claim, the story, put forward by followers of some religions, with no evidence. There are also the tales told by some people who say they had visions of it or actually went there and returned. Again, no evidence means *nothing but a story*. We know that people can have psychological experiences that seem real to them but did not happen, and we know that people can lie. What we do not know is if a heaven exists. While one certainly can understand the attraction of such a place, it is difficult to understand why so many people would think it is real.

According to a 2007 survey, 81 percent of American adults believe in heaven and 8 percent are not sure. Only 11 percent of Americans do not believe heaven exists. I'm not sure why—perhaps it is a sign of American optimism—but fewer people (69 percent) believe in hell than in heaven.[2] What does it really mean to "believe in heaven" anyway? I contend that this is all a bit more complex than it appears. My hunch is that for all the confident talk and pulpit pounding over it, heaven is more an *emotional hope* than a *known destination* in the minds of most believers. I feel this way because for every one person I have met who lectured me on the specifics of heaven as an exclusive club, I've met a hundred or so more who say their god would let all but the absolute worst of us into heaven—no matter who they are or even what religion they adhere to. Over the years I've observed that the notion of heaven is less divisive and more positive than most other aspects of religions. New research seems to support this. For example, the vast majority of Americans believe that good people who follow different religions can still go to heaven. This is fascinating because it seems to contradict directly the core claims of typical religions as well as the words of their leaders, many of whom are very clear about who will and will not go to heaven. The poll numbers, however, reveal a remarkable generosity of hope. Even a majority of normally rigid and by-the-book evangelical Protestants think that the gateway to heaven is wide enough for almost everyone—even non-Christians—to gain entry.[3] Yes, in the minds of most believers, heaven appears to be more about neighborly love and near-limitless hope than a real destination defined by narrow, tribal theology. Now, if only we can find a way to direct all that post-death goodwill toward the Earth and the living, we might get somewhere.

GO DEEPER . . .

- Boyer, Pascal. *Religion Explained.* New York: Basic Books, 2002.
- Dawkins, Richard. *The God Delusion.* New York: Mariner Books, 2008.
- Harris, Sam. *The End of Faith: Religion, Terror, and the Future of Reason.* New York: W. W. Norton, 2005.
- Hitchens, Christopher, ed. *The Portable Atheist: Essential Readings for the Nonbeliever.* New York: Da Capo Press, 2007.

Chapter 46

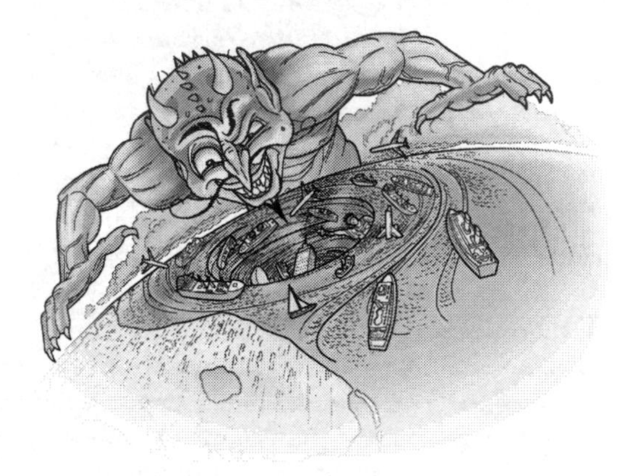

"SOMETHING VERY STRANGE IS GOING ON IN THE BERMUDA TRIANGLE."

The Bermuda Triangle is a manufactured mystery from start to finish.

—Terrence Hines, *Pseudoscience and the Paranormal*

The real mystery is how the Bermuda Triangle became a mystery.

—Robert Todd Carroll, *The Skeptic's Dictionary*

After growing up on the eastern coast of south Florida and then living twenty years of my adult life in the Caribbean, I know that the ocean can be as dangerous as it can be beautiful. Bad things often happen in the Atlantic Ocean and Caribbean Sea. People get lost. People vanish. People die. Incredibly powerful storms routinely wreak havoc in the seas and on coastal communities. Deadly creatures roam free and even currents just offshore sometimes kill weekend swimmers on sunny afternoons.

In 1992, I explored the rubble that was Homestead, Florida, and interviewed many of the newly homeless after category 5 Hurricane Andrew struck. I saw large boats stranded high and dry in fields far from the sea. I saw an eighteen-wheel truck left upside down by winds. In the Cayman Islands I watched a community mourn fishermen who went out to sea and were never heard from again. I've also seen waves taller than my house smash into land on a warm day with clear skies. It should be no surprise to anyone that things can and do go tragically wrong on the water. Not every ship and plane that travels over the ocean reaches its destinations as planned. If *nothing* unexpected and unfortunate ever happened out there, *then* we would have a real mystery.

During two decades spent swimming, snorkeling, diving, beachcombing, and working as a journalist in the Caribbean, I developed a heightened appreciation for how dangerous the ocean can be. Twice I even found myself in situations where I could have ended up a "missing at sea" statistic. While snorkeling one early morning just outside the reef along a lonely stretch of beach I had a face-to-face encounter with a seven-foot bull shark. It probably was attracted to the area by blood in the water from nearby spear fishermen and was only looking me over to see if I was towing a string of fish it could steal. I know that I'm more likely to be struck by lightning than eaten by a shark, but statistics offer little assurance when one is less than ten feet away from a shark and too far from shore for a quick escape. I felt virtually helpless—and was. Furthermore, watching National Geographic and Discovery Channel shark documentaries had not prepared me for the stunning size and power of this creature. Seven feet (an honest estimate, taking into account the way in which fear and water magnify objects) may not sound so big, but up close the shark seemed like a blue whale to me. Its rows of misaligned teeth, muscular bulk, and the way it sliced through the water sent me into a state of near panic. I have never had an irrational fear of sharks, but this was primal; my brain and my body went into full fight-or-flight mode. After a face-off that felt like forever and a brief game of peekaboo around a coral head, the shark surged away with a flick of its tail and faded into the deep. The encounter left me trembling and grateful to be alive.

On another occasion I ignored high winds and attempted a short but risky trip between two islands in a small inflatable boat. I grossly underestimated the strength of both the winds and the currents and my little boat rapidly gained speed—toward the open sea. What was supposed to be a brief crossing turned into a serious threat to my life. I was very fit at the time but struggled to make progress against the elements. With each stroke I fought back panic, fully aware that I could end up lost at sea. Swimming for shore was an option, of course, but it would mean abandoning the boat and all the supplies in it. It would also be a very long and difficult swim. I dug deep and paddled for what seemed like eternity before finally reaching shore, where I flopped down on the sand, exhausted. Memories like these left me with no doubt about the ever-present dangers of the ocean. Twice, on a sunny day and in sight of shore, I might have ended up missing at sea, another mystery open to speculation.

The area known as the Bermuda Triangle is fully capable of swallowing up the occasional ship or plane without the aid of anything supernatural or paranormal. This is why I find the wild claims and stories of the Bermuda Triangle not just unproven, but unnecessary as well. The beautiful but indifferent ocean sometimes kills the unskilled, the unprepared, and the just plain unlucky. That's the reality. Yes, boats, planes, and people go missing in a vast area that has a large volume of traffic. What's unusual about that? Why in the world would anyone think an explanation involving magic, Atlantis, extraterrestrials, black holes, or interdimensional time shifts are required? But some people do. However, the Bermuda Triangle may be one of the safest places in the world. That is what the evidence suggests. It turns out that the accident rate is actually *lower* there than in surrounding areas! For this reason, skeptic and science historian Michael Shermer suggests it might make more sense to call the area the "Non-Bermuda Triangle."[1]

The Bermuda Triangle, or Devil's Triangle as it is also known, is a classic case of how "nothing unusual" can be embellished and sold as "something very unusual." In figuring out the Bermuda Triangle, it helps to know something about the origin of the claim. The source is telling. The first use of the name for a specific triangular region that claims ships and planes by some mysterious means is widely believed to be an article in a 1960s pulp magazine. More than anything, Vincent Gaddis's story, "The Deadly Bermuda Triangle," in the February 1964 issue of *Argosy* magazine seems to have launched the myth that continues to tread water against all reason. *Argosy* would never be confused with *National Geographic* or *New Scientist*. It was a magazine that published *fiction* and *sensationalized stories* meant to appeal to adventure-minded young men. The late Gaddis was also the author of the books *Gold Rush Ghosts* and *Wide World of Magic*. He was not a scientist nor was he an expert on marine safety or marine history. None of this is meant to label Gaddis a con man or condemn *Argosy* for trying to publish interesting and fun stories that were loose with the truth. However, I think it's important to be aware of the original source because I have encountered people who are under the false impression that the Bermuda Triangle claim is somehow rooted in science, supported by scientists, and was/is endorsed by the US Navy and Coast Guard. None of that is true. The Bermuda Triangle myth was born out of imagination and has been kept alive all these years by sloppy reporting, exaggerations, and misrepresentation of facts—also known as lies.

If Gaddis and his *Argosy* article created a spark, writer Charles Berlitz fanned the flames to create a bonfire with his 1974 "nonfiction" bestseller, *The Bermuda Triangle*. It sold millions of copies and probably did more than anything else to entrench this weird claim into popular culture. I believe there is one primary reason the myth continues to appeal to so many people: great stories. The Bermuda Triangle was built not on statistics and evidence but on creepy stories of vanishing planes, missing people, strange sights in the skies, and mysterious ships being found adrift with no crews aboard. We are all suckers for a good tale. Thousands of years before films and novels, there was the campfire. Humans have always enjoyed sharing stories with each other. They are the foundation of religions, cultures, nations, and families. They hook us and set us up to learn and retain information—good or bad. So it should be no surprise that many people find the tale of Flight 19 irresistible.

THE FATAL JOURNEY OF FLIGHT 19

December 5, 1945. Flight 19, a group of five US Navy Avenger torpedo bombers based at Fort Lauderdale Naval Air Station vanished without a trace in ideal weather during daylight hours while on a routine training exercise. A frantic transmission from the pilots was received by the control tower back in Florida: "Everything is wrong . . . strange . . . we can't be sure of any direction. . . . Even the ocean doesn't look as it should!" Then silence—forever. Their fatal mistake had been to fly into the deadly and unforgiving Bermuda Triangle. Adding to the tragedy, a rescue plane sent to search for them disappeared as well. Despite an intensive search, no trace has ever been found of the planes. After an official investigation, one Navy officer said it was if the planes had "flown off to Mars."

Now that's the kind of story that sticks and gives a myth legs. But is it accurate? Fortunately, librarian and Bermuda Triangle skeptic Larry Kusche compiled credible research that suggests far more reasonable explanations for what may have happened to many of the ships and planes that went missing. His book, *The Bermuda Triangle Mystery—Solved* is fun to read and does a very effective job of dismantling the myth, piece by piece. Kusche conducted a detailed investigation of Flight 19, for example, and found numerous problems with the popular version of this story that has inspired so many Bermuda Triangle believers.

Kusche believes—just as the US Navy does—that Flight 19 simply got lost and the planes crashed into the sea and sank. The rescue plane that followed, a Martin PBM Mariner, blew up, probably due to a problem with the fuel system, something that model had a notorious reputation for. Kusche reveals that much of the popular version of the Flight 19 story is exaggerated, twisted, or entirely made up out of thin air. For example, there is no record in the transcripts or testimonies by navy personnel of dramatic radio transmissions from the pilots describing strange environmental conditions. In fact, transcripts of the radio transmissions read pretty much like one would expect from a lost group of planes and nothing more. Kusche points out that the flight's leader, Lieutenant Charles C. Taylor, had recently transferred to Florida and was not yet familiar with the area. And while Taylor was experienced, the other four pilots were not.[2]

One thing that seems to add to the legend is that five planes vanished rather than just one. This is not as strange or unlikely as it might seem, however. Flight 19 was lost and the pilots would have made every effort to stay together. Therefore they all would have run out of fuel at approximately the same time and ditched in the same general area. Kusche also found that the weather was not as ideal as it is almost always described in Bermuda Triangle lore. "Although [the weather] had been fair when they took off, it rapidly deteriorated," explains Kusche. "Search planes reported extreme turbulence and unsafe flying conditions, and one ship in the area reported 'high winds and tremendous seas.'" After flying around lost for some four hours, Kusche says most likely the Avengers ran out of fuel and went down somewhere in the Atlantic. Tragic yes, but nothing so odd given the fact that they were lost. Kusche adds: "Flight 19 was not a group of experienced pilots touching down on a calm sea in the middle of a sunny afternoon—it was one disoriented instructor and four student pilots attempting to ditch at sea on a dark and stormy night. It was a hopeless situation."[3]

Kusche also points out that, contrary to the myth, the US Navy Board of Investigation that looked into the incident was not "baffled" by it as is so often claimed by Triangle believers. The board's opinion after considering all the facts was that Flight 19 "made forced landings in darkness at sea east of the Florida peninsula" and that the conditions were "rough and unfavorable for a water landing."[4] There is nothing about this event that should have led anyone to imagine that paranormal forces are behind it.

While the story of Flight 19 is the most famous Bermuda Triangle tale, there are many more. Collectively they seem to make the case that *something* must be happening in the Triangle. Where there's smoke there's fire, right? One that often comes up is the disappearance of Joshua Slocum, one of history's great seamen. I interviewed one of his great grandsons in the early 1990s and became very interested in Slocum's story. The New England captain gained international fame as the first person to sail around the world alone. He did it in the 1890s aboard the thirty-six-foot *Spray* and wrote, *Sailing Alone around the World*, an entertaining book about the adventure. In 1909, however, the aging Slocum sailed for the Caribbean and was never heard from again. At the time, it was believed that he most likely had been run down at night by a freighter, or perhaps the *Spray* capsized in rough seas and he drowned. Only many decades later would his name be attached to the Bermuda Triangle. When Triangle believers tell the Slocum story today, it is sure to include emphasis upon his sailing skill: "He was too smart to have made a mistake at sea," goes the reasoning. "Something very mysterious must have happened to him." This is, of course, preposterous. Having elite skill at something does not make one infallible or invulnerable. If it did, Formula One racecar drivers and fighter pilots would never crash. And Tony Hawk would never need to wear a helmet and kneepads when he skateboards. But he does.

WHERE IS IT?

By the way, where exactly is the Bermuda Triangle anyway? It's a basic question, but the answer depends on whom you ask. No official Triangle boundaries have ever been designated by any person or organization with relevant credentials. There are no official military or government maps with the Bermuda Triangle marked on them. The most popular version tossed around today is close to the one popularized by the 1964 *Argosy* article. That triangle is formed by drawing a line from somewhere in south Florida, up to Bermuda, down to Puerto Rico, and then back to Florida. It should be not be surprising that there is ambiguity, however, because the entire claim is made up. Anyone can draw boundaries anywhere they want. Some believers claim that the danger zone extends down to cover the entire Caribbean Sea and west into the Gulf of Mexico. Yikes! According to those boundaries, I lived most of my entire life inside the Bermuda Triangle! I remember losing my car keys once, but apart from that mysterious incident, I guess I was lucky. Still others say the Triangle is nebulous and specific borders can't be drawn. That's convenient; let's just agree that every missing plane and ship anywhere should be considered a victim of the Bermuda Triangle. Don't laugh, that is precisely what Triangle believers have done in many cases. Skeptic Kusche points out that blame for Triangle vanishings has been given broadly and with little restraint. "If all the locations of 'Bermuda Triangle incidents' were plotted on a globe it would be found that they had taken place in an area that included the Caribbean Sea, the Gulf of Mexico, and most of the North Atlantic."[5] He also points out that many missing ships and planes should not be assumed to have been lost in the Triangle just because their course may have taken them through it. They were on long journeys and could have gone down outside the Triangle.

While working his way through numerous Bermuda Triangle stories, Kusche made a surprising discovery: "My research, which began as an attempt to find as much information as possible about the Bermuda Triangle, had an unexpected result. After examining all the evidence, I have reached the following conclusion: *there is no theory that solves the mystery.* It is no more logical to try to find a common cause for all the disappearances in the Triangle than, for example, to try to find one cause for all automobile accidents in Arizona. By abandoning the search for an overall theory and investigating each incident independently, the mystery began to unravel."[6]

For years, Kusche and other skeptics have made clear arguments against the Bermuda Triangle claim. The following is a compilation of key reasons why the myth has managed to thrive all these years:[7]

- Many of the incidents occurred late in the day or at night, which delayed visual searches and made it more unlikely to find debris.
- Many cases were not mysteries at all in light of known facts. For example, it is not mysterious to find a ship at sea without a crew soon after a hurricane struck the port where that ship had been moored.
- The incidents that are still unsolved remain so because there is insufficient evidence—not surprising considering that debris usually sinks at sea. The absence of evidence, however, does not justify reaching for the most unlikely explanations of all, such as aliens, black holes, and so on.
- Vehicles and people go missing on sea and on land. It happens everywhere in the world.
- Some missing vessel stories that are credited to the Bermuda Triangle did not happen there. Kusche describes the case of the *Freya*, for example, that left port in western Mexico, had troubles, and was discovered in the *Pacific Ocean*.
- Some planes and ships that went missing passed through the Triangle, but this does not justify claims that they went down there.
- Many incidents were not considered mysterious at all when they occurred, but many years later were described that way by writers promoting the Triangle myth.
- The weather was bad in many incidents, but often this is omitted or even contradicted in Triangle tales. Kusche also found that key information that points to an obvious likely explanation is often left out.

It should be clear by now that the Bermuda Triangle "mystery" does not hold up to scrutiny. It relies on exaggerated stories that omit key information and inject outright fabrications. Even if every story was factual—and actually took place inside some agreed-upon area called the Bermuda Triangle—there is still a problem. Pointing to a list of missing ships and planes in a vast expanse of water and then concluding that there must be some sinister paranormal reason behind it all can't be justified. Ships and planes go missing for natural and mostly well-understood reasons. If someone wants to make the claim that supernatural or paranormal forces are behind it, then he is obligated to prove it.

Still not convinced that there is nothing magical and sinister going down in the big triangle? Then how about the simple fact that the United States Navy—an organization that knows a lot about oceans, ships, and flying over water—thoroughly rejects the Bermuda Triangle claim. The navy does not believe in it or worry about it in the slightest, even as it operates multimillion-dollar ships, aircraft, and submarines, and sends thousands of sailors into the Bermuda Triangle every year.[7] But maybe the navy is too concerned with war and national defense to notice paranormal forces working against freight ships and weekend boaters. Therefore, to be thorough, I thought I should check with the people who do spend a lot of time specifically thinking about the safety of cargo ships and weekend boaters. I phoned the Miami, Florida, office of the US Coast Guard and asked if their organization has any concerns about unusual phenomena occurring in the Bermuda Triangle. "No," was the concise answer provided by a polite lieutenant who also referred me to the official US Coast Guard statement on the matter:

> The Coast Guard does not recognize the existence of the so-called Bermuda Triangle as a geographic area of specific hazard to ships or planes. In a review of many aircraft and vessel losses in the area over the years, there has been nothing discovered that would indicate that casualties were the result of anything other than physical causes. No extraordinary factors have ever been identified.[8]

It seems simple enough, if the US Navy and US Coast Guard don't believe in the Bermuda Triangle, why should anyone?

GO DEEPER . . .

- Earl, Sylvia, and Linda K. Glover. *Ocean: An Illustrated Atlas*. Washington, DC: National Geographic, 2008.

- Kusche, Larry. *The Bermuda Triangle Mystery—Solved.* Amherst, NY: Prometheus Books, 1995.
- Slocum, Joshua. *Sailing Alone around the World.* Memphis, TN: General Books, 2010.

Chapter 47

"AREA 51 IS WHERE THEY KEEP THE ALIENS."

You can't prove that aliens are visiting Earth by pointing to government secrecy. That establishes nothing.

—Seth Shostak, senior astronomer
at the SETI Institute

According to many UFO believers, a large area of land in southern Nevada is home to some very unusual activities involving extraterrestrials and the United States government. Some say debris and bodies from the alleged 1947 crash at Roswell, New Mexico, are kept there. Some believe the US military is reverse engineering alien spacecraft there. Many people claim to have seen alien spacecraft flying in the area. Also known by the names Paradise Ranch and Groom Lake, it is familiar to most by its now-legendary name, "Area 51." Although it's less than ninety miles from Las Vegas, Area 51 may as well be on far side of the Moon given all the secrecy and weird rumors surrounding it. The base is off-limits to the public and constantly guarded. Nonetheless, the US government has gone on record admitting that it is a facility where the US military and CIA develop and test new aircraft and new weapons systems. But that explanation has not satisfied many people who focus their UFO hopes and fears on this one specific air base.

Before considering whether or not aliens are on ice and flying saucers are housed in hangers at Area 51, it's important to understand that weird and mysterious flying vehicles *made by humans* are known to have been developed and flown there. This is well documented and widely known. For example, the U-2, A-12, and SR-71 jets were developed in extreme secrecy at least in part at Area 51. These reconnaissance planes flew at various times in the 1950s through the 1990s and were very different from conventional aircraft in both design and function. The SR-71 Blackbird was capable of Mach 3.3 speed (over 2,000 mph) and could operate at 85,000 feet. Even today, long after retirement, it looks like it belongs more on the set of the TV show *Battlestar Galactica* than on display in air and space museums as a relic of the Cold War. The F-117 Nighthawk and B-2 Spirit bomber were also secretly developed and test flown at Area 51. These revolutionary stealth planes still have a futuristic look to them more than two decades after their public debut. The F-117 was like nothing seen before in the history of aviation. Its swept wings, weird boxlike structure, tailless body, hybrid rudder-elevator design, and sharp angles could easily lead imaginations astray. The big B-2 bomber with its bat-wing design looks like it was straight out of a *Batman* comic book—or from another world. I saw one on the ground at an air show and I've seen video of one in flight. At certain angles the B-2 looks exactly like—surprise—the stereotypical flying saucer from outer space. Imagine what someone who happened to see one of these aircraft flying at night in the 1970s might have thought. An untrained and unexpecting person might easily assume that it was an alien spacecraft. The same probably still holds true today. Given the large sums of money poured into military attack and reconnaissance aircraft, it's safe to assume that much of whatever the US Air Force and others are up to out in the Nevada desert would surprise the average person. We, the public, are probably twenty years or more behind when it comes to knowing what's buzzing around above our heads. Unfortunately, however, the reasonable idea that strange sightings near a *secret military base* are probably nothing more than *secret mil-*

itary testing does not satisfy the cravings and suspicions of the more enthusiastic students of Area 51.

According to believers, Area 51 has secrets that extend far beyond the latest jet from Boeing or Lockheed Martin. One of the more popular claims is that the military took an alien spaceship that crashed at Roswell in 1947 to Area 51 in order to study it. Much of current aviation and space technology, they say, was obtained by reverse engineering that Roswell spacecraft. In addition to that, some people believe that dead extraterrestrials from the crash were taken to Area 51 in order to perform autopsies and store them away in secrecy. Those are big claims, but there's more. Some Area 51 enthusiasts also maintain that the government is developing teleportation and time travel capabilities there as well. The gargantuan, insurmountable, and fatal problem for all of this is that none of these claims are proven or even come with compelling evidence. The best we have to back up these claims are stories, UFO sightings, dubious documents, and, of course, that alien autopsy video that Fox Network made into a television special in 1995. Simply put, none of this is good enough. It doesn't come close to adding up to the sort of high-quality evidence required to back up such extraordinary claims. Stories can be based on lies, mistakes, or inaccurate memories. For these reasons, hard evidence is absolutely necessary to support stories as wild as the ones orbiting Area 51. UFO sightings may be interesting but they too fall short. Streaking lights in the night sky over Nevada are probably the latest bizarre, ultrasecret vehicle the US Air Force is testing that we won't learn about until many years from now. As for that "alien autopsy" film, it's just another silly dead end. The guy who made it confessed to the hoax in 2006.[1]

Why do some people insist on believing that a patch of Nevada desert is some kind of a spaceport or warehouse for dead aliens? Presumably the intense secrecy and security around Area 51 encourage these ideas. But I fail to see a big mystery here. Why wouldn't there be tight security? This is where the military prepares for wars that might have to be fought twenty, thirty, or fifty years from now. Pushing aviation engineering to the limits and then trying to reach beyond them is expected, and keeping new technologies a secret can be the difference between victory and defeat in war. As far as reports of seeing weird and unidentified flying objects around Area 51, isn't that to be expected as well? It's a *secret* base for developing *secret* technology and aircraft. If anyone could stroll around Area 51 anytime they wanted and if no weird shapes or lights were ever spotted in the southern Nevada sky, *then* I might be suspicious.

It is easy to underestimate just how bizarre top-secret aircraft probably are right now. A lack of imagination makes it more likely for a person to see something they can't make sense of and conclude that it must be otherworldly. Did you know, for example, that there are programs under way in the US right now to develop large unmanned vehicles capable of staying aloft at very high altitudes for five years or more without landing? Some are essentially giant, solar-powered wings like NASA's Helios. There are also huge, secret airships designed to hover close to the edge of space where they can serve as cheap alternates to surveillance satellites. Some of these aircraft don't fit the general public's preconceived notions of what flying machines are supposed to look like. I don't know if these kinds of aircraft are being developed at Area 51 specifically, but they could be.

It is also possible that the rapid rise of UAVs (unmanned aerial vehicles) is contributing to UFO sightings around Area 51 and elsewhere. I was at a conference for future technologies in Chicago a few years ago and was able to see a Global Hawk UAV and talk with a Northrop Grumman representative. I had seen this particular one before in photos and on TV but was surprised by how sleek and futuristic it looked up close. The winged robot seemed to project power and intelligence with its bulbous nose section and aesthetic lines. One detail I have noticed that is common to many UFO reports is that observers claim the light or object moved in ways no airplane ever could. It's important to remember that it's difficult and sometimes impossible to accurately determine distance, direction, and speed when viewing something you can't identify in the sky. This means it is relatively easy to misinterpret the movements of an unknown object. Aside from that, however, could it be that some of these descriptions of radical and "impossible" movements can be attributed to UAVs? UAVs are significantly different from traditional aircraft because with them there is no concern about a human pilot blacking out in the cockpit. UAVs are held back only by engine-to-weight ratio and structural strength. Undoubtedly advanced UAVs will be capable of flying in ways far beyond the limits of crewed military aircraft. It's no stretch to imagine that the latest generation of UAVs, such as the US Air Force's RQ-170 Sentinel and, of course, secret UAVs we don't yet know anything about, might confuse a layperson with their extraordinary maneuvering abilities.

Figure 6. This photo, taken in 1961 at an aviation and rocket research facility, is clear proof that the United States government was in possession of an extraterrestrial spacecraft during the Cold War. Or, more likely, it's just a photo of technicians cleaning the domed top of a vacuum tank made by humans for human use. As this innocent but potentially misleading photograph suggests, secretive cutting-edge research and testing of new aircraft over the last sixty years undoubtedly has contributed much to UFO belief. Photo: NASA

As mentioned in the UFO chapter, many strange sightings around Area 51 might be explained by something as simple as flares, a defensive countermeasure used by many military aircraft. In order to evade incoming missiles that seek a heat source (like a jet engine), many military aircraft, including helicopters, are equipped to fire multiple high-intensity flares at once or in rapid succession in hopes of fooling the missile into chasing a flare instead of the aircraft. Some of these flares fall passively, but many of them spiral away from the aircraft rapidly on their own unique course. It's not difficult to imagine that the possible testing of next-generation countermeasure systems around Area 51 could lead to UFO sightings.

Seth Shostak is the senior astronomer at the SETI Institute, which means he spends most of his waking hours not just thinking in general about the possible existence of intelligent extraterrestrials but thinking of ways in which we might detect them if indeed they are out there somewhere. He is open-minded about UFO claims but feels strongly that speculating about the government hiding aliens from us at Area 51 or anywhere else is a waste of time.

"I think that new and compelling evidence should always be investigated, but I would suggest to the UFO community that, despite more than six decades of claims, there's still no evidence of visitation good enough to stack up in the Smithsonian," he said. "I think that this circumstance should probably induce a degree of modesty on this subject and not—as all too often happens—an appeal to lack of interest by scientists or hidden evidence. You can't prove that aliens are visiting Earth by pointing to government secrecy. That establishes nothing."[2]

The fact is we don't know exactly what is going on at Area 51. But that's the whole point of Area 51. What we can safely assume is that there are cutting-edge aircraft hidden away in hangers there that come out at night and tear up the skies. Considering how bizarre and futuristic-looking some formerly secret aircraft look to most of us today, it's very likely that some experimental aircraft flying in secrecy today would blow our minds if we caught a glimpse of them. This base is no small-time operation. A report in Smithsonian's *Air and Space* magazine estimated annual expenditures at Area 51 to be more than $30 billion per year. That's almost $100 million per day being spent on new aviation technologies.[3] In 2011, an uncrewed US Air Force "plane" that was rocket launched and then reentered the atmosphere, crashed into the Pacific after achieving speeds twenty times the speed of sound. The Falcon Hypersonic Technology Vehicle 2 is part of a project to develop a bomber that can reach a target anywhere in the world within one hour. They told the public about this one. Just imagine all the aircraft that they don't tell us about. It is confirmed that the US military is sending aircraft into space and returning them. The potential for yet more UFO sightings should be obvious.

William Scott, an aviation reporter, believes that UFO excitement serves as a convenient distraction from the reality of what goes on in Area 51 that the US military and CIA find useful. "The UFO phenomenon is used to protect the base's deepest secrets. I once was advised that if I wanted clues about real-world classified aircraft projects, I should read the supermarket tabloids. . . . I once asked a Groom [Area 51] test pilot whether tainting classified-aircraft sightings with the UFO stench was ever done intentionally. He smiled and replied: 'It's worked for fifty years. Why would we change now?' Without question, black-world operators have become masters of such deception to protect their work. As a result, Groom Lake will likely retain its secrets for a very long time."[4]

During the height of the Cold War in the 1950s and 1960s, the CIA estimates that more than half of all reported UFO sightings were caused by flights of American U-2 and SR-71 spy planes.[5] *More than half.* The Cold War has ended, but military spending on new and exotic aircraft certainly has not. The "war on terror" has helped to spark an unprecedented rush to design and build robot spies and warriors for land, sea, and air. In his fascinating book *Wired for War: The Robotics Revolution and Conflict in the 21st Century,* P. W. Singer points out that the US Air Force now places orders for more UAVs than crewed aircraft. The skies above us are going to be increasingly filled with new and strange flying machines in the coming years. So the next time you see something odd in the sky, or someone tells you about a UFO sighting linked to Area 51, ask yourself if you can be sure it's not reflected light off the belly of a lumbering P-791 hybrid air vehicle,[6] an X-47B UAV[7] shredding the sky, or maybe even something unknown and unimaginable to the public at this time. If we can't be sure of those possibilities, then we can't justify leaping to the most extraordinary conclusion of all, that it must something from another world.

GO DEEPER . . .

- Burton, Robert A. *On Being Certain: Believing You Are Right Even When You're Not*. New York: St. Martin's Griffin, 2009.
- Merlin, Peter W., and Tony Moore. *X-Plane Crashes: Exploring Experimental, Rocket Plane & Spycraft Incidents, Accidents & Crash Sites*. North Branch, MN: Specialty Press, 2008.
- Patton, Phil. *Dreamland: Travels inside the Secret World of Roswell and Area 51*. New York: Villard, 1999.
- Rich, Ben R. *Skunk Works: A Personal Memoir of My Years of Lockheed*. New York: Back Bay Books, 1996.
- Singer, P. W. *Wired for War: The Robotics Revolution and Conflict in the 21st Century*. New York: Penguin, 2009.

DREAMING OF THE END

Chapter 48

"THE MAYANS WARNED US: IT'S ALL OVER ON DECEMBER 21, 2012."

[Two teenagers] were considering killing themselves, because they didn't want to be around when the world ends. Two women in the last two weeks said they were contemplating killing their children and themselves so they wouldn't have to suffer through the end of the world.

—David Morrison, NASA astronomer

If you wait for tomorrow, tomorrow comes. If you do not wait for tomorrow, tomorrow comes.

—West African proverb

The wild-eyed believer is worried because I'm not worried.

"Seriously, this is no joke," he tells me. "In late 2012, maybe early 2013, the Sun is going to fry the Earth."

"But how do you know this and NASA doesn't?"
"They know."

I have had more than a few encounters like that one recently. I have no right to complain, however. I usually seek them out to learn more for this book. But wait, what was I thinking? Why did I bother writing this book anyway? If the world is going to explode, implode, collide, flood, fry, freeze, or vanish on December 21, 2012? What's the point? I should be busy selling my possessions, hugging my children, and eating as much ice cream as possible before doomsday arrives. But, no, here I am writing away as if the world will still be here after that fateful date in 2012. Will there be a world for us on the morning of December 13? To save readers suspense, I checked to see if there is any good evidence or credible experts that support this extraordinary prediction. I also investigated the sources of 2012 apocalypse belief. What I found may terrify you, but I'll give it to you straight anyway, no holding back: While there are no guarantees, all indications are that the world is *not* going to end any time soon and we are all going to have to continue existing whether we like it or not. Sorry, you still have to pay those bills, save for retirement, and floss.

Although they could never have known it would come to this and shouldn't be blamed, the Maya of Central America and their calendar are the source of 2012 hysteria. Or, perhaps I should say misinterpretations, mistakes, exaggerations, and lies about their calendar are the sources. The Maya calendar is cyclical and resets to year zero every five thousand years or so. That's all there is to it, really. It's like the calendar on your fridge that eventually hits December 31 and ends. What's the big deal? Go buy another calendar and get on with your life. Experts on Maya culture—people who have dedicated their lives to studying Maya culture—disagree with the claim that the Maya predicted the end of the world in 2012.

"There is nothing in the Maya or Aztec or ancient Mesoamerican prophecy to suggest that they prophesied a sudden or major change of any sort in 2012," explains Mark Van Stone, author of *2012: Science and Prophecy of the Ancient Maya*. "The notion of a 'Great Cycle' coming to an end is completely a modern invention. Maya inscriptions that predict the future consistently show that they expected life to go on pretty much the same forever. At Palenque, for instance, they predicted that people in the year 4772 AD would be celebrating the anniversary of the coronation of their great king Pakal."[1]

So here we have the odd situation where a past culture *didn't* predict a December 21, 2012, doomsday date but millions of people are choosing to believe they did anyway. More to the point, even if the Maya had clearly made this prediction, it still shouldn't cause anyone to worry. After all, how likely is it that a preindustrial culture was able to identify a specific date for a global catastrophe *centuries in advance* while NASA and all the world's scientists today can't see it coming mere months in advance? As absurd as this belief may be, it's not unique in its lack of a credible origin. This failure to consider and assess the source of extraordinary claims is a widespread problem that reaches far beyond 2012. Many irrational beliefs depend on people failing to question how the source knows what it says it knows. When it comes to the age of the Earth, for example, many people choose to trust the word of a preacher with no science education over the conclusions of professional geologists with doctorates. On some matters of science, millions give more credence to radio talk show hosts and politicians than they do the world's leading scientists. The source of a claim may not prove or disprove anything, but it is often a good starting point for someone who sincerely wants to figure out if something is more or less likely to be true.

The popularity of the 2012 claim is remarkable, even for a species long-obsessed with bizarre and baseless end-of-the-world predictions. In most cases, these predictions are conceived and promoted by organized religious groups. The 2012 doomsday claim sprang from one weird notion about an old Mesoamerican calendar and then snowballed to lure in believers all over the world. It also has joined forces with a long list of paranormal and pseudoscience mainstays, including Nostradamus, astrology, aliens, Atlantis, psychics, reincarnation, and the Bermuda Triangle. It has been surprising, even to me as one who has observed unusual beliefs for many years, to see the connections people have made with 2012. Some say the Maya prophecy even involves Hitler somehow! It was, of course, a no-brainer that Nostradamus would be tied to 2012. *Ah, yes, Nostradamus knew centuries ago that 2012 would be doomsday. Yes, it's all so clear in his writings.* Of course, after 2012 passes without anything happening Nostradamus's name will be dropped from any further association with the Maya. Then, soon enough, another doomsday prediction or surprise catastrophe will come along. And we will be told that Nostradamus knew all about it centuries ago and warned of us, of course.

Belief in 2012 has been everywhere in recent years, from books to websites to blockbuster movies. It's even in the Bible, according to *The Bible Code* author Michael Drosnin. He decoded "2012," "comet" and "Earth annihilated" from a passage in the Bible.[2] Of course, it has to be mentioned that Drosnin is the same guy who also found an encoded biblical prediction of a nuclear holocaust that was supposed to have occurred in 2006.[3] Silly as all this may seem to some, I think 2012 is worthy of study and should not be forgotten after the date comes and goes without global doom. This is a textbook-worthy demonstration of how a slim scrap of an idea can be inflated with bluster, tied to other hollow claims, and then sweep up millions of imaginative people. If more psychologists, anthropologists, historians, and journalists analyzed these recurring group fantasies and spoke about them more, we might figure out ways to make people less vulnerable to them.

There are many claims about what exactly is supposed to happen on December 21, 2012. One of the more creative ones is the approaching "alien invasion fleet," an image of which was captured for all to see by a powerful telescope. It turned out to be nothing more than an "image defect on the observation plate," however.[4] In fact, none of the specific claims for how our planet will be destroyed are worth worrying about, at least not if the world's astronomers can be trusted to know anything about planets and space. For example, the popular 2012 claim of "planets aligning" or a "galactic alignment" may sound important, maybe even dangerous, but we can relax, say astronomers. Astrologers and 2012 believers may talk about alignments a lot, but "the reality is that alignments are of no interest to science. They mean nothing," explains David Morrison, senior scientist at the NASA Astrobiology Institute.[5]

Far scarier than the gravitational effect of distant planets aligning is the claim that one mysterious planet is heading directly our way and will cause unimaginable destruction by striking Earth or at least passing close by us. This is not totally crazy because collisions like that do happen in space. In fact, it happened to us once already. Astronomers believe that a Mars-sized planet struck Earth shortly after Earth formed. This probably created the debris that eventually became the Moon. Fortunately, we don't have to worry about this happening again anytime soon because there is no such planet heading our way. Keep in mind: this is coming from the world's astronomers—the people who know more about planets than anyone else. Believers who fear a collision with "Planet X," also known as "Nibiru," claim it was originally discovered by the ancient Sumerians. I have spoken to believers who say this planet is "hiding" behind the Sun right now. Many unconvincing photos of Planet X peeking out from behind the Sun are posted on the Web. Apparently these images impress people who don't understand what often happens when cameras are pointed at the Sun. (It's called a lens flare.) Originally it was predicted that this planet would hit the Earth in May 2003.[6] When it failed to show, many Nibiru believers revised their doomsday date and jumped on the 2012 bandwagon. Changing dates, by the way, is nothing new. Prophets of doom have been playing that game for thousands of years. I'm no Nostradamus, but after December 21, 2012, fails, I boldly predict somebody will come up with a new and "correct" interpretation of the Maya calendar that pushes the date forward by several years or so. And many people will believe it!

A "polar shift" or "pole shift" will bring about the 2012 catastrophe, say some believers. Depending on whom you listen to, this either refers to a flip of magnetic polarity, or that the Earth's rotational spin will stop or reverse. Again, don't stop flossing. You are probably going to be alive in 2013. A pole shift does not threaten Boy Scouts making their way through the woods with a compass nor the rest of us. NASA astronomer Morrison explains: "A reversal in the rotation of Earth is impossible. It has never happened and never will. There are slow movements of the continents (for example Antarctica was near the equator hundreds of millions of years ago), but that is irrelevant to claims of reversal of the rotational poles. However, many of the disaster websites pull a bait-and-switch to fool people. They claim a relationship between the rotation and the magnetic field of Earth, which does change irregularly with a magnetic reversal taking place, on average, every four hundred thousand years. As far as we know, such a magnetic reversal does not cause any harm to life on Earth. A magnetic reversal is very unlikely to happen in the next few millennia, anyway."[7]

WHAT IS THE APPEAL?

Why has this hollow claim attracted and entranced so many people? Perhaps 2012 belief swept around the world seducing believers in the early twenty-first century because it took the common desire to glance back at our past and combined that sentiment with the understandable concern one has about being incinerated by the Sun and crushed by falling buildings. It mixes in astronomical pseudoscience to give it a modern, high-tech flavor as well. It's natural for us to want to hear whispers from the ancients, to know their wisdom and glean a bit of parental knowledge as we face the stress of modern life and an uncertain future. Furthermore, there is no denying our obsession with the end. The doomsday theme often emerges in everything from songs to comic books, from films to religions. I'm certainly not immune to it. Some of my favorite fictional books and films center on the end of the world. I've always been drawn to nonfiction science books that address real doomsday possibilities, such as asteroid strikes, plagues, and nuclear war. It's exciting stuff, no denying it. There is an "apocalyptic porn" market out there that seems to feed the reptilian brain of a wide range of people, including those who like to think of themselves as rational skeptics. But mentally toying with such ultimate horror is tolerable, I suspect, only because deep down very few of us *really* believe any of these things will come to pass in our lifetimes. It's like riding a roller coaster. You feel like your life is in danger, but even while screaming you still trust that the car will stay on the rails.

GO DEEPER . . .

- Coe, Michael D. *The Maya*. London: Thames and Hudson, 2011.
- Sharer, Robert. *The Ancient Maya*. Palo Alto, CA: Stanford University Press, 2005.

Chapter 49

"THE END IS NEAR!"

I don't even think about the Rapture not happening on May 21 [2011] because I know that it will. You will see.
—Rita, May 7, 2011

Only after disaster can we be resurrected.
—Brad Pitt as Tyler Durden, *Fight Club*

I grew up in a small coastal town in south Florida, which means I was fed a diet rich in oranges, fish, and strange warnings about the end of the world. Judgment Day, I often heard, would start with "good Christians" being swept up to heaven while everybody else was left on Earth to riot, loot, and be eaten by a dragon . . . or something like that. Eventually Jesus would return on horseback and fix the world. I can't recall my parents ever mentioning the "End Times," but I do remember hearing kids at school, TV preachers, and a friend's mother talk about it. I also remember sitting through a few church sermons in which a preacher warned about the Rapture, the moment when born-again Christians vanish and avoid all the chaos and suffering to come. They always said it was almost here, and might even come that day. Earthquakes and wars were clear signs, they said, and when I was a child there were earthquakes and wars—just like today. I specifically remember one preacher warning me that if the Rapture happened before I accepted Jesus into my heart, I would burn in hell forever with no second chances—nice thing to tell a ten-year-old. But a funny thing happened on the way to adulthood. The Four Horsemen of the Apocalypse never showed up. But, no matter, the Rapture is still coming, *very soon*, of course. Century after century, it's always just around the corner.

According to the book of Revelation in the Bible, which 31 percent of Americans take to be a literal prediction of the immediate future, the world will be struck with epidemics, famines, natural disasters, and wars.[1] People who don't know much about history tend to be very impressed by this and point to current headlines as proof that the Apocalypse is drawing near. However, the reality is that bad news is nothing new. Many people were certain the world was going to end when the Black Death raged in Europe in the 1300s. World War II looked a lot like the end to some. And the Rapture was very near during some of the more tense days of the Cold War, many believers said with confidence fifty years ago. Not much has changed. The Left Behind series of Rapture-themed novels has sold tens of millions of copies. According to a 2002 *Time*/CNN poll, 59 percent of Americans believe that prophecies contained in the book of Revelation will come true, even if not every detail is accurate.

If all goes according to plan, earthly problems will mount for us and then the Antichrist will rise up to direct our descent into enslavement, destruction, and eternal suffering. Thanks to the *Omen* movies, we all know to be very afraid of the creepy but charismatic Antichrist. For half my life I heard people declare with great confidence that he was Pope John Paul II. That didn't work out, of course, so it must be someone else. A significant number of Americans think it might be President Barack Obama. But, don't forget, nearly 20 percent of Americans also think the Sun revolves around the Earth, so I wouldn't suggest relying on popular opinion polls to find out who the Antichrist is.[2] Still, belief about the Antichrist is worth noting for what it reveals about people and their ideas about the end of the world. According to a 2010 study by Harris Interactive, 25 percent of Tea Party members suspect that Obama is the great evil one.[3] Among Republicans in general, 24 percent think he may be the Antichrist. Thirteen percent of independents and six percent of Democrats also worry that Obama may be hiding the "mark of the beast" somewhere on his body. Overall, 14 percent of Americans think that the US president could be the "seed of Satan."[4]

Did I mention there is also a red seven-headed dragon wearing crowns and a giant sea monster involved in all this? Dragons are always fun, but my favorite Apocalyptic creatures as promised in Revelation are the locusts: "The appearance of the locusts was like horses prepared for battle; and on their heads appeared to be crowns like gold, and their faces were like the faces of men. They had hair like the hair of women, and their teeth were like the teeth of lions. They had breastplates like breastplates of iron; and the sound of their wings was like the sound of chariots, of many horses rushing to battle."[5]

The selective nature of the Rapture is extreme to the say the least. According to Rapture believers it's not enough to be a Christian. One must be specific *type* of Christian. For example, Catholics and Mormons are in trouble, according to many confident Rapture-ready people I have spoken to. They also tell me that a "born-again" experience is required to clear the bar on Judgment Day. A key point many Rapture believers seem to miss in all of this is how incredibly cruel and barbaric it would be if it actually happened. It is unconscionable that believers often seem thrilled about the horrors of doomsday in order to scare people—including young children—but never seem to reflect on the apparent madness and evil of it all. Why is it necessary for some five or six billion people to suffer and die? Why is so much destruction required? One would think that an all-powerful god could come up with a smoother transition plan.

In the end, Satan and the Antichrist fail, of course, which raises the question, Can't they read? The story is laid out in the Bible, so why would Satan bother trying when he knows he is going to lose? The climax of the whole affair, by the way, is Jesus returning as a warrior with a sword and riding on a white horse to smash nations and establish a new kingdom of God. This image seems to contradict the popular notion that Jesus is concerned only with love and forgiveness.

I met the late evangelist Sir Lionel Luckhoo several years ago while reporting on a faith-healing service in the Caribbean. The Guyanese lawyer was best known for being listed in the *Guinness Book of World Records* as the world's most successful lawyer with 245 successive murder acquittals. More interesting to me than courtroom conquests was his ability to know the future. He told me that he was in a race against time to spread the "word of Jesus." It was November 1993 when we spoke, and I asked him about the end of the world. He said he did not know the exact date but did know that it would be soon. "I am sure of one thing," he said. "We [Christians] will not see December 31, 1999, because we will be in heaven! The return of the master is imminent. Jesus is at the door."[6] Luckhoo was just one of countless millions of Christians over the centuries who were confident that they would see the Rapture occur in their lifetime only to go to their graves disappointed. It's a safe bet that every one of the preachers today who repeatedly promise that the end is almost here will one day die and the world will keep on spinning. The only question is, How many more generations must pass before doubts about the Rapture finally take hold and believers accept that it's just never going to happen?

ENCOUNTER WITH A DOOMSDAY BELIEVER

"I would estimate that about 97 percent of the people in the world will not make it," the woman behind the sunglasses said. "That's sad, of course. That's why I'm out here trying to let people know."

Her name was Rita and she told me that the world as I knew it would end on May 21, 2011. That's only about two weeks from the day we spoke in southern California. I was confident that the world would not end—and obviously it did not, as her fatal date came and went. Doomsday believers like Rita are nothing new, of course. They have been around for thousands of years. For me, however, every time I meet one it's like the first time all over again. My fascination rises with their sincerity—and Rita was as sincere as they came.

Here was this soft-spoken and polite middle-aged woman walking back and forth on a sunny day warning the heathens that they were almost out of time. Her tiny body was draped in a heavy sandwich board with "End of the World—May 21, 2011," written in large letters, front and back. Rita clearly cared. She wasn't signing up new members or asking for money. She only wanted people to know that her loving God was going to kill billions of people in a couple of weeks and some heavy duty praying was in order. I walked away from our conversation sad for her. Strange as it may seem, I almost wanted the world to end just so this sweet misguided woman wouldn't have to face her own internal psychological Armageddon.

For readers who may be wondering where the May 21 date came from, I did the mind-numbing research to find out so you don't have to. It was Harold Camping, head of a Christian radio network, who came up with the date based primarily on the timing of the Noah's ark story. Camping believes the flood occurred in the year 4990 BCE (seven thousand years ago). Of course any high school kid with a half-decent education would know to tune out right there. For the entire Earth to have been flooded a mere seven thousand years ago would mean that the entire fields of geology, biology, zoology, anthropology, marine science, and archaeology are all wrong. Obviously a global flood did not happen seven thousand years ago, so no extrapolated date from that point is going to work. Camping's prediction never had a chance because it was a mistake based on a mistake. While talking with Rita, I knew doomsday wasn't coming but I couldn't help caring about her.

Me: "Are you confident that you will be Raptured if the world ends May 21?"

Rita: "No, not at all. I hope that I will be, but only God knows."

"Does walking around in the hot sun warning people about May 21 help your chances?"

"I hope so, but I really don't know. It's not about what you do or who you are. It's all been predetermined who will be and who won't be."

"So what you do from now until then doesn't really matter because it's already decided?"

"Yes."

"So it's possible that you could be left behind to suffer while some killer who is in prison right now could be Raptured?"

"Yes. Well, I don't think Osama bin Laden would go to heaven because he didn't believe in the God of the Bible. But we can't really know."

"How do strangers react to you when you tell them the world is going to end later this month?"

"I get every kind of response you might imagine. Some are nice, some are not."

"What happens if May 21 comes and goes without earthquakes, tsunamis, or anything unusual? How will you feel?"

"I don't think about that because I know it will happen. I made the decision months ago to disconnect my computer and stop thinking about it. It's going to happen."

"But there have been countless predictions of the end of the world throughout history, and all of them failed to come true. Doesn't that concern you a little bit?"

"But they were all based on misreadings of the Bible and misunderstandings. Many people make the mistake of looking too closely at political happenings, things that don't really matter. It's all in the Bible."

"You won't be tempted to kill yourself or anything like that if nothing happens on May 21, will you?"

"No, that's not a concern at all. I don't even think about the Rapture not happening on May 21 because I know that it will. You will see."

After May 21, 2011, passed Raptureless, Harold Camping announced to his followers and to the world that Judgment Day had indeed occurred but no one realized it because it was an "invisible Judgment Day." He then said the actual date of destruction had been revised to October 21, 2011. A few days after the May 21 disappointment I listened to Camping take phone calls from listeners on his radio station. Some callers were upset and some mocked him. Many others, however, assured him that they still believed in him and trusted that his new prophecy was accurate.

I think about the lady with the doomsday sign sometimes. I hope Rita fares well with whatever psychological stress Camping's failed predictions might have caused her. I wish her well, wherever she is today. She and I are not so different, really. Rita believed she had an important warning for all humankind and was willing to face criticism and ridicule to share it. I write essays and books to share my warning about irrational beliefs. Sometimes I am criticized and mocked for my efforts. But I do it anyway because I care about others—just like Rita who saw the mirage of yet another apocalypse on the horizon and was compelled to warn strangers about it. People like her think the world will end in supernatural destruction very soon with a few special people escaping to heaven. I believe the world will keep on spinning and life will endure. I suppose we have to wait and see to find out who is right. In the meantime, however, I suggest that my outlook is the healthiest way to live.

WHO GETS TO MAKE THE RULES?

Warnings about the Rapture and global destruction are not as simple as most preachers describe because there is tremendous disagreement about it within Christianity. Various churches and denominations have conflicting views about precisely how it will happen, where everybody goes, who gets saved and who doesn't, and which parts of Revelation are meant to be interpreted metaphorically and which parts are meant to be taken literally. How is a Christian supposed to know which one of the many versions of Christianity has it right on the End Times? What matters? Are the Catholics right or are the Baptists right? What about the Rastafarians, Mormons, Jehovah's Witnesses, and so on? Key areas are viewed very differently within the many different versions of Christianity. Points of disagreement include Holy Communion, baptism, the Book of Mormon, Saturday or Sunday Sabbath, the return of black people to Africa, pants for women, blood transfusions, tattoos, dancing, divorce, modern science, homosexuality, and many more. Depending on whom you ask, those points and many others are the difference between salvation and damnation. It's a tough call to choose which way to go because no one denomination can make the case that it has the weight of evidence or logic in its favor. But it's even more difficult than that for those who might want to play it safe because Christianity is not the only religion that comes with a doomsday warning.

How can we know which religion's apocalypse is the one we should be concerned about? There are many belief systems that have their own unique end-of-the-world scenarios. And most of them require one to be correctly aligned with their demands in order to survive or at least end up in a happy place. Islam, for example, is the world's second most popular religion with more than one billion followers, and it has its own distinct final chapter. Are you up to speed on the predicted appearance of the Mahdi in the final days? Some say he's due to arrive any day now. Should we all become Muslims in order to play it safe? But wait, should we be Sunni or Shia Muslims? What about the great flood to come, as promised by Norse religion? Should we all start building arks? With so many to choose from, how do we decide? Can we vote on our preferred apocalypse? If so, I think I'd go with the Zoroastrians' more inclusive version of doomsday. Theirs involves a bit of suffering followed by forgiveness for everyone and peace on Earth. And no dragons or creepy, long-haired locusts to worry about.

END GAMES

During the buildup of hype for Y2K, the digital Armageddon that was supposed to happen on New Year's Day, 2000, I remember chucking an extra jar of peanut butter and a few more cans of green beans than usual into my shopping cart at the grocery store—just in case. I think I may have felt a momentary twinge of that weird rush of excitement the apocalyptic faithful must feel when she stores up freeze-dried food and ammunition in preparation for her supernatural doomsday. Finally, I thought to myself, a rational, thinking skeptic can enjoy a small taste of apocalyptic paranoia. I have to confess, it was fun to imagine not having to go to work on Monday or wash my car anymore. It was perversely exhilarating to imagine a world without rules, fishing for dinner, growing vegetables, plenty of time to read, setting up a solar-powered DVD player, coming up with creative ways to defend my fortress from looters and zombies. Perhaps this is what is behind all the end-of-the-world and human-extinction buzz. Maybe it is nothing more than the fantasy of freedom from daily drudgery and a chance to escape from responsibility that draws us toward belief in a magical finish. In fundamentalist churches one does not have to sit in a pew for long before hearing about the End Times. According to a Baylor study, approximately one-fifth of all Americans have read at least one book in the *Left Behind* series of novels about the Rapture.[7] Many millions of people worldwide view most natural disasters as signs of Judgment Day's approach. It's assumed that all of this interest is fueled by religion, but maybe it has as much or more to do with *Mad Max* fantasies as belief in destructive gods. It's possible. After all, one does not even need religion to be seduced by the apocalypse.

GEEK RAPTURE

Not to be left out, secular nerds finally have their own Rapture. It's called the Technological Singularity and—you guessed it—it will be here very soon! Lead prophet Ray Kurzweil claims the Singularity will arrive around the year 2040. The prediction is that exponential growth will endow computers with godlike powers, and then machines will saturate the universe at a rate that will be incomprehensibly fast. We will merge with the machines and become gods too. It sounds crazy, of course, but there are intriguing data, projections, and logical arguments that are difficult to dismiss.[8] I admit to being partially seduced by the Singularity. I find it irresistible to imagine being swept up in my old age by a digital miracle that will give me immortality and infinite opportunities for fun and discovery, while simultaneously seeing the elimination of poverty, disease, and war from the planet. But, of course, there's always a catch. The Singularity's heaven scenario is shadowed ominously by a hell scenario that envisions terrorism, global war, mass slavery, and even the extermination of humankind. I don't know about you, but I prefer the heaven scenario.

No matter where my Singularity fantasies may lead, my mind always ends up back in the real world. Hoping may be fun, but it's not the same as knowing, not even close. Sure, I hope the Singularity arrives just in time before I die. I hope that one day I can download every book in the world into my enhanced brain and move into a nice new robot body so that I can trek across the Arctic and break ten seconds for the 100 meters. In the meantime, however, I'll keep working from the assumption that my existence is finite and precious so I had better make the best of every day. Maybe that's the most practical outlook for all of us.

GO DEEPER . . .

- Binford, Gregory, and Elsabeth Malartre. *Beyond Human: Living with Robots and Cyborgs*. New York: Forge Books, 2008.
- Bostrom, Nick, and Milan M. Cirkvoic, eds. *Global Catastrophic Risks*. New York: Oxford University Press, 2008.
- Brockman, John, ed. *The Next Fifty Years: Science in the First Half of the Twenty-first Century*. New York: Vintage Books, 2002.
- Brockman, John. *Science at the Edge: Conversations with the Leading Scientific Thinkers of Today*. New York: Union Square Press, 2008.
- Diamond, Jared. *Collapse: How Societies Choose to Fail or Succeed*. New York: Penguin, 2011.
- Ehrman, Bart D. *Jesus: Apocalyptic Prophet of the New Millennium*. New York: Oxford University Press, 2001.
- Guyatt, Nicholas. *Have a Nice Doomsday: Why Millions of Americans Are Looking Forward to the End of the World*. United Kingdom: Ebury Press, 2007.
- Halpern, Paul. *Countdown to the Apocalypse: A Scientific Exploration of the End of the World*. New York: Basic Books, 2000.
- Impey, Chris. *How It Ends: From You to the Universe*. New York: W. W. Norton, 2011.
- Kurzweil, Ray. *The Singularity Is Near: When Humans Transcend Biology*. New York: Penguin, 2006.
- Lewis, James R. *Doomsday Prophecies: A Complete Guide to the End of the World*. Amherst, NY: Prometheus Books, 2000.

- Taylor, Justin. *The Apocalypse Reader* (fiction). Philadelphia, PA: Running Press, 2007.
- Ward, Peter D., and Donald Brownlee. *The Life and Death of Planet Earth*. New York: Henry Holt, 2004.

Chapter 50

"WE'RE ALL GONNA DIE!"

I don't think the human race will survive the next one thousand years.

—Stephen Hawking

Why does human extinction fascinate us? Probably because we are humans and we will be the ones who go away, of course. As we will see in this chapter, however, our final exit is unlikely in the short term. There are no guarantees, of course; it could happen tomorrow. But talk of our ultimate end seems like it is everywhere these days and I believe it misleads many people about how vulnerable we really are. Humans are smart, numerous, and spread around much of the world. We won't be easy to eliminate. The avalanche of apocalyptic books, comics, movies, songs, and magazine articles in recent years suggests that we have become downright obsessive about the subject. Interestingly, these warnings and scary scenarios are not all coming from the same place. These days, one is as likely to encounter the idea of global destruction and the end of civilization in a respectable science publication as in a blockbuster film or comic book. There obviously is just something about the end of humanity that has grabbed us and won't let go. Such thoughts are not new, of course. They have been with us for a very long time. But once they were based on supernatural concepts of gods and magic and were the exclusive centerpiece of religions. These days, however, the grisly possibility of human extinction is just as likely to be brought up by a scientist as by a preacher. So, if the wrath of angry gods or the fulfillment of some sacred plan of global destruction doesn't appeal or seem rational enough, then one can always opt for the secular asteroid strike, gamma ray burst, or killer virus. Yes, no doubt about it, these are the golden days of doomsday dreamers, as there is bound to be at least one disaster scenario to suit anyone's taste. I contend, however, that none of the popular threats, whether religious or evidence based, will do us in any time soon.

WE'RE (PROBABLY) NOT GOING ANYWHERE

Before exploring reasons why it is very unlikely that humans will become extinct any time soon, let's be clear about what extinction really means. The words *extinction* and *end of the world* are often used inaccurately and far too casually in popular culture. I suspect many people do not have a clear understanding about what extinction actually means. For example, a nuclear war of unprecedented savagery that leaves two-thirds of the population dead wouldn't qualify as a human extinction event. Not even a global disaster that completely destroys civilization and reduces us to a few random clans of scavengers would do it. Even a lethal virus that wipes out 99 percent of humanity wouldn't be enough because a 99 percent kill rate still leaves tens of millions of people. Extinction, keep in mind, means *everyone*. To bring about ours, an event would have to be so thoroughly catastrophic that it killed everyone outright or so altered the environment that every single person would succumb eventually.

Most religious doomsday scenarios don't even qualify because a supposed extinction event fails if some people are holding a supernatural escape ticket. This eliminates virtually every religious take on human extinction. No matter what they may say, most people who believe in a religion-based end-of-the-world event do not see it as an extinction event. Their scenarios almost always include safe passage for people who are favored within their particular belief system. If people existed in a heaven or a hell after doomsday and could still be considered people, then the human species wouldn't be extinct. So this class of doomsday scenarios does not belong in an analysis of possible human extinction. Nobody gets saved, Raptured, reincarnated, sent somewhere else. Extinction means everyone and forever.

While I think that extinction is very unlikely in the near future, I do not believe that *Homo sapiens* will last forever. No matter how smart and lucky we may turn out to be, it's virtually impossible that we can hold on indefinitely because, sooner or later, every species either dies out or evolves into something else. Even if we survive, hundreds of thousands if not millions of years from now, our descendants will be so physically and genetically different that they would have to be considered a different species. Either way, "we" won't be around.

EXPIRATION DATE

This chapter is concerned only with the popular belief that humans are likely to become extinct within a century or so. Given enough time, predictions of our extinction will almost certainly be proven correct one way or another eventually. The glaring example provided by our planet's natural history is that extinction is the norm. Every species checks out sooner or later, at least that's what the evolutionary history of this planet shows. Approximately 99 percent of all species that have ever existed are now gone. Based on what we know today, it is clear that extinction for us is far more likely than eternal survival is. Even if we manage to avoid the many dangers we face here on Earth, the Sun will die four billion years or so from now. And even if we survive the death spasms of our star by colonizing other worlds, constructing habitats in space, or somehow controlling the Sun and extending its duration, we still would have to confront the eventual "end" of life in an expanding universe. Trillions of years from now it will grow too cold and dark for life. But even then there might be a chance. In 2011, astronomers went public with the intriguing idea that dark matter might warm planets and make them hospitable to life.[1] Then there is the possibility that our descendants will have become so smart that we would be able to figure out how to slip into another universe or dimension. Or maybe we could even time travel a few billion years in reverse to place us back in a more agreeable universe. But this chapter is not about the deep future; it's about the popular belief or fear that human extinction will happen sooner rather than later. It's possible, yes, but how likely is it? First, let's review some of the more serious and scientifically reasonable potential disasters we face, then we will consider what our chances of surviving them are.

Asteroid Strike

What is the chance that a large asteroid or comet will hit Earth in the future? According to astronomers, the odds are pretty high. How high? Try 100 percent! I'm not sure about you, but I don't like those odds. Earth has been pounded by large objects in the past and it will be hit again. Some of the greatest mass extinctions in Earth history are believed to have been triggered by asteroid strikes, most famously the one that wiped out many dinosaur species sixty-five million years ago. If a big enough object, say one three or four miles in diameter, hit Earth, we might be finished. It's not the impact itself that does the most damage, however. It's all the dust, smoke, and debris that blocks the Sun and chokes out life. It's mind-boggling to imagine such destruction and terror. Unfortunately, this is our reality. The good news is that impacts are few and far between from our human perspective. We are also smart enough as a species to do something about it when one is heading our way in order to avoid extinction. We don't even have to play space cowboys and send Bruce Willis to intercept them with nukes. All it takes is a slight nudge far enough out and they would miss us. But we can't be complacent about it because it requires setting up a reliable warning system and having a solid plan in place that we can execute immediately when an object is spotted heading our way.

I met former Apollo astronaut Rusty Schweickart at a space conference in Arizona a few years ago. While he signed autographs for fans I took the opportunity to chat with the friendly space hero about his work on asteroid and comet defense. Schweickart is a cofounder of the B612 Foundation and cochaired the 2010 NASA Advisory Council Ad-Hoc Task Force on Planetary Defense. He told me that he is concerned and believes we are not doing nearly enough to identify, track, and deal with objects that will threaten us in the future. I agree. To be smashed by some random rock from space would be a tragic end after all the effort we made to move out of caves and clean ourselves up.

Nuclear War

Everybody cheered when the Cold War ended (I'm still waiting for my peace dividend check to arrive in the mail) but thousands of nuclear weapons are still ready to fly should World War III ignite. Most of them are in the arsenals of the United States and Russia. Currently tensions between the two nations are nowhere near as high as they were during the Cold War. Still, the threat is there so long as the missiles are. Most people have no idea about the true power of modern nuclear weapons. They assume they are similar to the atomic bombs dropped on Hiroshima and Nagasaki at the close of World War II. Not even close. Today's nukes are *thousands of times* more powerful than those.

As destructive as these weapons are, however, humankind could and probably would survive a nuclear war. Given the fact that people are spread out so far and wide on the planet, most will no doubt be spared direct hits from missiles and bombs. Many big cities will be hit, of course, but the United States, Russia, China, or whichever nations are fighting would not waste their missiles on uninvolved cities such as Mogadishu (population more than one million) and Reykjavik (population more than one hundred thousand), for example. The real challenge for species survival would come when radioactive particles blow around the world and the sky turns dark from all the smoke, dust, and ash. Still, there is an excellent chance that some pockets of humans could survive if they figure out ways to insulate themselves from the radiation, stay warm, and procure enough water and food. It's not impossible by any means. Humans are very good at figuring out solutions to difficult challenges. Remember, we already know how to live in harsh environments such as the Antarctic and in space.

Killer Germs

We are always in danger of some germ evolving into a lethal monster capable of spreading from human to human and killing millions, maybe even billions. Massive epidemics have already happened many times in the past. Killer germs have had profound impacts on European and Asian history, and they wiped out perhaps as much as 90 percent of the Native American population who died from smallpox and other diseases soon after the first Europeans arrived in the New World. The movies and history books often downplay it or leave it out entirely, but more people were killed by germs than bullets in the Civil War and in World War I. Bacteria and viruses can be wicked. They could even bring about our extinction.

To make matters worse, it's not only nature we have to worry about. Deadly engineered microbes could be created and released from a laboratory either by accident or on purpose. It's unlikely but still possible that a killer species of tiny viruslike invaders may descend upon us from space or hitch a ride on the hull of one of our returning space vehicles. But while all these scenarios could happen, I don't see any of them leading to our extinction. Again, if our smarts couldn't save us, our sheer numbers and geographical spread probably would.

I remember in the days following the 2004 Indian Ocean tsunami reading about an Indian Coast Guard helicopter that was sent to assess damaged coastlines. While flying over one island, a naked man defiantly shot an arrow at the helicopter. He was a member of an isolated tribe of people called the Sentinelese. Such groups of people who live in near or total isolation from the rest of humankind would probably do very well if biological catastrophe strikes the part of the world the rest of us live in. Their isolation might not be enough in the event of an asteroid strike, nuclear war, or supervolcano eruption. But if the threat is a virus, they have an excellent chance of survival and continuing the species. Their separation would leave them well positioned to inherit the Earth.

Climate Change

Of all the extinction scenarios discussed in popular culture these days, global warming is my pick for the least likely to do us in. Yes, the climate is changing rapidly. Yes, the evidence indicates that humankind is causing or contributing to it. Yes, it is extremely serious and the toll is likely going to be immense in both lives and money from the impacts on sea level, weather, agriculture, and disease. But extinction? No way. There is no global warming outcome that I have heard described by credible scientists that is likely to wipe out all of humanity within the next one hundred to two hundred years. Could a worst-case scenario cripple or even bring down modern civilization and set us back a thousand years? Sure, that's possible, but millions of people somewhere would find a way to survive. Never forget, *Neanderthals* with prehistoric technology lived successfully for at least one hundred thousand years in Europe *during an ice age*. We know that human populations—even with minimal technology—can survive in desert, rainforest, wetland, snow, and mountain environments. It's a safe bet that at least some of us could manage to hang on somewhere.

Rise of the Machines

"Do we build gods, or do we build our potential exterminators?" asks artificial intelligence researcher Hugo de Garis in his book *The Artilect War*.[2] Yes, we may all end up at the end of a leash held by intelligent machines one day. And, if they grow bored with us, we might be eliminated. But I don't think so. I have no problem believing in the likelihood of superpowerful computers and robotics bringing profound changes to our world. However, I don't believe it would result in our demise. Radical evolution maybe, but not abrupt extinction. What is going to happen is that we will merge with machines and computers. There is no doubt about this because it's already happening. It will be difficult if not impossible for machines to wage war against us in the future because they will be us and we will be them. There is also the likelihood that some genius somewhere will think to include an off switch when designing a computer capable of conquering the world.

Supervolcanoes

The first time I ever heard about supervolcanoes was many years ago during an interview with Apollo Moonwalker John Young. The veteran astronaut was making the point that science isn't only fun and exciting, it also may be the key to saving our species from extinction. Intrigued, I began researching supervolcanoes. To sum up what I learned: they are horrifyingly destructive and one will erupt sooner or later, perhaps destroying civilization. Don't mistake a supervolcano with the run-of-the-mill variety of volcano we are all most familiar with. Supervolcanoes are *thousands of times more powerful* than those. The problem for modern civilization is that, like nuclear war, supervolcanoes eject tons of smoke, ash, and dust up into the atmosphere, enough to rapidly change the world's climate and cause mass extinctions. Some scientists think a supervolcano almost killed us off at least once already. Approximately seventy thousand years ago the Toba supervolcano erupted in what is now in the region of Indonesia. It triggered a long, cold, dark winter that may have cut human numbers down to less than a few thousand people. That was a scary brush with extinction. But it makes the point that if our ancestors, with nothing more than stone and wooden tools, could survive a global supervolcano catastrophe, so could we.

Nanobots Eat the World

One of the more bizarre possible human extinction events is the one in which we and everything else become food for microscopic robots. Crazy as it sounds, it's possible. Nanotechnology is the science and engineering of incredibly small machines that are scaled down to the atomic level. It's exciting, almost intoxicating, to dream about the possibilities. The applications seem infinite, from stronger, lighter building materials to nanobots that swim around inside of us monitoring our bodies and keeping us healthy to tiny probes that scour the universe and send back information about everything they discover.

Some experts envision miniature machines that are self-sufficient and self-replicating. They would consume available atoms in their environment and reconstruct them to make more nanobots. Such machines could have many practical and positive applications. But there is also a possible downside. Self-replicating nanobots might run amok and become unstoppable eating machines, converting all matter they encounter into more hungry nanobots. Scientists call this the "gray goo" scenario. There was a depiction of something vaguely similar to this near the end of the 2008 remake of the film, *The Day the Earth Stood Still*. I think we could avoid the gray goo doomsday, however, by programming a kill switch or simply letting loose bigger and badder nanobots that are programed to eat renegade nanobots.

WE'RE ALL *NOT* GONNA DIE!

Wow, what a horrifying list. And keep in mind that these are all things that really could happen. None of them are based on supernatural or paranormal claims. Given enough time, some of them are inevitable.

Now here's the good news: None of these events, bad as they could be, are likely to bring about our extinction. A 2009 scientific study looked at various disaster scenarios and concluded that no one disaster is likely to wipe us out. "The human race is unlikely to become extinct without a combination of difficult, severe, and catastrophic events," said Tobin Lopes, the study's project leader. "[We] were very surprised about how difficult it was to come up with plausible scenarios in which the entire human race would become extinct."[3]

We have faced ice ages, supervolcanoes, and deadly epidemics in the past and survived. We can handle a lot, thanks to language, culture, and the creative power of the human brain. Short of being hit with a series of major events in succession—say the impact of an asteroid that is covered with lethal space viruses and accidently triggers a full-scale nuclear war—we should be able to survive. Groups of people in one or more places would likely find a way to cope. We may suffer and struggle, but the odds are very good that humankind will endure.

GO DEEPER . . .

Books

- Adams, John Joseph, ed. *Wastelands: Stories of the Apocalypse* (fiction). San Francisco, CA: Night Shade Books, 2008.
- Anders, Lou, ed. *Future Shocks: What Terror Does Tomorrow Hold?* (fiction). New York: New American Library, 2006.
- Benton, Michael. *When Life Nearly Died: The Greatest Mass Extinction of All Time*. London: Thames and Hudson, 2005.
- Brockman, John, ed. *What Are You Optimistic About? Today's Leading Thinkers Lighten Up*. New York: Harper Perennial, 2007.
- Clegg, Brian. *Armageddon Science*. New York: St. Martin's Press, 2010.
- Hanlon, Michael. *Eternity: Our Next Billion Years*. Hampshire, UK: Palgrave Macmillan, 2008.
- McCarthy, Cormac. *The Road* (fiction). New York: Vintage, 2009.
- McGuire, Bill. *Global Catastrophes: A Very Short Introduction*. New York: Oxford University Press, 2009.

- McGuire, Bill. *A Guide to the End of the World*. Oxford: Oxford University Press, 2002.
- McGuire, Bill. *Surviving Armageddon: Solutions for a Threatened Planet*. New York: Oxford University Press, 2005.
- Miller, Walter M., Jr., and Martin H. Greenberg, eds. *Beyond Armageddon* (fiction). Lincoln, NE: Bison Books, 2006.
- Plait, Philip. *Death from the Skies!* New York: Penguin Books, 2008.
- Rees, Martin. *Our Final Hour*. New York: Basic Books, 2003.
- Stevenson, Mark. *An Optimist's Tour of the Future*. New York: Avery, 2011.
- Stewart, George R. *Earth Abides* (fiction). New York: Del Ray, 2006.
- Wilson, Daniel H. *Robopocalypse* (fiction). New York: Doubleday, 2011.

Other Sources

- Earth Impacts Effects Program (website that allows visitors to calculate damage from asteroid strikes), http://impact.ese.ic.ac.uk/ImpactEffects/.

FAREWELL AND GOOD LUCK

The whole of life is but a moment of time. It is our duty, therefore to use it, not to misuse it.
—Plutarch

It scares me to imagine how my life might have turned out if I hadn't adopted a skeptical outlook early on. I might have wasted countless hours worrying about psychic readings, horoscopes, and whether or not one or more of a million gods approved of me. That would have meant less time for reading books, exploring the world, flirting with girls, watching sci-fi movies, and eating chocolate-covered peanuts. A diminished life, for sure. I also could have wasted thousands of dollars and possibly harmed my health with medical quackery. Worse, I might have dedicated my life to spreading and selling dangerous irrational beliefs to others. I feel fortunate and proud that I am willing and able to think about what I think. The habit of thinking before believing has served me well.

No one should think of skepticism and science as too complicated or demanding. You don't have to be a professional scientist to use the principles of science. You don't have to be a card-carrying member of some skeptic society in order to be able to think your way around all the steamy piles of nonsense lying everywhere. It's not difficult to be a skeptic. Just imagine having a skeptical filter or screen wrapped around your brain. Any weird claim that wants to worm its way into your life must first squeeze through that filter. If it can't, then it's not worth your time. The sooner one begins thinking like this, the better. When my children were very young, for example, I explained to them the value of approaching life as amateur scientists. I encouraged them to question everything, consider the source, ask for evidence, look for flaws in a claim, consider alternative explanations, and so on. Today, they routinely impress me with their insightful questions and ability to see through things they see in a TV ad or hear in everyday conversations. I didn't teach them to think like me. I taught them to think for themselves, and their lives will surely be better for it.

The skeptical life is for anyone and everyone. It's not dependent on extensive education or exceptional intelligence. At its core, being a skeptic means nothing more than recognizing and understanding something about the natural processes, frailties, and vulnerabilities of the human brain—and deciding not to surrender to them without a fight. It's about knowing that we all can and will be fooled, over and over, throughout our lives. It's inevitable. No one is immune to the infection of bad ideas and crazy beliefs. I'm the author of three skeptic-themed books and I can still fall victim to junk thinking. A few years ago, for example, I was habitually pouring expensive protein shakes down my gullet after workouts. It seemed like the smart thing to do at the time, but I was a victim of advertising and peer influence. I have since realized that a post-workout protein-rich meal is more sensible and better for me. I also briefly bought into the Mozart-makes-you-smarter craze back in the 1990s. I listened while working, hoping it might help me come up with a way to finally prove the existence of parallel universes or maybe invent a better light bulb. Now I know better, however, and I just listen to Mozart's music because I like it. You live and learn. You fall down and get back up.

This world we have made for ourselves is a maze of madness. Around every corner there is someone waiting to tell you a story about things that never happened or make a promise that will never come true. The best we can do is to fight back against the illusions, delusions, and scams with critical thinking and the scientific method. In my view, skepticism is necessary for everyone everywhere. Who in their right mind doesn't want to avoid being tricked and fooled, wasting time and money? Don't fall for the lie that skepticism equates to cynicism, a negative world view, or requires some great sacrifice. Only those who feel threatened by honest inquiry tend to make such charges. As millions of skeptical and freethinking people prove every day, it is possible to be productive, creative, and content without clinging to irrational beliefs. Besides, once skepticism has done its work of weeding out so much nonsense, there is a lot more time and energy available for family, friends, fun, exercise, self-improvement, and creative pursuits—the very things that can add up to a happy life. There is nothing negative about wanting to live in the real world and deal with life as it really is. Isn't that part of what being an adult is all about? Misery, misfortune, and death are terrible things none of us can avoid. But I contend that tuning out—via fantasy, willful ignorance, or embracing lies—is not the best defense against the hardships of life. Hope is found in reality as well. Inspiration exists in truth, too. We may be momentary specks of existence in an incomprehensibly vast universe, but the amazing human brain has the potential to find some measure of confidence, comfort, and contentment in reality, if we give it a chance. Between the endless gifts of scientific discovery and the warmth of a few fellow humans, what need do we have for the fool's gold of myth and make-believe? Fiction is wonderful—right up to the point where it is confused for reality—then it becomes a liability and a burden.

We are not necessarily doomed to be forever burdened by irrational belief. Never forget that the brain that so often leads us into traps is the same brain that can free us. We only have to make the necessary effort to be vigilant skeptical thinkers. If we choose to, we can give our lives great meaning and purpose, based on things that are known to be real. Nobody has to take this route, of course, but do not doubt that you can. There is no need to rely on superstition for strength or to let fantasy define your existence. If you want to be a nice person, then be nice. If you want to leave your mark on the world, then go do something. If you want to be a slacker, then stop reading this book and go take a nap. You don't need unproven claims of the supernatural and paranormal guiding you and reassuring you every step of the way. You are a human being with an immensely powerful brain. Don't underestimate its ability to separate you from much of the nonsense and danger out there. Your life is yours, live it wisely.

NOTES

INTRODUCTION

1. James Randi, "The Amazing Randi Wants You to Think," *Caymanian Compass*, June 1, 2000, p. A12.

CHAPTER 1: "I BELIEVE IN THE PARANORMAL AND THE SUPERNATURAL."

1. "Sasquatch: Legend Meets Science," *Talk of the Nation*, NPR, November 10, 2006, http://www.npr.org/templates/story/story.php?storyId=6469070 (accessed December 22, 2010).

2. Andrew A. Skolnick, "Natasha Demkina: The Girl with Normal Eyes," *Skeptical Inquirer*, May/June 2005, http://www.csicop.org/si/show/natasha_demkina_the_girl_with_normal_eyes/ (accessed March 23, 2011).

3. David W. Moore, "Three in Four Americans Believe in Paranormal," Gallup News Service, June 16, 2005, http://www.gallup.com/poll/16915/three -four-americans-believe-paranormal.aspx (accessed March 11, 2011).

4. Linda Lyons, "Paranormal Beliefs Come (Super) Naturally to Some," Gallup News Service, November 1, 2005, http://www.gallup.com/poll/19558/paranormal-beliefs-come-supernaturally-some.aspx (accessed March 15, 2011).

5. Tauriq Moosa, interview with the author, April 26, 2011.

6. Hank Davis, *Caveman Logic: The Persistence of Primitive Thinking in a Modern World* (Amherst, NY: Prometheus Books, 2009), pp. 183–84.

7. Michael Shermer, *The Believing Brain* (New York: Times Books, 2011), p. 5.

CHAPTER 2: "I KNOW THERE IS AN AFTERLIFE BECAUSE OF ALL THE NEAR-DEATH EXPERIENCES."

1. Frank Ward, interview with the author, January 10, 2011.

2. Kevin Nelson, *The Spiritual Doorway in the Brain* (New York: Dutton, 2010), p. 95.

3. Vilayanur S. Ramachandran and Diane Rogers Ramachandran, "Reflections on the Mind," *Scientific American Mind*, July/August 2011, pp. 18–22.

4. Nelson, *Spiritual Doorway*, p. 130.

5. Ibid., p. 137.

6. Ibid., pp. 142–43.

7. Susan Blackmore, *Dying to Live: Near-Death Experiences* (Amherst, NY: Prometheus Books, 1993), p. 263.

8. Sean M. Carroll, "Physics and the Immortality of the Soul," *Scientific American*, May 23, 2011, http://www.scientificamerican.com/blog/post.cfm?id =physics-and-the-immortality-of-the-2011-05-23 (accessed June 22, 2011).

9. Kevin Nelson, quoted in Amanda Gefter, "The Light in the Tunnel," *New Scientist* 208, nos. 2792–93 (December 2010–January 2011): 81.

CHAPTER 3: "A PSYCHIC READ MY MIND."

1. Tauriq Moosa, interview with the author, April 26, 2011.

2. James van Praagh, *Heaven and Earth: Making the Psychic Connection* (New York: Pocket Books, 2006), p. 188.

3. Bootie Cosgrove-Mather, "Poll: Most Believe in Psychic Phenomena," CBS News, February 11, 2009, www.cbsnews.com/stories/2002/04/29/opinion/polls/main507515.shtml (accessed July 1, 2011).

4. BBC News, "Britons Report 'Psychic Powers,'" *BBC News*, May 26, 2006, http://news.bbc.co.uk/2/hi/uk_news/5017910.stm (accessed December 11, 2010).

CHAPTER 4: "YOU'RE EITHER BORN SMART OR YOU'RE NOT."

1. Guy P. Harrison, *Race and Reality: What Everyone Should Know about Our Biological Diversity* (Amherst, NY: Prometheus Books, 2009), p. 253.

2. David Shenk, *The Genius in All of Us* (New York: Doubleday, 2010), p. 16.

CHAPTER 5: "THE BIBLE CODE REVEALS THE FUTURE."

1. David E. Thomas, "Beyond the Bible Code: Hidden Messages Everywhere!" in *Skeptical Odysseys*, ed. Paul Kurtz (Amherst, NY: Prometheus Books, 2001), p. 389.

2. David E. Thomas, "Hidden Messages and the Bible Code," *Skeptical Inquirer*, November/December 1997, http://www.csicop.org/si/show/hidden_messages_and_the_bible_code/ (accessed February 22, 2011).

3. Brendan McKay, "Assassinations Foretold in *Moby Dick*!" 1997, http://cs.anu.edu.au/~bdm/dilugim/moby.html (accessed March 17, 2011).

4. Michael Drosnin, *Bible Code II* (New York: Viking, 2002), p. 186.

5. Ibid., p. 101.

6. Submission.org, "Bible Code vs. Quran Code," http://www.submission.org/quran/biblecode.html (accessed March 11, 2011).

CHAPTER 6: "STORIES OF PAST LIVES PROVE REINCARNATION IS REAL."

1. David W. Moore, "Three in Four Americans Believe in Paranormal," Gallup News Service, June 16, 2005, http://www.gallup.com/poll/16915/three -four-americans-believe-paranormal.aspx (accessed March 11, 2011).

2. Carl Haub, "How Many People Have Ever Lived on Earth?" Population Reference Bureau, February 1995, www.prb.org/Articles/2002/HowManyPeopleHaveEverLivedonEarth.aspx (accessed January 12, 2011).

3. ABC News, "Parents Think Boy Is Reincarnated Pilot," *Primetime*, June 30, 2005, http://abcnews.go.com/Primetime/Technology/story?id=894217 (accessed February 12, 2011).

4. Judy Kroeger, "About Past Lives . . . Uniontown WWII Flyer's Memories in Louisiana Boy," *Daily Courier*, April 15, 2004, http://www.pittsburghlive.com/x/dailycourier/news/s_189477.html (accessed October 19, 2011).

5. J. Allen Danelek, *The Case for Reincarnation* (Woodbury, MN: Llewellyn Publications, 2010), p. 19.

6. Terence Hines, *Pseudoscience and the Paranormal* (Amherst, NY: Prometheus Books, 2003), pp. 109–10.

7. Danelek, *Reincarnation*, p. 188.

CHAPTER 7: "ESP IS THE REAL DEAL."

1. Richard Wiseman, "'Heads I Win, Tails You Lose,' How Parapsycholo-

gists Nullify Null Results," *Skeptical Inquirer*, February 2010, http://www.csicop.org/si/show/heads_i_win_tails_you_loser_how_parapsychologists_nullify_null_results (accessed March 2, 2011).

2. David W. Moore, "Three in Four Americans Believe in Paranormal," Gallup News Service, June 16, 2005, http://www.gallup.com/poll/16915/three-four-americans-believe-paranormal.aspx (accessed March 11, 2011).

3. Jonathan C. Smith, *Pseudoscience and Extraordinary Claims* (West Sussex, UK: Wiley-Blackwell, 2010), p. 251.

4. Michael D. Mumford, Andrew M. Rose, David A. Goslin, "An Evaluation of Remote Viewing: Research and Applications," American Institutes for Research, September 29, 1995, http://www.fas.org/irp/program/collect/air1995.pdf (accessed March 1, 2011).

5. Michael Shermer, "Freeman Dyson, Miracles, and the Belief in the Paranormal," *eSkeptic*, May 4, 2004, http://www.skeptic.com/eskeptic/04-05-04/ (accessed February 2, 2011).

6. Ibid.

7. Jefferson M. Fish, interview with the author, March 26, 2011.

8. Terrence Hines, *Pseudoscience and the Paranormal* (Amherst, NY: Prometheus Books, 2003), p. 150.

CHAPTER 8: "NOSTRADAMUS SAW IT ALL COMING."

1. James Randi, *The Mask of Nostradamus* (Amherst, NY: Prometheus Books, 1993), pp. 11–12.

2. Ibid., pp. 6–7.

3. Ibid., p. 149.

4. Ibid., p. 233.

5. Ibid., pp. 212–13.

6. Janelle Brown, "Nostradamus Called It! Internet Conspiracy Theorists Are Having a Field Day after the Attacks," *Salon*, September 17, 2001, http://www.salon.com/technology/feature/2001/09/17/kooks/index.html (accessed November 1, 2010).

7. Randi, *Mask of Nostradamus,* p. 223.

CHAPTER 9: "I BELIEVE IN MIRACLES."

1. Harris Poll, "What People Do and Do Not Believe In," Harris Interactive, December 15, 2009, http://www.harrisinteractive.com/vault/Harris_Poll_2009_12_15.pdf (accessed March 11, 2011).

2. Jonathan C. Smith, *Pseudoscience and Extraordinary Claims of the Paranormal* (West Sussex, UK: Wiley-Blackwell, 2010), p. 130.

CHAPTER 10: "NASA FAKED THE MOON LANDINGS."

1. Guy P. Harrison, "The Last Moonwalker," *Caymanian Compass*, August 9, 2002, pp. A12–13.

2. Ibid.

3. Frank Newport, "Landing a Man on the Moon: The Public's View," July 20, 1999, Gallup News Service, http://www.gallup.com/poll/3712/landing-man-moon-publics-view.aspx (accessed November 13, 2010).

4. Mary Lynne Dittmar, "Engaging the 18–25 Generation: Educational Outreach, Interactive Technologies, and Space," Dittmar Associates, 2006, http://www.dittmar-associates.com/Publications/Engaging%20the%2018-25%20 pdate~web.pdf (accessed January 26, 2011).

5. "*Apollo 11* Hoax: One in Four People Do Not Believe in Moon Landing," *Telegraph*, July 17, 2009, http://www.telegraph.co.uk/science/space/5851435/Apollo-11-hoax-one-in-four-people-do-not-believe-in-moon-landing.html (accessed January 3, 2011).

6. California Academy of Sciences, "American Adults Flunk Basic Science," *ScienceDaily*, March 13, 2009, http://www.sciencedaily.com/releases/2009/03/090312115133.htm (accessed March 11, 2010).

7. Steve Crabtree, "New Poll Gauges Americans' General Knowledge Levels," Gallup News Service, July 6, 1999, http://www.gallup.com/poll/3742/new-poll-gauges-americans-general-knowledge-levels.aspx (accessed January 1, 2011).

CHAPTER 11: "ANCIENT ASTRONAUTS WERE HERE."

1. Nancy White, interview with the author, March 21, 2011.

2. Carl Sagan, *The Varieties of Scientific Experience: A Personal View of the Search for God* (New York: Penguin, 2007), p. 129.

3. Penn State University, "How Were the Egyptian Pyramids Built?" *ScienceDaily*, March 29, 2008, http://www.sciencedaily.com/releases/2008/03/080328104302.htm (accessed March 28, 2011).

4. Erich von Däniken, *Chariots of the Gods* (Berkley, CA: Berkley Trade, 1999), p. 87.

5. Ibid., p. 96.

6. Ibid., p. 65.

7. Ibid., p. 73.

CHAPTER 12: "UFOS ARE VISITORS FROM OTHER WORLDS."

1. Read about the Drake equation on SETI's website: http://www.seti.org/drakeequation.

2. Linda Lyons, "Paranormal Beliefs Come (Super) Naturally to Some," Gallup News Service, November 1, 2005, http://www.gallup.com/poll/19558/paranormal-beliefs-come-supernaturally-some.aspx (accessed January 3, 2011).

3. Antonio Regalado, "Poll: Mexicans Express Belief in Spirits, Not Science," January 5, 2011, http://news.sciencemag.org/scienceinsider/2011/01/poll-mexicans-express-belief-in.html?ref=hp (accessed June 9, 2011).

4. Steve Crabtree, "New Poll Gauges Americans' General Knowledge Levels," Gallup News Service, July 6, 1999, http://www.gallup.com/poll/3742/new-poll-gauges-americans-general-knowledge-levels.aspx (accessed January 1, 2011).

5. Mark Lewis, interview with the author, September 7, 2011.

6. Stephen Webb, *If the Universe Is Teeming with Aliens . . . Where Is Everybody?* (New York: Copernicus Books, 2002), p. 30.

7. Stephen L. Macknic and Susana Martinez-Conde, *Sleights of Mind* (New York: Henry Holt, 2010), pp. 11–12.

8. D. J. Simons, C. F. Chabris, "What People Believe about How Memory Works: A Representative Survey of the U.S. Population," *PLoS ONE* 6, no. 8: e22757.doi:10.1371/journal.pone.0022757, http://www.plosone.org/article/info:doi%2F10.1371%2Fjournal.pone.0022757 (accessed August 11, 2011).

9. The video is available at www.theinvisiblegorilla.com/videos.html.

10. Philip Plait, *Bad Astronomy* (New York: John Wiley and Sons, 2002), pp. 202–204.

11. Christopher Chabris and Daniel Simons, *The Invisible Gorilla and Other Ways Our Intuitions Deceive Us* (New York: Crown, 2010), pp. 8–10.

12. Paul Davies, *The Eerie Silence: Renewing Our Search for Alien Intelligence* (Boston: Houghton Mifflin Harcourt, 2010), p. 19.

13. Andrew Fraknoi, "An Astronomer Looks at UFOs: A Lot Less Than Meets the Eye," *Skeptical Inquirer* 33, no. 1, January/February 2009, http://www.csicop.org/si/show/astronomer_looks_at_ufos_a_lot_less_than_meets_the_eye (accessed January 5, 2011).

14. Seth Shostak, interview with the author, April 6, 2011.

CHAPTER 13: "A FLYING SAUCER CRASHED NEAR ROSWELL, NEW MEXICO, IN 1947 AND THE GOVERNMENT KNOWS ALL ABOUT IT."

1. "RAAF Captures Flying Saucer in Roswell Region," *Roswell Daily Record*, July 8, 1947.

2. B. D. Gildenberg, "A Roswell Requiem," *Skeptic* 10, no. 1 (2003): 60–61.

3. Ibid., p. 61.

4. Ibid., p. 62.

5. Ibid., p. 63.

6. Charles Berlitz and William L. Moore, *The Roswell Incident* (New York: Berkley Books, 1980).

7. Joe Kittinger, interview with the author. Quoted in Guy P. Harrison, "I Was the First Man in Space," *Caymanian Compass*, October 26, 2001. The complete interview with Joe Kittinger can be read at http://www.spaceguy.8k.com/custom.html, under the heading "I Was the First Man in Space."

8. Ibid.

9. Gildenberg, "Roswell Requiem."

10. Air Force Web Information Service, "The Roswell Report: Case Closed," US Air Force, June 24, 1997, http://www.af.mil/information/roswell/index.asp (accessed February 22, 2011).

11. Kittinger, interview.

12. Jonah Lehrer, "Ads Implant False Memories," *Wired*, May 25, 2011, http://www.wired.com/wiredscience/2011/05/ads-implant-false-memories/ (accessed July 4, 2011).

13. Frank Newport, "What If Government Really Listened to the People?" Gallup News Service, October 15, 1997, http://www.gallup.com/poll/4594/What-Government-Really-Listened-People.aspx (accessed January 11, 2011).

14. Linda Lyons, "Paranormal Beliefs Come (Super) Naturally to Some," Gallup News Service, November 1, 2005, http://www.gallup.com/poll/19558/paranormal-beliefs-come-supernaturally-some.aspx (accessed March 15, 2011).

CHAPTER 14: "ALIENS HAVE VISITED EARTH AND ABDUCTED MANY PEOPLE."

1. Frank Newport, "Americans More Likely to Believe in God Than the Devil, Heaven More Than Hell," Gallup News Service, June 13, 2007, http://www.gallup.com/poll/27877/Americans-More-Likely-Believe-God-Than-Devil-Heaven-More-Than-Hell.aspx (accessed November 3, 2010).

2. Susan A. Clancy, *Abducted: How People Come to Believe They Were Kidnapped by Aliens* (Cambridge, MA: Harvard University Press, 2005), pp. 28–29.

3. Ibid., p. 33.

4. Kat McGowan, "Past Perfect," in "The Brain," special issue, *Discover* (Fall 2010): 69–70.

5. Clancy, *Abducted*, p. 59.

6. Elizabeth Loftus, interview with the author, March 27, 2011.

7. Clancy, *Abducted*, p. 35.

CHAPTER 15: "ASTROLOGY IS SCIENTIFIC."

1. Pew Forum on Religion and Public Life, "Many Americans Mix Multiple Faiths," Pew Forum, December 9, 2009, http://pewforum.org/Other -Beliefs-and-Practices/Many-Americans-Mix-Multiple-Faiths.aspx (accessed March 10, 2011).

2. Time Staff, "Good Heavens! An Astrologer Dictating the President's Schedule?" *Time*, May 16, 1988, http://www.time.com/time/magazine/article/0,9171,967389-1,00.html (accessed March 10, 2011).

CHAPTER 16: "ALL SCIENTISTS ARE GENIUSES AND SCIENCE IS ALWAYS RIGHT."

1. These examples are taken from "Not Even Wrong," in "Genius," special issue, *Discover* (Winter 2011): 94–95.

2. "VP: US Conducted 17 Types of Experiments on Guatemalans," *Latin American Herald Tribune*, January 15, 2011, http://www.laht.com/article.asp?ArticleId=373723&CategoryId=23558 (accessed January 15, 2011).

CHAPTER 17: "THE HOLOCAUST NEVER HAPPENED."

1. Guy P. Harrison, "Band of Brothers," *Caymanian Compass*, September 7, 2001, p. A19.

2. Guy P. Harrison, "Embraced by Evil," *Caymanian Compass*, December 5, 2002, pp. 15–16.

3. Guy P. Harrison, "Defying Hitler's Evil," *Caymanian Compass*, August 8, 2003, p. A16.

4. Guy P. Harrison, "Stronger Than Evil," *Caymanian Compass*, March 5, 2004, pp. A25–26.

5. Michael Shermer, *Why People Believe Weird Things* (New York: MJF Books, 1997), p. 190.

6. Ibid., p. 212.

7. Ibid., p. 214.

8. Nick Wynne, interview with the author, March 27, 2011.

CHAPTER 18: "GLOBAL WARMING IS A POLITICAL ISSUE AND NOTHING MORE."

1. Nathanial Gronewold and Christa Marshall, "Rising Partisanship Sharply Erodes US Public's Belief in Global Warming," *New York Times*, December 3, 2009, http://www.nytimes.com/cwire/2009/12/03/03climatewire -rising-partisanship-sharply-erodes-us-public-47381.html (accessed October 19, 2011).

2. Ibid.

CHAPTER 19: "TELEVISION NEWS GIVES ME AN ACCURATE VIEW OF THE WORLD."

1. Benjamin Radford, *Media Mythmakers: How Journalists, Activists, and Advertisers Mislead Us* (Amherst, NY: Prometheus Books, 2003), p. 69.

2. Ibid.

3. Daniel Gardner, *The Science of Fear* (New York: Dutton, 2008), p. 250.

4. Ibid., pp. 250–51.

5. Chris Hedges, *Empire of Illusion: The End of Literacy and the Triumph of Spectacle* (New York: Nation Books, 2010), p. 44.

CHAPTER 21: "BIOLOGICAL RACE DETERMINES SUCCESS IN SPORTS."

1. Michael Jordan, *Driven from Within* (New York: Atria, 2006), p. 33.

CHAPTER 22: "MOST CONSPIRACY THEORIES ARE TRUE."

1. Thomas Hargrove, "Third of Americans Suspect 9-11 Government Conspiracy," *Scripps News*, August 11, 2006, www.scrippsnews.com/911poll (accessed July 22, 2011).

2. Jon Hamilton, "Psst! The Human Brain Is Wired for Gossip," *Morning Edition*, May 20, 2011, http://www.npr.org/2011/05/20/136465083/psst-the-human-brain-is-wired-for-gossip (accessed June 28, 2011).

CHAPTER 23: "ALTERNATIVE MEDICINE IS BETTER."

1. Pride Chigwedere, George R. Seage III, Sofia Gruskin, Tun-Hou Lee, and M. Essex, "Estimating the Lost Benefits of Antiretroviral Drug Use in South Africa," *Journal of Acquired Immune Deficiency Syndrome* 49, no. 4 (December 1, 2008): 410, http://www.aids.harvard.edu/Lost_Benefits.pdf (accessed December 1, 2008).

2. Ibid., p. 414.

3. Richard L. Nahin, Patricia M. Barnes, Barbara J. Stussman, and Barbara Bloom, "Costs of Complementary and Alternative Medicine (CAM) and Frequency of Visits to CAM Practitioners: United States, 2007," *National Health Statistics Report* 18 (July 30, 2009), http://www.cdc.gov/NCHS/data/nhsr/nhsr018.pdf (accessed March 3, 2011).

4. Ibid., p. 3.

5. A. Malik and S. Gopalan, "Use of CAM Results in Delay in Seeking Medical Advice for Breast Cancer," *European Journal of Epidemiology* (August 18, 2003), http://www.ncbi.nlm.nih.gov/sites/entrez?cmd=Retrieve&db=pubmed&dopt=AbstractPlus &list_uids=12974558 (accessed March 6, 2011).

6. V. A. Luyckx, V. Steenkamp, and M. J. Stewart, "Acute Renal Failure Associated with the Use of Traditional Folk Remedies in South Africa," *Ren Fail* (January 27, 2005), http://www.ncbi.nlm.nih.gov/pubmed/15717633 (accessed March 6, 2011).

7. Elizabeth Mendes, "In US, More Than 8 in 10 Rate Nurses, Doctors Highly," Gallup News Service, December 13, 2010, http://www.gallup.com/poll/145214/rate-nurses-doctors-highly.aspx (accessed January 12, 2011).

8. Nany Shute, "Desperate for an Autism Cure," *Scientific American*, October 2010, p. 81.

CHAPTER 24: "HOMEOPATHY REALLY WORKS, AND NO SIDE EFFECTS!"

1. University of Abertay Dundee, "Homeopathy Is 'Dangerous and Wasteful,' Bioethics Expert Argues," *ScienceDaily*, May 9, 2011, www.sciencedaily.com/releases/2011/05/110509065749.htm (accessed July 7, 2011).

2. Bruce Hood, *The Science of Superstition: How the Developing Brain Creates Supernatural Beliefs* (New York: HarperCollins Paperback, 2010), p. 157.

3. Ben Goldacre, *Bad Science* (New York: Faber and Faber, 2010), p. 35.

4. Simon Singh and Edzard Ernst, *Trick or Treatment: The Undeniable Facts about Alternative Medicine* (New York: W. W. Norton, 2008), p. 93.

5. Laura Donnelly, "Homeopathy Is Witchcraft, Say Doctors," *Telegraph*, May 15, 2010, http://www.telegraph.co.uk/health/alternativemedicine/7728281/Homeopathy-is-witchcraft-say-doctors.html (accessed December 10, 2010).

6. Katelyn Catanzariti, "Homeopath, Wife Jailed over Baby's Death," *Sydney Morning Herald*, September 28, 2009, http://news.smh.com.au/breaking-news-national/homeopath-wife-jailed-over-babys-death-20090928 -g8w4.html (accessed January 22, 2011).

7. Richard Oakley, "Call for Stricter Checks on Therapists," *Sunday Times*, June 19, 2005, http://www.timesonline.co.uk/tol/news/world/ireland/article535014.ece (accessed January 1, 2011).

8. "Healer Dies after Letting Cut Foot Rot," *Metro*, November 17, 2008, http:// www.metro.co.uk/news/405720-healer-dies-after-letting-cut-foot-rot (accessed March 15, 2011).

9. Yusuke Fukui and Akiko Okazaki, "Homeopathy under Scrutiny after Lawsuit over Death of Infant," *Asahi Shimbun*, September 6, 2010, http://www.asahi.com/english/TKY201008050254.html (accessed January 20, 2011).

10. Pallab Ghosh, "Homeopathic Practices 'Risk Lives,'" *BBC News*, July 13, 2006, http://news.bbc.co.uk/2/hi/uk_news/5178488.stm (accessed January 22, 2011).

CHAPTER 25: "FAITH HEALING CURES THE SICK AND SAVES LIVES."

1. The problem of unhealed amputees has its own site at http://whywontgodhealamputees.com/god5.htm.

2. Guy P. Harrison, "God Is in This Place," *Caymanian Compass*, November 19, 1993, pp. 10–11.

3. Terrence Hines, *Pseudoscience and the Paranormal* (Amherst, NY: Prometheus Books, 2003), pp. 346–48.

CHAPTER 26: "RACE-BASED MEDICINE IS A GREAT IDEA."

1. American Anthropological Association, "Statement on 'Race,'" May 17, 1998, http://www.aaanet.org/stmts/racepp.htm (accessed February 1, 2011).

2. Kenan Malik, "Is This the Future We Really Want? Different Drugs for Different Races," *TimesOnline*, June 18, 2005, www.timesonline.co.uk/tol/comment/columnists/guest_contributors/article534565.ece (accessed January 11, 2011).

3. Charles N. Rotimi, "Are Medical and Nonmedical Uses of Large-Scale Genomic Markers Conflating Genetics and 'Race'?" *Nature Genetics*, October 2004.

4. Jonathon Marks, interview by the author, February 2009.

5. National Human Genome Research Institute, "The Human Genome Project Completion," April 14, 2003, updated October 30, 2010, http://www.genome.gov/11006943 (accessed March 3, 2011).

6. Emily Singer, "The $30 Genome?" *Technology Review*, June 7, 2010, http://www.technologyreview.com/biomedicine/25481/ (accessed November 10, 2010).

7. "Backgrounders from the Unnatural Causes Health Equity Database," from *Unnatural Causes . . . Is Inequality Making Us Sick?* pp. 9–10, www.unnaturalcauses.org (companion website for documentary; accessed November 6, 2010).

8. Ibid., p. 10.

CHAPTER 27: "NO VACCINES FOR MY BABY!"

1. Paul A. Offit, *Book TV*, C-Span2, January 27, 2011.

2. Centers for Disease Control and Prevention, "Update: Measles," August 28, 2008, http://www.cdc.gov/mmwr/preview/mmwrhtml/mm5733a1.htm (accessed February 12, 2011).

3. Paul A. Offit, *Deadly Choices: How the Anti-Vaccine Movement Threatens Us All* (New York: Basic Books, 2010), p. 92.

4. Paul A. Offit, *Autism's False Prophets: Bad Science, Risky Medicine, and the Search for a Cure* (New York: Columbia University Press, 2010), p. 24.

5. Offit, *Deadly Choices*, pp. 94–96.

6. Michael Specter, *Denialism: How Irrational Thinking Hinders Scientific Progress, Harms the Planet, and Threatens Our Lives* (New York: Penguin Press, 2009), p. 72.

7. Andy Coghlan, "Autism Rises Despite MMR Ban in Japan," *New Scientist,* March 5, 2005, www.newscientist.com/article/mg18524895.300-autism-rises-despite-mmr-ban-in-japan.html (accessed January 15, 2011).

8. Specter, *Denialism*, p. 71.

9. United Nations Children's Fund (UNICEF), "Vaccines Bring Seven Diseases under Control," http://www.unicef.org/pon96/hevaccin.htm (accessed January 17, 2011).

10. Offit, *Deadly Choices*, p. ix.
11. Paul A. Offit, interview with the author, April 14, 2011.
12. Offit, *Deadly Choices*, p. ix.
13. Ibid., p. xviii.
14. Ibid., p. 37.
15. World Health Organization, "Influenza," March 2003, http://www.who.int/mediacentre/factsheets/2003/fs211/en/ (accessed April 2, 2011).
16. Shawn R. Browning, interview with the author, March 3, 2011.
17. Offit, *Autism's False Prophets*, p. 247.
18. Specter, *Denialism*, p. 97.
19. Sarah Bruyn Jones, "Whooping Cough Outbreak in Floyd County Blamed on Lax Vaccinations," *Roanoke Times*, April 6, 2011, http://www.roanoke.com/news/roanoke/wb/282419 (accessed April 11, 2011).

CHAPTER 29: "MY RELIGION IS THE ONE THAT'S TRUE."

1. Steven Prothero, *Religious Literacy: What Every American Needs to Know—And Doesn't* (New York: HarperOne, 2007), p. 5.
2. Pew Forum on Religion and Public Life, "US Religious Knowledge Survey," September 28, 2010, http://pewforum.org/other-beliefs-and-practices/u-s-religious-knowledge -survey.aspx (accessed November 27, 2010).

CHAPTER 30: "CREATIONISM IS TRUE AND EVOLUTION IS NOT."

1. Guy P. Harrison, "The Dinosaur Hunter," *Caymanian Compass*, September 14, 2001, p. B6.
2. Guy P. Harrison, "Lucy in the Sky," *Caymanian Compass*, August 23, 2002, p. A11.
3. Frank Newport, "Four in 10 Americans Believe in Strict Creationism," Gallup News Service, December 17, 2010, http://www.gallup.com/poll/145286/Four-Americans-Believe-Strict-Creationism.aspx (accessed January 3, 2011).
4. Tim White, interview with the author. Quoted in Guy P. Harrison, "Who's Your Daddy?" *Caymanian Compass*, October 17, 2003, p. A16.
5. National Academy of Sciences and Institute of Medicine, *Science, Evolution, and Creationism* (Washington, DC: National Academies Press, 2008), p. 7.

CHAPTER 31: "INTELLIGENT DESIGN IS REAL SCIENCE."

1. Jonathan Marks, *Why I Am Not a Scientist* (Berkley: University of California Press, 2009), pp. 118–19.

2. John Brockman, ed., *Intelligent Thought: Science versus the Intelligent Design Movement* (New York: Vintage Books, 2006), p. 22.

3. Michael Shermer, *Science Friction* (New York: Times Books, 2004), p. 199.

CHAPTER 33: "MANY PROPHECIES HAVE COME TO PASS."

1. Koran 54:1.

CHAPTER 34: "PRAYER WORKS!"

1. "How Monks Find Their Happy Groove," CNN, November 19, 2008, http://articles.cnn.com/2008-11-19/health/brain.meditation_1_buddhist-monks-meditation-brain?_s=PM:HEALTH (accessed February 19, 2011).

2. Save the Children, "State of the World's Mothers," http://www.savethechildren.org/site/c.8rKLIXMGIpI4E/b.6748295/k.BE47/State_of_the_Worlds_Mothers_2011_Statistics_and_Facts.htm (accessed May 11, 2011).

CHAPTER 35: "RELIGIONS ARE SENSIBLE AND SAFE. CULTS ARE SILLY AND DANGEROUS."

1. All quotes from Deborah Layton can be found in Guy P. Harrison, "Facing the Darkness," *Caymanian Compass*, October 25, 2002, p. A10.

CHAPTER 36: "THEY FOUND NOAH'S ARK!"

1. "Most Americans Take Bible Stories Literally," *Washington Times*, February 16, 2004, http://www.washingtontimes.com/news/2004/feb/16/20040216-113955-2061r/ (accessed February 18, 2011).

2. For more details about the Noah's Ark replica tourist attraction, see Ark Encounter, http://arkencounter.com/faq/.

CHAPTER 38: "HOLY RELICS POSSESS SUPERNATURAL POWERS."

1. David R. Arnott, "Thousands Gather to See a Hair from the Prophet Mohammad's Beard," MSNBC, February 16, 2011, http://photoblog.msnbc.msn.com/_news/2011/02/16/6064520-thousands-gather-to-see-a-hair-from -the-prophet-muhammads-beard (accessed March 3, 2011).

2. Joe Nickell, *Relics of the Christ* (Lexington: University Press of Kentucky, 2007), pp. 18–19.

3. Ibid.

4. Ibid., p. 21.

5. Ibid., p. 190.

CHAPTER 39: "A TV PREACHER NEEDS MY MONEY."

1. Hanna Rosin, "Televangelist Jim Bakker's Road to Redemption," *Washington Post*, August 11, 1999, http://www.washingtonpost.com/wp-srv/style/daily/aug99/bakker11.htm (accessed March 23, 2011).

2. "Track Owner Pledges $1.3 Million to Roberts," *New York Times*, March 22, 1987, http://www.nytimes.com/1987/03/22/us/track-owner-pledges -1.3-million-to-roberts.html?src=pm (accessed March 23, 2011).

3. Wayne King, "Swaggart Says He Has Sinned; Will Step Down," *New York Times*, February 22, 1988, http://www.nytimes.com/1988/02/22/us/swaggart -says-he-has-sinned-will-step-down.html (accessed September 12, 2011).

4. Dan Harris, "Benny Hinn: 'I Would Not Do This for Money,'" ABC News, October 19, 2009, http://abcnews.go.com/Nightline/benny-hinn -evangelical-leader-senate-investigation-speaks/story?id=8862027 (accessed March 23, 2011).

5. Ric Romero, "Preacher Sells Debt Removal through Prayer," KABC-TV, February 28, 2011, http://abclocal.go.com/kabc/story?section=news/consumer&id=7984766 (accessed March 2, 2011).

6. James Randi, "The Amazing Randi Wants You to Think," *Caymanian Compass*, June 1, 2000, p. A12.

CHAPTER 40: "GHOSTS ARE REAL AND THEY LIVE IN HAUNTED HOUSES."

1. Harris Poll, "What People Do and Do Not Believe In," Harris Interactive, December 15, 2009, http://www.harrisinteractive.com/vault/Harris _Poll_2009_12_15.pdf (accessed March 11, 2011).

2. Linda Lyons, "Paranormal Beliefs Come (Super) Naturally to Some," Gallup News Service, November 1, 2005, http://www.gallup.com/poll/19558/paranormal-beliefs-come-supernaturally-some.aspx (accessed March 15, 2011).

3. Pew Forum on Religion and Public Life, "Many Americans Mix Multiple Faiths," Pew Forum, December 9, 2009, http://pewforum.org/Other -Beliefs-and-Practices/Many-Americans-Mix-Multiple-Faiths.aspx (accessed March 10, 2011).

4. Paul Parsons, "Just an Illusion?" *Focus* (BBC), November 2010, p. 31.

5. Terrence Hines, *Pseudoscience and the Paranormal* (Amherst, NY: Prometheus Books, 2003), p. 92.

6. Parsons, "Just an Illusion?" p. 32.

7. Lyons, "Paranormal Beliefs."

8. Christopher Bader, Carson Mencken, and Joseph Baker, *Paranormal America* (New York: New York University Press, 2010), p. 73.

9. Joe Nickell, *The Mystery Chronicles: More Real-Life X-Files* (Lexington: University Press of Kentucky, 2004), p. 138.

CHAPTER 41: "BIGFOOT LIVES AND CRYPTOZOOLOGY IS REAL SCIENCE!"

1. Fernando Carbayo and Antonio C. Marques, "The Costs of Describing the Entire Animal Kingdom," *Trends in Ecology and Evolution* 26, no. 4 (February 7, 2011): 154–55, http://www.cell.com/trends/ecology-evolution/fulltext/S0169-5347%2811%2900017-6#bib0015 (accessed February 14, 2011).

2. C. Mora, D. P. Tittensor, S. Adl, A. G. B. Simpson, and B. Worm, "How Many Species Are There on Earth and in the Ocean?" *PLoS Biol* 9, no. 8 (2011): e1001127. doi:10.1371/journal.pbio.1001127.

3. Robert Kunzig, "20,000 Microbes under the Sea," *Discover*, March 2004, http://discovermagazine.com/2004/mar/cover (accessed April 8, 2011).

4. Christopher D. Bader, Carson F. Mencken, and Joseph O. Baker, *Paranormal America* (New York: New York University Press, 2010), p. 106.

5. Curtis Wienker, interview with the author, March 21, 2011.

6. Bob Young, "Lovable Trickster Created a Monster with Bigfoot Hoax," *Seattle Times*, December 5, 2002, http://community.seattletimes.nwsource.com/archive/?date=20021205&slug=raywallaceobit05m (accessed July 1, 2011).

7. Timothy Egan, "Search for Bigfoot Outlives the Man Who Created Him," *New York Times*, January 03, 2003, http://www.nytimes.com/2003/01/03/us/search-for-bigfoot-outlives-the-man-who-created-him.html (accessed July 3, 2011).

8. Cameron M. Smith, interview with the author, March 21, 2011.

9. "Sasquatch: Legend Meets Science," *Talk of the Nation*, NPR, Novem-

ber 10, 2006, http://www.npr.org/templates/story/story.php?storyId=6469070 (accessed December 22, 2010).

10. Greg Long, *The Making of Bigfoot: The Inside Story* (Amherst, NY: Prometheus Books, 2004), pp. 443–51.

11. Ibid., p. 336.

12. Jeff Meldrum, *Sasquatch: Legend Meets Science* (New York: Tom Doherty Associates, 2006), p. 44.

CHAPTER 42: "ANGELS WATCH OVER ME."

1. Hebrews 13:2.

2. Frank Newport, "Americans More Likely to Believe in God Than the Devil, Heaven More Than Hell," Gallup News Service, June 13, 2007, http://www.gallup.com/poll/27877/Americans-More-Likely-Believe-God-Than-Devil-Heaven-More-Than-Hell.aspx (accessed November 3, 2010).

3. David Kinnaman, "New Research Explores Teenage Views and Behavior Regarding the Supernatural," Barna Group, January 23, 2006, http://www.barna.org/barna-update/article/5-barna-update/164-new-research -explores-teenage-views-and-behavior-regarding-the-supernatural?q=angels (accessed February 23, 2010).

4. Heather Mason Kiefer, "Divine Subjects: Canadians Believe, Britons Skeptical," Gallup News Service, November 16, 2004, http://www.gallup.com/poll/14083/Divine-Subjects-Canadians-Believe-Britons-Skeptical.aspx (accessed March 11, 2011).

5. James Randi, *An Encyclopedia of Claims, Frauds, and Hoaxes of the Occult and Supernatural* (New York: St. Martin's Griffin, 1995), p. 11.

6. Christopher D. Bader, Carson F. Mencken, and Joseph O. Baker, *Paranormal America* (New York: New York University Press, 2010), p. 184.

7. Ibid., pp. 185–86.

8. Ibid., p. 185.

CHAPTER 43: "MAGIC IS REAL AND WITCHES ARE DANGEROUS."

1. BBC News, "Nigeria 'Child Witch Killer' Held," *BBC News*, December 4, 2008, http://news.bbc.co.uk/2/hi/africa/7764575.stm (accessed February 12, 2011).

2. Salman Ravi, "Village 'Witches' Beaten," *BBC News*, October 20, 2009, http://news.bbc.co.uk/2/hi/south_asia/8315980.stm (accessed February 12, 2011).

3. BBC News, "Indian 'Witchcraft' Family Killed," *BBC News*, March 19,

2006, http://news.bbc.co.uk/2/hi/south_asia/4822750.stm (accessed February 13, 2011).

4. James Uribarri, "Mobs in Haiti Kill 'Witches' Accused of Intentionally Spreading Cholera," *New York Daily News*, December 3, 2010, http://www.nydailynews.com/news/world/2010/12/03/2010-12-03_mobs_in_haiti_kill_suspected_witches_accused_of_intentionally_spreading_cholera_.html (accessed February 1, 2011).

5. "African Churches Denounce Children as Witches," MSNBC/Associated Press, October 17, 2009.

6. Ibid.

7. Graham Wood, "Hex Appeal," *Atlantic*, June 2010.

8. Watch the video of Sarah Palin being protected from witches and witchcraft on YouTube: "Sarah Palin Gets Protection from Witches," YouTube video, 9:47, from ceremony where Thomas Muthee of the Assembly of God provides "special supernatural protection from witchcraft" to Sarah Palin on September 23, 2008, posted by "nyprogressive," September 23, 2008, http://www.youtube.com/watch?v=jl4HIc-yfgM (accessed February 10, 2011).

9. David W. Moore, "Three in Four Americans Believe in Paranormal," Gallup News Service, June 16, 2005, http://www.gallup.com/poll/16915/three -four-americans-believe-paranormal.aspx (accessed March 11, 2011).

10. Michelle Mead, interview with the author, March 2, 2011.

CHAPTER 44: "ATLANTIS IS DOWN THERE SOMEWHERE."

1. Paul Rincon, "Satellite Images 'Show Atlantis,'" *BBC News*, June 6, 2004, http://news.bbc.co.uk/2/hi/science/nature/3766863.stm (accessed February 20, 2011).

2. BBC News, "Atlantis 'Obviously Near Gibraltar,'" *BBC News*, September 20, 2001, http://news.bbc.co.uk/2/hi/science/nature/1554594.stm (accessed February 11, 2011).

3. BBC News, "Tsunami Clue to 'Atlantis' Found," *BBC News*, August 15, 2005, http://news.bbc.co.uk/2/hi/science/nature/4153008.stm (accessed February 20, 2011.)

4. Baylor Institute for Studies of Religion, "American Piety in the 21st Century," Baylor University, September 2006, p. 45, http:// www.baylor.edu/content/services/document.php/33304.pdf (accessed December 12, 2010).

5. Theodore Schick and Lewis Vaughn, *How to Think about Weird Things* (New York: McGraw-Hill, 2011), p. 7.

6. Bob Ballard, interview with the author. Quoted in Guy P. Harrison, "Deep Secrets," *Caymanian Compass*, June 13, 2002, p. 14.

7. Plato describes Atlantis in two of his dialogues: the *Timaeus* and the

Critias.

8. Kenneth L. Feder, *Encyclopedia of Dubious Archaeology: From Atlantis to the Walam Olum* (Santa Barbara, CA: Greenwood, 2010), p. 33.

CHAPTER 45: "I'M GOING TO HEAVEN WHEN I DIE."

1. Guy P. Harrison, "God Is in This Place," *Caymanian Compass*, November 19, 1993, pp. 10–11.

2. Frank Newport, "Americans More Likely to Believe in God Than the Devil, Heaven More Than Hell," Gallup News Service, June 13, 2007, http://www.gallup.com/poll/27877/Americans-More-Likely-Believe-God-Than-Devil-Heaven-More-Than-Hell.aspx (accessed March 22, 2011).

3. Pew Forum on Religion and Public Life, "American Grace: How Religion Divides and Unites Us," Pew Research Center Publications, January 7, 2011, http://pewresearch.org/pubs/1847/how-religion-divides-and-unites-us -david-campbell-conversation-transcript (accessed March 22, 2011).

CHAPTER 46: "SOMETHING VERY STRANGE IS GOING ON IN THE BERMUDA TRIANGLE."

1. Michael Shermer, *Why People Believe Weird Things* (New York: MJF Books, 1997), pp. 54–55.

2. Larry Kusche, *The Bermuda Triangle Mystery—Solved* (Amherst, NY: Prometheus Books, 1995), pp. 117–18.

3. Ibid., p. 118.

4. Ibid., p. 120.

5. Ibid., p. 276.

6. Ibid., pp. 275–77.

7. Ibid.

8. Naval Heritage and History Command, "The Bermuda Triangle," http://www.history.navy.mil/faqs/faq8-1.htm (accessed March 7, 2011).

CHAPTER 47: "AREA 51 IS WHERE THEY KEEP THE ALIENS."

1. Joe Nickell, "The Story behind the 'Alien Autopsy' Hoax," *Science Daily*, May 7, 2006, http://www.livescience.com/742-story-alien-autopsy -hoax.html (accessed February 22, 2011).

2. Seth Shostak, interview with the author, April 6, 2011.

3. William B. Scott, "The Truth Is out There: A Veteran Reporter

Describes His Search for the Aircraft of Area 51," *Air and Space*, September 1, 2010, http://www.airspacemag.com/military-aviation/The-Truth-is-Out-There.html?c=y&page=1 (accessed November 12, 2010).

4. Ibid.

5. Phil Patton, "6 Top-Secret Aircraft That Are Mistaken for UFOs," *Popular Mechanics*, February 18, 2009, http://www.popularmechanics.com/technology/aviation/ufo/4304207 (accessed March 1, 2011).

6. Depending on the angle from which it is observed, Lockheed's P-791 can look like an ordinary blimp or an alien spacecraft straight out of Hollywood casting: "Hybrid Air Vehicle (P-791)," Lockheed Martin, http://www.lockheedmartin.com/products/p-791/ (accessed February 22, 2011).

7. Northrop Grumman has information and photos of the X-47B: X-47B UCAS, Northrop Grumman, http:// www.as.northropgrumman.com/products/nucasx47b/index.html (accessed February 17, 2011).

CHAPTER 48: "THE MAYANS WARNED US: IT'S ALL OVER ON DECEMBER 21, 2012."

1. Mark Van Stone, "2012 FAQ," Foundation for the Advancement of Mesoamerican Studies, http://www.famsi.org/research/vanstone/2012/faq.html (accessed November 2, 2012).

2. Michael Drosnin, *Bible Code II* (New York: Touchstone, 1998), p. 230.

3. Ibid., p. 186.

4. Ian Oneil, "2012 Alien Invasion? Um, No," *Discovery News*, December 27, 2010, http://news.discovery.com/space/the-2012-alien-invasion-um-no.html (accessed March 20, 2011).

5. Brian Handwerk, "2012: Six End-of-the-World Myths Debunked," *National Geographic*, November 6, 2010, http://news.nationalgeographic.com/news/2009/11/091106-2012-end-of-world-myths.html (accessed March 2, 2011).

6. David Morrison, "2012 and Counting," *Skeptic*, http://www.skeptic.com/reading_room/2012-and-counting/ (accessed October 19, 2011).

7. Ibid.

CHAPTER 49: "THE END IS NEAR!"

1. Frank Newport, "One-Third of Americans Believe the Bible Is Literally True," Gallup News Service, May 25, 2007, http://www.gallup.com/poll/27682/OneThird-Americans-Believe-Bible-Literally-True.aspx (accessed January 30, 2011).

2. Steve Crabtree, "New Poll Gauges Americans' General Knowledge Levels," Gallup News Service, July 6, 1999, http://www.gallup.com/poll/3742/new-poll-gauges-americans-general-knowledge-levels.aspx (accessed March 24, 2011).

3. Harris Poll, "A Third of Public, Including Three in Five Republicans, Support the Tea Party Movement and about a Quarter Oppose It," Harris Interactive, March 31, 2010, http://www.harrisinteractive.com/vault/Harris_Interactive_Poll_Tea_Party_Opposition_2010_03.pdf (accessed March 23, 2011).

4. LiveScience staff, "Quarter of Republicans Think Obama May Be the Anti-Christ," LiveScience, March 25, 2010, http://www.livescience.com/8160 -quarter-republicans-obama-anti-christ.html (accessed March 23, 2011).

5. Revelation 9.

6. Guy P. Harrison, "God Is in This Place," *Caymanian Compass*, November 19, 1993, p. 11.

7. Baylor Institute for Studies of Religion, "American Piety in the 21st Century," Baylor University, September 2006, p. 45, http://www.baylor.edu/content/services/document.php/33304.pdf (accessed December 12, 2010).

8. Ray Kurzweil, *The Singularity Is Near* (New York: Penguin, 2006). See also Joel Garreau, *Radical Evolution: The Promise and Peril of Enhancing Our Minds, Our Bodies—And What It Means to Be Human* (New York: Broadway, 2006).

CHAPTER 50: "WE'RE ALL GONNA DIE!"

1. "Dark Matter, the Bringer of Life," *New Scientist*, April 9, 2011, p. 16.

2. Hugo de Garis, *The Artilect War* (Palm Springs, CA: ETC, 2005), p. 25.

3. Jennifer Viegas, "Human Extinction: How Could It Happen?" *Discovery News*, http://news.discovery.com/human/human-extinction-doomsday.html (accessed September 5, 2011).

BIBLIOGRAPHY

Aaronovitch, David. *Voodoo Histories: The Role of the Conspiracy Theory in Shaping Modern History*. New York: Riverhead, 2010.

Bader, Christopher, F. Carson Mencken, Joesph Baker. *Paranormal America*. New York: New York University Press, 2010.

Barrett, Stephen, and William T. Jarvis, eds. *The Health Robbers: A Close Look at Quackery in America*. Amherst, NY: Prometheus Books, 1993.

Bausell, R. Barker. *Snake Oil Science: The Truth about Complementary and Alternative Medicine*. Oxford: Oxford University Press, 2009.

Belanger, Jeff, ed. *Encyclopedia of Haunted Places*. Franklin Lakes, NJ: Castle Books, 2008.

Bennett, Jeffrey. *Beyond UFOs: The Search for Extraterrestrial Life and Its Astonishing Implications for Our Future*. Princeton, NJ: Princeton University Press, 2008.

Berlitz, Charles. *Atlantis: The Lost Continent Revealed*. London: Fontana/Collins, 1985.

Berlitz, Charles, and William L. Moore. *The Roswell Incident*. New York: Berkley Books, 1980.

Blackmore, Susan. *Beyond the Body*. Chicago: Academy Chicago Publishers, 1992.

———. *Dying to Live: Near-Death Experiences*. Amherst, NY: Prometheus Books, 1993.

———. *In Search of the Light: The Adventures of a Parapsychologist*. Amherst, NY: Prometheus Books, 1996.

Bostrom, Nick, and Milan M. Cirkvoic, eds. *Global Catastrophic Risks*. New York: Oxford University Press, 2008.

Brockman, John, ed. *Intelligent Thought: Science versus the Intelligent Design Movement*. New York: Vintage Books, 2006.

Brockman, Max, ed. *What's Next? Dispatches from the Future of Science*. New York: Vintage, 2009.

Buh, Joshua Blu. *Bigfoot: The Life and Times of a Legend*. Chicago: University of Chicago Press, 2010.

Calder, Nigel. *Magic Universe: A Grand Tour of Modern Science*. New York: Oxford University Press, 2003.

Carroll, Robert Todd, ed. *The Skeptic's Dictionary*. Hoboken, NJ: John Wiley and Sons, 2003.

Cayce, Edgar. *Edgar Cayce on Atlantis*. New York: Paperback Library, 1968.

Chabris, Christopher, and Daniel Simons. *The Invisible Gorilla and Other Ways Our Intuitions Deceive Us*. New York: Crown, 2010.

Chaffe, John. *Thinking Critically*. Boston: Houghton Mifflin, 2000.

Charpak, Georges, and Henri Broch. *Debunked! ESP, Telekinesis, and Other Pseudoscience*. Baltimore: Johns Hopkins University Press, 2004.

Clancy, Susan A. *Abducted: How People Come to Believe They Were Kidnapped by Aliens*. Cambridge, MA: Harvard University Press, 2005.

Clegg, Brian. *Armageddon Science*. New York: St. Martin's Press, 2010.

———. *The God Effect: Quantum Entanglement, Science's Strangest Phenomenon*. New York: St. Martin's Griffin, 2006.

Darling, David. *Life Everywhere: The Maverick Science of Astrobiology*. New York: Basic Books, 2001.

Davies, Paul. *The Eerie Silence: Renewing Our Search for Alien Intelligence*. Boston: Houghton Mifflin Harcourt, 2010.

Davis, Hank. *Caveman Logic: The Persistence of Primitive Thinking in a Modern World*. Amherst, NY: Prometheus Books, 2009.

Dawkins, Richard. *The Ancestor's Tale*. Boston: Houghlin Mifflin, 2004.

———. *The Blind Watchmaker: Why the Evidence of Evolution Reveals a Universe without Design*. New York: W. W. Norton, 1996.

———. *Climbing Mount Improbable*. New York: W. W. Norton, 1997.

———. *The Greatest Show on Earth: The Evidence for Evolution*. New York: Free Press, 2009.

Dunning, Brian. *Skeptoid: A Critical Analysis of Pop Phenomena*. Seattle, WA: Thunderwood Press, 2007.

———. *Skeptoid 2: More Critical Analysis of Pop Phenomena*. Seattle, WA: Skeptoid Media, 2008.

Ellis, Richard. *Imagining Atlantis*. New York: Vintage, 1999.

Epstein, Greg. *Good without God: What a Billion Nonreligious People Do Believe*. New York: Harper Paperbacks, 2010.

Feder, Kenneth L. *Frauds, Myths, and Mysteries: Science and Pseudoscience in Archaeology*. Boston: McGraw Hill, 2008.

Ferris, Timothy. *The Whole Shebang: A State-of-the-Universe(s) Report*. New York: Simon and Schuster, 1998.

Fine, Cordelia. *A Mind of Its Own: How Your Brain Distorts and Deceives*. New York: W. W. Norton, 2006.

Fish, Jefferson, ed. *Race and Intelligence: Separating Science from Myth*. New York: Routledge, 2001.

Forrest, Barbara, and Paul R. Gross. *Creationism's Trojan Horse: The Wedge of Intelligent Design*. Oxford: Oxford University Press, 2004.

Frazier, Kendrick. *Science under Siege: Defending Science, Exposing Pseudoscience*. Amherst, NY: Prometheus Books, 2009.

Freedman, Carl. *Conversations with Isaac Asimov*. Jackson: University Press of Mississippi, 2005.

Gardner, Martin. *Did Adam and Eve Have Navels: Debunking Pseudoscience*. New York: W. W. Norton, 2001.

———. *The New Age: Notes of a Fringe-Watcher.* Amherst, NY: Prometheus Books, 1991.

———. *Science: Good, Bad, and Bogus*. Amherst, NY: Prometheus Books, 1989.

Garis, Hugo de. *The Artilect War*. Palm Springs, CA: ETC Publication, 2005.

Goldacre, Ben. *Bad Science: Quacks, Hacks, and Big Pharma Flacks*. New York: Faber and Faber, 2010.

Graves, Joseph L. *The Emperor's New Clothes: Biological Theories of Race at the Millennium*. Piscataway, NJ: Rutgers University Press, 2003.

———. *The Race Myth: Why We Pretend Race Exists in America*. New York: Plume, 2005.

Greene, Brian. *The Elegant Universe: Superstrings, Hidden Dimensions, and the Quest for the Ultimate Theory.* New York: Vintage Books, 2000.

Guyatt, Nicholas. *Have a Nice Doomsday: Why Millions of Americans Are Looking Forward to the End of the World*. United Kingdom: Ebury Press, 2007.

Halpern, Paul. *Countdown to Apocalypse: A Scientific Exploration of the End of the World*. Cambridge, MA: Perseus Publishing, 1990.

Hanlon, Michael. *Eternity: Our Next One Billion Years*. London: MacMillan, 2009.

Hansen, James. *Storms of My Grandchildren: The Truth about the Coming Climate Catastrophe and Our Last Chance to Save Humanity*. New York: Bloomsbury USA, 2010.

Harrison, Guy P. *50 Reasons People Give for Believing in a God*. Amherst, NY: Prometheus Books, 2008.

———. *Race and Reality: What Everyone Should Know about Our Biological Diversity*. Amherst, NY: Prometheus Books, 2009.

Hazen, Robert M. *Genesis: The Scientific Quest for Life's Origins*. Joseph Henry Press, 2007.

Head, Tom, ed. *Conversations with Carl Sagan*. Jackson: University Press of Mississippi, 2006.

Hedges, Chris. *Empire of Illusion: The End of Literacy and the Triumph of Spectacle*. New York: Nation Books, 2010.

Hemenway, Priya. *Hindu Gods*. San Francisco, CA: Chronicle Books, 2003.

Henry, Lewis, ed. *Five Thousand Quotations for All Occasions*. New York: Doubleday, 1945.

Hines, Terrence. *Pseudoscience and the Paranormal*. Amherst, NY: Prometheus Books, 2003.

Hood, Bruce. *The Science of Superstition: How the Developing Brain Creates Supernatural Beliefs*. New York: HarperCollins, 2010.

Horstman, Judith. *The Scientific American: Brave New Brain*. San Francisco, CA: Jossey-Bass, 2010.

———. *The Scientific American: Day in the Life of Your Brain*. San Francisco, CA: Jossey-Bass, 2009.

Humes, Edward. *Monkey Girl: Education, Education, Religion, and the Battle for America's Soul*. New York: HarperCollins, 2007.

Jack, Albert. *Loch Ness Monsters and Raining Frogs: The World's Most Puzzling Mysteries Solved.* New York: Random House, 2007.

Jayawardhana, Ray. *Strange New Worlds: The Search for Alien Planets and Life beyond Our Solar System*. Princeton, NJ: Princeton University Press, 2011.

Jordan, Michael. *Encyclopedia of Gods*. London: Kyle Cathie, 2002.

Kaku, Michio. *Physics of the Future: How Science Will Shape Human Destiny and Our Daily Lives by the Year 2100*. New York: Doubleday, 2011.

Kaminer, Wendy. *Sleeping with Extra-terrestrials*. New York: Pantheon Books, 1999.

Kaufman, Marc. *First Contact: Scientific Breakthroughs in the Hunt for Life beyond Earth*. New York: Simon and Schuster, 2011.

Keegan, John. *The Second World War.* New York: Penguin, 2005.

Kelly, Lynne. *The Skeptic's Guide to the Paranormal*. New York: Avalon, 2004.

Kida, Thomas. *Don't Believe Everything You Think*. Amherst, NY: Prometheus Books, 2006.

Klass, Philip J. *The Real Roswell Crashed-Saucer Coverup*. Amherst, NY: Prometheus Books, 1997.

Kurtz, Paul. *Affirmations: Joyful and Creative Exuberance*. Amherst, NY: Prometheus Books, 2004.

———. *The New Skepticism: Inquiry and Reliable Knowledge*. Amherst, NY: Prometheus Books, 1992.

———, ed. *Science and Religion: Are They Compatible?* Amherst, NY: Prometheus Books, 2003.

———, ed. *Skeptical Odysseys*. Amherst, NY: Prometheus Books, 2001.

———. *The Transcendental Temptation: A Critique of Religion and the Paranormal*. Amherst, NY: Prometheus Books, 1991.

Kurzweil, Ray. *The Singularity Is Near.* New York: Penguin, 2006.

Kusche, Larry. *The Bermuda Triangle Mystery—Solved*. Amherst, NY: Prometheus Books, 1995.

Leeming, David. *A Dictionary of Creation Myths.* New York: Oxford University Press, 1994.

Lewis, James R. *Doomsday Prophecies: A Complete Guide to the End of the World*. Amherst, NY: Prometheus Books, 2000.

Lipstadt, Deborah. *Denying the Holocaust: The Growing Assault on Truth and Memory*. New York: Plume, 1994.

Long, Greg. *The Making of Bigfoot: The Inside Story*. Amherst, NY: Prometheus Books, 2004.

Loxton, Daniel. *Evolution: How We and All Living Things Came to Be*. Tonawanda, NY: Kids Can Press, 2010.

Macknic, Stephen L., and Susana Martinez-Conde. *Sleights of Mind*. New York: Henry Holt, 2010.

Margulis, Lynn, and Dorian Sagan. *Microcosmos: Four Billion Years of Microbial Evolution*. Berkeley: University of California Press, 1997.

Marks, Jonathan. *Why I Am Not a Scientist*. Berkley: University of California Press, 2009.

Mayr, Ernst. *What Evolution Is*. London: Weidenfeld and Nicolson, 2002.

McAndrew, James. *Roswell Report: Case Closed*. Texas: Books Express Publishing, 2011.

McGuire, Bill. *A Guide to the End of the World*. Oxford: Oxford University Press, 2002.

Medina, John. *Brain Rules: 12 Principles for Surviving and Thriving at Work, Home, and School*. Seattle, WA: Pear Press, 2008.

Mnookin, Seth. *The Panic Virus: A True Story of Medicine, Science, and Fear*. New York: Simon and Schuster, 2011.

Montagu, Ashley. *Man's Most Dangerous Myth: The Fallacy of Race*. Lanham, MD: AltaMira Press, 1997.

Mooney, Chris, and Sheril Kirshenbaum. *Unscientific America: How Scientific Illiteracy Threatens Our Future*. New York: Basic Books, 2009.

Murdoch, Stephen. *IQ: A Smart History of a Failed Idea*. Hoboken, NJ: Wiley, 2007.

National Academy of Sciences. *Science, Evolution, and Creationism*. Washington, DC: National Academies Press, 2008.

Nelson, Kevin. *The Spiritual Doorway in the Brain*. New York: Dutton, 2011.

Nickell, Joe. *Adventures in Paranormal Investigation*. Lexington: University Press of Kentucky, 2007.

———. *Looking for a Miracle: Weeping Icons, Relics, Stigmata, Visions & Healing Cures*. Amherst, NY: Prometheus Books, 1999.

———. *The Mystery Chronicles: More Real-Life X-Files*. Lexington: University Press of Kentucky, 2004.

———. *Psychic Sleuths: ESP and Sensational Cases*. Amherst, NY: Prometheus Books, 1994.

———. *Relics of the Christ*. Lexington: University Press of Kentucky, 2007.

———. *Tracking the Man-Beasts: Sasquatch, Vampires, Zombies, and More*. Amherst, NY: Prometheus Books, 2011.

Nisbett, Richard E. *Intelligence and How to Get It*. New York: W. W. Norton, 2009.

Offit, Paul A. *Autism's False Prophets: Bad Science, Risky Medicine, and the Search for a Cure*. New York: Columbia University Press, 2010.

———. *Deadly Choices: How the Anti-Vaccine Movement Threatens Us All*. New York: Basic Books, 2010.

Offit, Paul A., and Charlotte A. Moser. *Vaccines and Your Child: Separating Fact from Fiction*. New York: Columbia University Press, 2011.

Olson, Steve. *Mapping Human History: Genes, Race, and Our Common Origins*, New York: Mariner Books, 2003.

Palmer, Douglas. *Origins: Human Evolution Revealed*. New York: Mitchell Beazley, 2010.

Park, Robert. *Voodoo Science: The Road from Fraud to Foolishness*. New York: Oxford University Press, 2000.

Pigliucci, Massimo. *Denying Evolution: Creationism, Scientism, and the Nature of Science*. Sunderland, MA: Sinauer Associates, 2002.

———. *Nonsense on Stilts: How to Tell Science from Bunk*. Chicago: University of Chicago Press, 2010.

Piper, Don. *90 Minutes in Heaven*. Grand Rapids, MI: Revell, 2004.

Plait, Philip. *Bad Astronomy*. New York: John Wiley and Sons, 2002.

———. *Death from the Skies!* New York: Penguin Books, 2008.

Prothero, Stephen. *God Is Not One: The Eight Rival Religions That Run the World—And Why Their Differences Matter*. New York: HarperOne, 2007.

———. *Religious Literacy: What Every American Needs to Know—And Doesn't*. New York: HarperOne, 2007.

Radford, Benjamin. *Media Mythmakers: How Journalists, Activists and Advertisers Mislead Us*. Amherst, NY: Prometheus Books, 2003.

———. *Scientific Paranormal Investigation: How to Solve the Unexplained Mysteries*. Corrales, NM: Rhombus, 2010.

Randi, James. *An Encyclopedia of Claims, Frauds, and Hoaxes of the Occult and Supernatural*. New York: St. Martin's Griffin, 1995.

———. *The Faith Healers*. Amherst, NY: Prometheus Books, 1989.

———. *Flim-Flam!* Amherst, NY: Prometheus Books, 1982.

———. *The Mask of Nostradamus*. Amherst, NY: Prometheus Books, 1993.

Rees, Martin. *Our Final Hour*. New York: Basic Books, 2003.

Ryan, Craig. *Pre-Astronauts: Manned Ballooning on the Threshold of Space*. Annapolis, MD: US Naval Institute Press, 2003.

Sagan, Carl. *The Demon-Haunted World: Science as a Candle in the Dark*. New York: Random House, 1995.

———. *The Varieties of Scientific Experience: A Personal View of the Search for God*. New York: Penguin, 2007.

Saler, Benson, Charles A. Ziegler, and Charles B. Moore. *UFO Crash at Roswell: The Genesis of a Modern Myth*. Washington, DC: Smithsonian Books, 2010.

Schick, Theodore, and Lewis Vaughn. *How to Think about Weird Things*. New York: McGraw-Hill, 2011.

Scott, Eugenie C. *Evolution vs. Creationism: An Introduction*. Berkeley: University of California Press, 2009.

Sharpiro, Rose. *Suckers: How Alternative Medicine Makes Fools of Us All*. London: Harvill Secker, 2008.

Shenk, David. *The Genius in All of Us*. New York: Doubleday, 2010.

Shermer, Michael. *The Believing Brain: From Ghosts and Gods to Politics and Conspiracies—How We Construct Beliefs and Reinforce Them as Truths*. New York: Times Books, 2011.

———. *The Borderlands of Science: Where Sense Meets Nonsense*. New York: Oxford University Press, 2002.

———. *Science Friction: Where the Known Meets the Unknown*. New York: Times Books, 2005.

———. *Why Darwin Matters: The Case against Intelligent Design*. New York: Times Books, 2006.

———. *Why People Believe Weird Things*. New York: MJF Books, 1997.

Shermer, Michael, and Alex Grobman. *Denying History: Who Says the Holocaust Never Happened and Why Do They Say It?* Berkley: University of California Press, 2009.

Shostak, Seth. *Confessions of an Alien Hunter: A Scientist's Search for Extraterrestrial Intelligence*. Washington, DC: National Geographic, 2009.

Singh, Simon, and Edzard Ernst. *Trick or Treatment: The Undeniable Facts about Alternative Medicine*. New York: W. W. Norton, 2000.

Smith, Cameron M. *The Fact of Evolution*. Amherst, NY: Prometheus Books, 2011.

Smith, Cameron M., and Charles Sullivan. *The Top 10 Myths about Evolution*. Amherst, NY: Prometheus Books, 2006.

Smith, Jonathan C. *Pseudoscience and Extraordinary Claims of the Paranormal: A Critical Thinker's Toolkit*. West Sussex, UK: Wiley-Blackwell, 2010.

Soutwood, Richard. *The Story of Life*. New York: Oxford University Press, 2004.

Specter, Michael. *Denialism: How Irrational Thinking Hinders Scientific Progress, Harms the Planet, and Threatens Our Lives*. New York: Penguin Press, 2009.

Stanovich, Keith E. *How to Think Straight about Psychology*. New York: HarperCollins, 1996.

Stavropaulos, Steven. *The Beginning of All Wisdom*. Jackson, TN: Da Capo Press, 2003.

Stenger, Victor J. *The Fallacy of Fine-Tuning: Why the Universe Is Not Designed for Us*. Amherst, NY: Prometheus Books, 2011.

———. *The New Atheism: Taking a Stand for Science and Reason*. Amherst, NY: Prometheus Books, 2009.

Striber, Whitley. *Communion*. New York: Beech Tree Books, 1987.

Stringer, Chris, and Peter Andrews. *The Complete World of Human Evolution*. New York: Thames and Hudson, 2005.

Stringer, Lauren, and Peters Westberg. *Our Family Tree: An Evolution Story*. New York: Harcourt Children's Books, 2003.

Tattersall, Ian. *Extinct Humans*. New York: Basic Books, 2001.

Van Hecke, Madeleine. *Blind Spots: Why Smart People Do Dumb Things*. Amherst, NY: Prometheus Books, 1997.

Vankin, Jonathan, and John Whalen. *The 80 Greatest Conspiracies of All Time*. New York: Citadel Press Books, 2004.

Van Praagh, James. *Ghosts among Us: Uncovering the Truth about the Other Side*. New York: HarperOne, 2009.

———. *Heaven and Earth: Making the Psychic Connection*. New York: Pocket, 2006.

Vyse, Stuart A. *Believing in Magic: The Psychology of Superstition*. New York: Oxford University Press, 1997.

Wanjek, Christopher. *Bad Medicine: Misconceptions and Misuses Revealed, from Distance Healing to Vitamin O*. New York: Wiley, 2002.

Ward, Peter D., and Donald Brownlee. *The Life and Death of Planet Earth*. New York: Henry Holt, 2004.

Webb, Stephen. *If the Universe Is Teeming with Aliens . . . Where Is Everybody?* New York: Copernicus Books, 2002.

Wheen, Francis. *How Mumbo Jumbo Conquered the World: A Short History of Modern Delusions*. New York: Public Affairs, 2004.

Wiseman, Richard. *Paranormality: Why We See What Isn't There*. London: Macmillan, 2011.

Woerlee, G. M. *Mortal Minds: The Biology of Near-Death Experiences*. Amherst, NY: Prometheus Books, 2005.

Wood, Michael. *In Search of the Trojan War*. Los Angeles: University of California Press, 1998.

Young, Matt, and Taner Edis. *Why Intelligent Design Fails: A Scientific Critique of the New Creationism*. Piscataway, NJ: Rutgers University Press, 2006.

Zimmer, Carl. *The Tangled Bank: An Introduction to Evolution*. New York: Roberts, 2009.

Zuckerman, Phil. *Society without God: What the Least Religious Nations Can Tell Us about Contentment*. New York: NYU Press, 2010.

INDEX

Abbey, Edward, 81
Adler Planetarium, 140
Africa, 26, 168, 173, 184, 190, 196, 201–202, 204, 228, 229, 231, 267, 293, 355, 356, 403
AIDS, 201–202, 220–22, 347
Air and Space (magazine), 387
Aldrin, Buzz, 91
alien abductions, 132–38
aliens, 100–104, 111, 119–20, 122, 125, 127, 130, 132–38, 365, 379, 382. *See also* alien abductions
al Qaeda, 174, 193, 196
alternative medicine, 14, 26, 33, 134, 201–10
American Anthropological Association (AAA) statement on race, 226–27
American Institutes for Research, 70
angels, 132, 350–53
Antichrist, 78
Apollo (space program), 89–92
Arafat, Yasser, 60
Aratat (mountain), 300
archaeology, 305–309
Area 51, 382–88
Aristotle, 81, 150
Armstrong, Neil, 89, 90, 95
Arnold, Kenneth, 123, 124
Artilect War, The (De Garis), 413
asteroid strike, 410–11
astrology, 18, 19, 24, 25, 28, 29, 33, 76, 139–43, 276, 318, 324, 357
Atlantis, 28, 352, 374, 393, 363–67
Atlantis: The Lost Continent Revealed (Berlitz), 364
Auschwitz, 157
autism and vaccines, 206, 234–41
Ayatollah Khomeini, 78

Babylonians, 141
Bader, Christopher, 350
Baker, Joseph, 350
Bakker, Jim, 315
Bakker, Tammy Faye, 315
Ballard, Bob, 365
Band of Brothers (television miniseries), 156
Bean, Alan, 91
Beck, Glenn, 165
Behrens, Ron, 217
Belgium, 213
Bergen, Peter, 196
Berlitz, Charles, 125, 364, 366, 375
Bermuda Triangle, 364, 367, 372–81, 393
Bermuda Triangle, The (Berlitz), 375
Bermuda Triangle Mystery—Solved, The (Kusche), 376
Bible code, 58–61, 394
Bible Code, The (Drosnin), 59, 76, 394

Bible Code II (Drosnin), 60
Bieber, Justin, 109
Bigfoot, 14, 24, 26, 30, 116, 333, 339–49
Binet, Alfred, 52, 53
birth weight, 231
Blackmore, Susan, 40
Blasingame, Loretta, 368–70
Bodnath Stupa, 312
Borman, Frank, 91
Brazel, Mack, 123
Brinkley, David, 167
British Medical Association, 214
Bronson, Charles, 62
Browning, Shawn R., 239
Buddha, 201
Bush, George W., 164, 193, 195, 255, 275–76

CAM (complimentary and alternative medicine), 201–209
Carpenter, Scott, 91
Carroll, Robert Todd, 372
Carroll, Sean M., 40
Case for Reincarnation, The (Danelek), 66
Case of the Ancient Astronauts, The (NOVA television program), 101
Casey Anthony murder trial, 168
Caveman Logic: The Persistence of Primitive Thinking in a Modern World (Davis), 31, 201
Cayman Islands, 84, 184, 223–25, 368, 373
CDC. *See* Centers for Disease Control and Prevention
Census of Marine Life, 336
Centers for Disease Control and Prevention (CDC), 240
Central African Republic (witchcraft court cases in), 356
Central Intelligence Agency. *See* CIA
Cernan, Gene, 91–93
Chabris, Christopher, 116, 192
Chaikin, Andrew, 89
Chariots of the Gods? (Von Däniken), 100
Cheney, Dick, 195
Cherne, Jack, 91
"child witches," 355–56
Chopra, Deepak, 28
Christians, 63, 82–83
Christopher, Kevin, 363
Church of the Holy Sepulchre, 285, 310–11
CIA (Central Intelligence Agency), 383
Clancy, Susan A., 134–35
Clarke, Arthur C., 167
Clay, Bryan, 188
climate change (global warming), 161–66, 337, 412–13
Coast Guard (US), 375, 380
cold reading, 42–46
Cold War, 70, 89, 94, 122 124, 193, 386, 388, 411
Colosseum, 106
Colquhoun, David, 215
Committee for Skeptical Inquiry, 333
complimentary and alternative medicine. *See* CAM
confirmation bias, 29–31, 43, 71, 165, 194–95, 252, 261, 291
conspiracy theories, 192–97
Conspiracy Theory: Did We Land on the Moon? (Fox television documentary), 97
countermeasures (as UFOs), 113–14
Cousteau, Jacques, 334
Cowherd, Colin, 128
Coyne, Jerry, 276
creationism, 175, 259–72
Creation Museum, 303, 158
cryptozoology, 333–49

cults, 296–99
CVS (pharmacy), 214

Danelek, Alan J., 66, 67
Danube River, 78
Darwin, Charles, 261, 262, 266, 276–77
Davis, Hank, 31, 201, 323
Dawkins, Richard, 266
Day the Earth Stood Still, The (film), 109, 414
Deadly Choices: How the Anti-Vaccine Movement Threatens Us All (Offit), 237
Dead Sea Scrolls, 308
Deep: The Extraordinary Creatures of the Deep, The (Nouvian), 337
Deer, Brian, 236
De Garis, Hugo, 413
Democratic Republic of Congo, 173
Demon-Haunted World, The (Sagan), 178
Denialism: How Irrational Thinking Hinders Scientific Progress, Harms the Planet, and Threatens Our Lives (Specter), 236
diabetes, 230–31
Diamond, Jared, 180
Dick, Philip K., 259
DiMaggio, Joe, 128
Discover (magazine), 177
Dollar, Creflo A., 315
Drosnin, Michael, 59, 60, 394
DTP (vaccine), 238
Duke, Charlie, 91, 92

Easter Island, 106
Edwards v. Aguillard (US Supreme Court case), 273–74
Egypt, 101
Einstein, Albert, 53
Emperor's New Clothes, The (Graves), 180
ESP (extrasensory perception), 25, 33, 68–73
Ethiopians (as elite runners), 188
evolution, 149, 158, 175, 259–72
extinction (of humans), 407–16
extrasensory perception. *See* ESP

faith healing, 218–25, 248
famine, 168
Feder, Kenneth, 365–66
Feynman, Richard, 147
50 Reasons People Give for Believing in a God (Harrison), 248
Fish, Jefferson M., 72
flares (as UFOs), 113–14
Flight 19, 375–77
Fort Lauderdale Naval Air Station, 375
Fort Worth Army Air Base, 124
Frank, Anne, 157
Frauds, Myths, and Mysteries: Science and Pseudoscience in Archaeology (Feder), 365
Friedman, Stanton, 125

Gaddafi, Muamar, 78
Gaddis, Vincent, 374
Galapagos Rift, 365
Gardner, Daniel, 174
General Medical Council, 236
Genius in All of Us, The (Shenk), 51, 56
genome, 230
germs, 411–12
ghosts, 323–32
Gigantopithecus, 343, 346
Gildenberg, B. D., 121, 123, 124, 127
Gimlin, Bob, 345
global warming, 161–66, 337, 412–13
Gobekli Tepe, 104, 106
gods (existence of), 245–49

Goldacre, Ben, 212
Goodall, Jane, 24, 343
Gore, Al, 163–64
gossip, 196
Goya, Francisco, 333
Graves, Joseph L., Jr., 180, 186
Great Rift Valley, 190
Groom Lake, 382

Hahnemann, Samuel, 211–13
Hammonds, Evelyn, 180
Hannity, Sean, 165
Hansen, James, 161
haunted houses, 328–32
Hawk, Tony, 377
Hawking, Stephen, 140, 407
heaven, 368–71
Heironimus, Bob, 346
Herzl, Theodor, 286
Heston, Charlton, 109
Hinduism, 63, 246, 247, 252, 256, 292
Hines, Terrence, 225, 372
Hinn, Benny, 218–22, 315–19, 347
Hister, 78
History Channel, 75
Hitchens, Christopher, 100
Hitler, Adolf, 77, 78, 154, 156, 158
Hitler Youth, 156
Holocaust, 59, 154–60, 175, 286
homeopathic medicine, 23, 26, 201, 203–205, 210–17
Homestead, Florida, 172, 373
Horner, Jack, 261
House of Commons Science and Technology Committee, 215
Hurricane Andrew, 172, 373
Hussein, Saddam, 78

Idaltu Man, 265
inattentional blindness, 116–17
intelligence, 51–57
intelligent design, 28, 112, 158, 175, 273–78, 280
Invisible Gorilla (Chabris and Simons), 116, 192

Jacobi, Walter, 91
Jenkins, Russel, 217
Jerusalem, 284–87, 308, 310–11
Jesus, 223, 224, 247, 248, 285, 287, 292, 307, 310–13, 318, 369–70, 397, 399
Johanson, Donald, 263, 265
Johnson, Magic, 189
Jones, Jim, 297
Jonestown, 296–98
Jordan, Michael, 189

Kamper, Mineke, 216–17
Kean, Leslie, 118
Keegan, John, 154
Kelenjin, 189
Kennedy, John F., 59, 60, 194
Kenyans (as elite runners), 188–91
Kida, Thomas, 68
Kirshenbaum, Sheril, 167
Kittinger, Joe, 126
KKK. *See* Ku Klux Klan
Kor, Eva Mozes, 157–58
Koran code, 60
Kranz, Gene, 91
Ku Klux Klan (KKK), 158
Kurzweil, Ray, 404
Kusche, Larry, 376–79

Laden, Osama bin, 78
Lancet (journal), 236
Layton, Deborah, 296–99
LeDermann, Barbara, 157
Left Behind (book series), 403
Lehmann, Armin, 156
Leininger, James, 65–66
Lewis, Mark, 113–14

Limbaugh, Rush, 165
Lipton, Carwood, 156
Littlewood, John, 83–84
Loch Ness monster, 343
Loftus, Elizabeth, 135
Lohan, Lindsay, 176
Long, Greg, 346
Lowell, Percival, 150
Luckhoo, Lionel, 399
lunar module, 96

Macknic, Stephen L., 116
magic, 149, 354–59
Making of Bigfoot, The (Long), 346
Malik, Kenan, 228–29
Marcel, Jesse, 124, 125
Marks, Jonathan, 229, 275
Marley, Bob, 223
Martinez-Conde, Susana, 116
Mask of Nostradamus, The (Randi), 75, 77
Mayan 2012 doomsday prediction, 391–96
McDivitt, Jim, 91
measles, 233, 234, 236–38, 240
Media Mythmakers: How Journalists, Activists, and Advertisers Mislead Us (Radford), 170
Medina, John, 108
Meldrum, Jeff, 347
memory, 31–32, 115–16, 127–29
Mencken, Carson, 350
Mengele, Joseph, 157–58
Mercury Seven, 91
miracles, 81–86, 104, 218, 257, 302, 307, 316, 357
MMR (vaccine), 237, 239, 240
Moby Dick (Melville), 60
Mohammed, 247, 311
Mooney, Chris, 167
Moosa, Tauriq, 29, 42, 46–47
Morris, Phillip, 346
Morris Costumes, 346
Morrison, David, 391, 394, 395
Morrison, Jim, 314
Mothers' Index, 294
Mumbai, India, 62
Mystery Chronicles, The (Nickell), 332

Napoleon, 78
NASA (National Aeronautics and Space Administration), 89–94, 175, 342, 394, 410
National Aeronautics and Space Administration. *See* NASA
National Football League. *See* NFL
National Geographic (magazine), 177
Nature (journal), 184
Nature Genetics (journal), 229
Navy (US), 239, 380
Navy SEALS (US), 78
near-death experience, 36–41
Nelson, Kevin, 39, 41
nematode worm, 000
Nevada, 382–85
New Scientist (magazine), 177, 342
Newton, Isaac, 53, 148
Nibiru, 395
Nickell, Joe, 310, 312, 313, 331–32
9/11 attacks, 78–79, 174, 193
Nixon, Richard, 90
NFL (National Football League), 51, 187
Noah's ark, 300–304, 400
Noelle, David C., 250
Nostradamus, 288, 393–94
Nostradamus: The Complete Prophecies (Hogue), 79
Nouvian, Claire, 337

Obama, Barack, 227–28
O'Brien, Dan, 188

Offit, Paul, 237–38, 240, 244
O'Kane, James, 91
Old Town, San Diego, 328
Olympics, 187–91, 224, 345
"one-drop rule," 184
optical illusions, 32
out-of-body experiences, 37–40

Pakistan, 204
Palin, Sarah, 164, 357
Paradise Ranch, 382
Paranormal America (Bader, Mencken, and Baker), 350
Park, Robert L., 279
Parthenon, 106
Patterson, Roger, 345
Patterson Bigfoot film, 345–47
Pauling, Linus, 151
Pearl Harbor, 77
Peoples Temple, 297
Pima (Native Americans), 230–31
Plait, Phil, 13, 117, 139
Planet of the Apes (1968 film), 109, 346
Planet X, 395
Plato, 273, 364
Plutarch, 417
Poe, Edgar Allen, 36
Pope John Paul II, 262, 398
Popoff, Peter, 318
postdiction, 60
poverty and prayer, 293–95
prayer, 290–95
  and poverty, 293–95
Prince Charles, 211
Project Mogul, 121–27, 193
prophecy, 283–89
Prothero, Stephen, 255
Pseudoscience and the Paranormal (Hines), 372
psychic detectives, 71–72
psychics, 14, 24, 25, 28, 29, 42–50, 70, 134, 142, 150, 175, 288, 292, 352, 417. *See also* psychic detectives
Pyramid of Djoser, 103–104
pyramids (Egyptian), 101–103, 105–106

Rabin, Yitzhak, 59
race
  American Anthropological Association statement on, 226–27
  and medicine, 226–32
  races as cultural categories, 180–85
  and sports, 186–91
Race and Reality: What Everyone Should Know about Our Biological Diversity (Harrison), 55–56, 103
Race Myth, The (Graves), 186
Radford, Benjamin, 170–71, 333
RAND-MIPT terrorism database, 174
Randi, James, 17–18, 75–77, 79–80, 139, 318, 354
Rapture, 397–406
Rather, Dan, 170
Reagan, Ronald, 140
Redford, Donald, 104
reincarnation, 62–67, 324, 331
relics, 300–14
Relics of the Christ (Nickell), 310, 312
religion, 250–58
Religious Literacy: What Every American Needs to Know—And Doesn't (Prothero), 255
remote viewing, 70
Roberts, Oral, 315–16
Rose, Pete, 128
Roswell, 59, 98, 121–31, 193, 382, 384
Roswell Daily Record (newspaper), 123

"Roswell Report: Case Closed, The" (US Air Force), 127
Rotimi, Charles N., 229
Rovelli, Carlo, 147

Sagan, Carl, 28, 103, 132, 140, 178, 218
Sailing Alone around the World (Slocum), 377
Sam, Gloria Mary, 216
Saqqara, Egypt, 104
Sasquatch: Legend Meets Science (Meldrum), 347
Schmitt, Harrison, 92
Science of Fear, The (Gardner), 174
Scientific American (magazine), 177
scientists, 147–53
Scott, Dave, 91
Scott, William, 387–88
Sentinelese, 412
SETI Institute, 119, 382
Shelley, Percy Bysshe, 245
Shenk, David, 51, 56
Shermer, Michael, 32, 70–71, 154, 158–59, 277, 368
Shostak, Seth, 108, 109, 119, 382, 386–87
sickle-cell disease, 228–29
Simons, Daniel, 116, 192
Singer, P. W., 388
Singularity, 404–405
Skeptic (magazine), 49, 70, 277
Skeptical Inquirer (magazine), 49
Skeptic's Dictionary, The (Carroll), 372
Slader, John, 108
sleep paralysis, 136–38
Sleights of Mind (Macknic and Martinez-Conde), 116
Slocum, Joshua, 377
Smith, Cameron M., 340–41
Smith, Kevin, 210
South Africa, 201–202, 204
Specter, Michael, 236
Stafford, Thomas, 91
Star Trek (television series), 109, 151
Stone of Unction, 310–11
supervolcano, 413
  Toba supervolcano, 414
Swaggart, Jimmy, 316
Sweickart, Rusty, 91, 410

tardigrade, 337–38
Tay-Sachs disease, 229
televangelists, 315–19
television news, 167–79
Teller, Edward, 150
terrorism, 174–75
thimerosal, 237, 240
Thomas, Dave, 59
Thompson, Daley, 188
Titanic, 365
Toba supervolcano, 414
Turkey, 104
Tuskegee experiment, 151, 193
Twilight Zone, The (television series), 109
2001: A Space Odyssey (1968 film), 346
2012 doomsday prediction, 391–96
2012: Science and Prophecy of the Ancient Maya (Van Stone), 392
Tyson, Neil deGrasse, 273

UAVs (unmanned aerial vehicles), 119–20, 382, 388
UFOs (unidentified flying objects), 28, 31, 108–23, 125, 148, 325, 383–88
UFOs: Generals, Pilots, and Government Officials Go on Record (Kean), 118
UNICEF, 233, 293, 319
unidentified flying objects. *See* UFOs
unmanned aerial vehicles. *See* UAVs

Vaccine Education Center at the Children's Hospital of Philadelphia, 237
vaccines, 208, 233–41, 293
Van Praagh, James, 48, 62, 323
Van Stone, Mark, 392
Venter, Craig, 337
Von Däniken, Erich, 100–107

Wakefield, Andrew, 236
Wallace, Michael, 340
Wallace, Ray, 340
Walsh, Don, 335
Ward, Frank, 36, 41
War of the Worlds (Wells), 121
Watergate, 90
Wicca, 357
Wienker, Curtis, 339–40
witches, 354–59
Wells, H. G., 121
Whaley House, 328–30
White, Nancy, 103
White, Tim, 259, 265
Why Darwin Matters (Shermer), 277
Wilson, Edward O., 300
Wired for War: The Robotics Revolution and Conflict in the 21st Century (Singer), 388
Wiseman, Richard, 70
World Trade Center, 78, 194–96
World War II, 65
Wynne, Nick, 159

Yad Veshem Holocaust History Museum, 155
Young, John, 90, 91, 95

Zundel, Ernst, 159